CAMBRIDGE LIBRARY COLLECTION

Books of enduring scholarly value

Earth Sciences

In the nineteenth century, geology emerged as a distinct academic discipline. It pointed the way towards the theory of evolution, as scientists including Gideon Mantell, Adam Sedgwick, Charles Lyell and Roderick Murchison began to use the evidence of minerals, rock formations and fossils to demonstrate that the earth was older by millions of years than the conventional, Bible-based wisdom had supposed. They argued convincingly that the climate, flora and fauna of the distant past could be deduced from geological evidence. Volcanic activity, the formation of mountains, and the action of glaciers and rivers, tides and ocean currents also became better understood. This series includes landmark publications by pioneers of the modern earth sciences, who advanced the scientific understanding of our planet and the processes by which it is constantly re-shaped.

The Great Ice Age and Its Relation to the Antiquity of Man

James Geikie (1839–1915) was born in Edinburgh, and his work from 1861 as a field geologist for the Geological Survey in Scotland provided the evidence for the theories he proposes in this work, first published in 1874 (revised editions appeared in 1877 and 1894). Geikie brought together his own research and the findings of other geologists in Scotland to support his main thesis of 'drift' being evidence of the action not of sea ice but of land ice. He was influenced by James Croll's theory that changes in the Earth's orbit led to epochs of cold climate in one hemisphere and warm in the other, and Geikie believed that the geological record provided evidence for inter-glacial periods. The book was hailed as a breakthrough at the time, and brought the author international recognition. With intricate scientific theories explained in clear uncluttered language, this remains a classic text.

Cambridge University Press has long been a pioneer in the reissuing of out-of-print titles from its own backlist, producing digital reprints of books that are still sought after by scholars and students but could not be reprinted economically using traditional technology. The Cambridge Library Collection extends this activity to a wider range of books which are still of importance to researchers and professionals, either for the source material they contain, or as landmarks in the history of their academic discipline.

Drawing from the world-renowned collections in the Cambridge University Library and other partner libraries, and guided by the advice of experts in each subject area, Cambridge University Press is using state-of-the-art scanning machines in its own Printing House to capture the content of each book selected for inclusion. The files are processed to give a consistently clear, crisp image, and the books finished to the high quality standard for which the Press is recognised around the world. The latest print-on-demand technology ensures that the books will remain available indefinitely, and that orders for single or multiple copies can quickly be supplied.

The Cambridge Library Collection brings back to life books of enduring scholarly value (including out-of-copyright works originally issued by other publishers) across a wide range of disciplines in the humanities and social sciences and in science and technology.

The Great Ice Age and Its Relation to the Antiquity of Man

JAMES GEIKIE

CAMBRIDGE UNIVERSITY PRESS

Cambridge, New York, Melbourne, Madrid, Cape Town,
Singapore, São Paolo, Delhi, Mexico City

Published in the United States of America by Cambridge University Press, New York

www.cambridge.org
Information on this title: www.cambridge.org/9781108050081

This edition first published 1874
This digitally printed version 2012

ISBN 978-1-108-05008-1 Paperback

The original edition of this book contains a number of colour plates, which have been reproduced in black and white. Colour versions of these images can be found online at www.cambridge.org/9781108050081

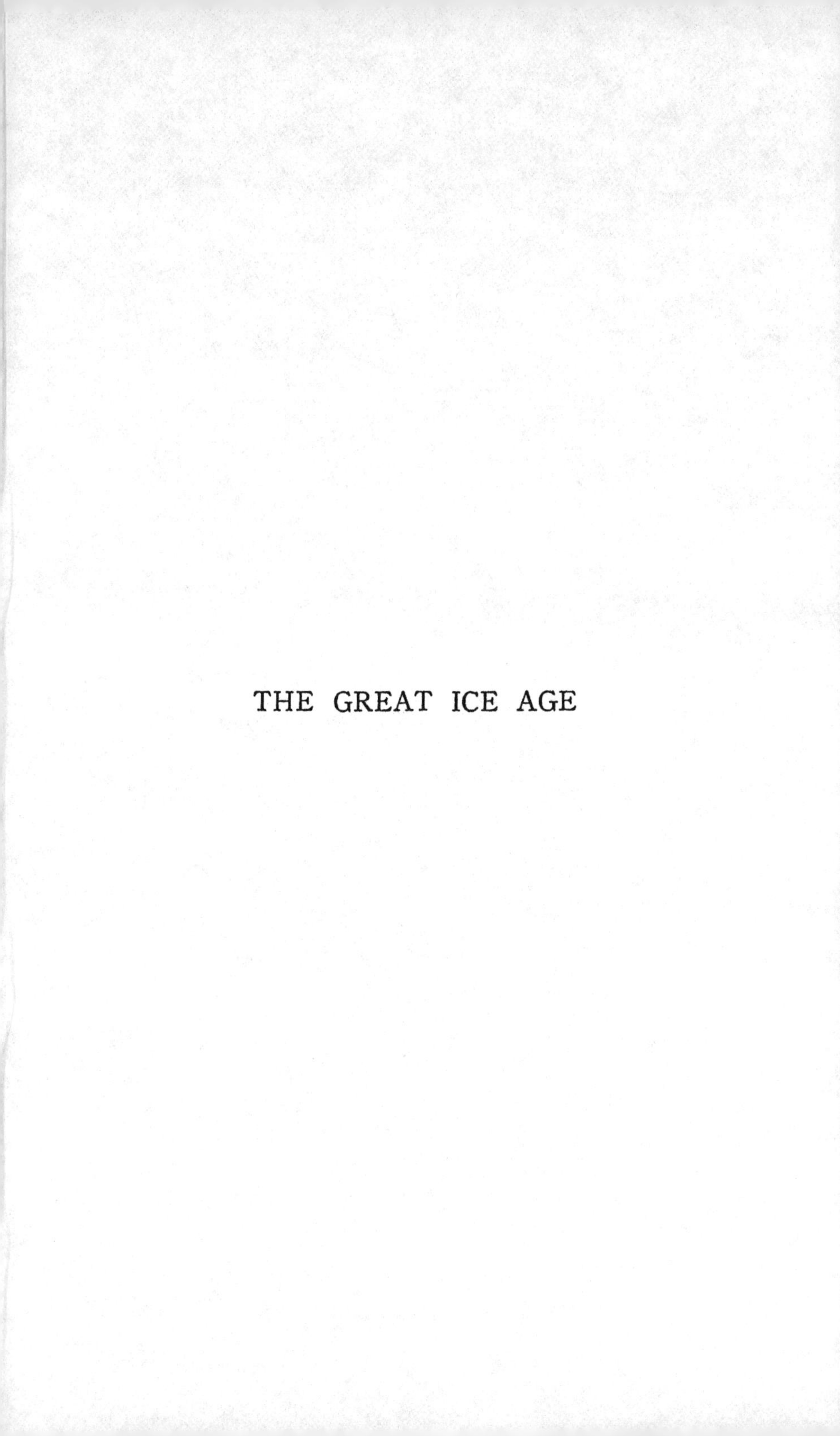

THE GREAT ICE AGE

THE GREAT ICE AGE

And its Relation to the Antiquity of Man

BY JAMES GEIKIE, F.R.S.E., F.G.S.
OF H. M. GEOLOGICAL SURVEY OF SCOTLAND

W. ISBISTER & CO.
56, LUDGATE HILL, LONDON
1874

TO

ANDREW CROMBIE RAMSAY, LL.D., F.R.S.

DIRECTOR-GENERAL OF THE GEOLOGICAL SURVEYS OF THE UNITED KINGDOM.

MY DEAR RAMSAY,

There is no one from whom in the course of my geological studies I have received more help and encouragement than you. Not only have your scientific writings been of essential service to me, but I have gathered much also in conversation from your wide experience, and feel that I have gained a great deal by having come in contact with one possessed of such originality of conception and independence of thought as yourself. The lessons that stand us most in stead, are not always got out of books, and I am glad of this opportunity to acknowledge my indebtedness to you for many such lessons. It gives me much pleasure to inscribe this volume with your name, in token of my regard and affection for you as my teacher and friend.

Believe me,

My dear RAMSAY,

Very sincerely yours,

JAMES GEIKIE.

EDINBURGH : *December*, 1873.

PREFACE.

IN the following pages I have endeavoured to give a systematic account of the Glacial Epoch, with special reference to its changes of climate. My intention at first was to restrict myself to a brief description of British glacial and post-glacial deposits, for the purpose of pointing out the general succession established, or in process of being established, by geologists in this country; and, thereafter, of drawing some conclusions as to the position in the series of the palæolithic gravels, &c., of southern England. But I eventually found, that in order to make my argument intelligible to non-specialists, it would be necessary to describe in detail some sufficiently large area, in which glacial deposits might be considered as typically developed. It is for this reason that I have entered so fully into the geological history of glacial and post-glacial Scotland. In delineating the post-tertiary geology of that country, I have been enabled to discuss many elementary matters, which are, no doubt, sufficiently familiar to my fellow-hammerers, but which a general reader could hardly be expected to know. In short, while treating of the Scottish deposits, I have endeavoured to explain the mode of investigating, and the principles of interpreting glacial phenomena. I also thought that by confining detailed sketching to the glacial history of a well-defined region, it would be possible to convey to the reader's mind a more vivid impression of what the Glacial Epoch really was, than if I ventured to take a wider canvas. But my chief aim throughout has been to indicate the succession of climatal changes that obtained during the Glacial Epoch—not in Scotland alone, but in every glaciated region which has been carefully studied by geologists. For I have long been of opinion that until this has been done, until we clearly understand what the succession of changes during the Ice Age was, it is premature to speculate upon the geological age of those deposits which hold the earliest traces of man in Britain. The great difference that obtains between the fauna of undoubted post-glacial and recent beds in Scotland, north of England, Wales, and Ireland, on the one hand, and the cave-deposits and palæolithic gravels of Southern England on the other, has long been a puzzle to me, as it has, no doubt, been to other geologists. But it was not until years had been spent in the study of the glacial deposits that what I conceive to be the true explanation of the difficulty dawned upon me.

Geologists are aware that the post-glacial age of the cave-deposits has not infrequently been called in question. Dr. Buckland, Mr. Godwin-Austen, and Professor Ramsay, have each expressed a belief that some of our cave-deposits may date back to pre-glacial times; and Mr. Godwin-Austen long ago pointed out that the "sub-aerial beds" of the English Channel districts were the equivalents of the glacial deposits elsewhere; and that the broad alluvia of our more southerly rivers, such as the Severn, the Fal, the Dart, and the Thames, belong to a period prior to the great submergence, during which the high-level marine drifts of Wales were accumulated. In other words he showed that these river-gravels could not be referred to post-glacial times. Within more recent years a modification of Mr. Godwin-Austen's view has been energetically put forward by Mr. W. Boyd Dawkins, who is of opinion that our palæolithic deposits belong to a time subsequent to the great submergence just referred to. He holds that man and the extinct mammalia lived in the south of England at a period when Scotland, Wales, and the northern districts of England were covered with ice and snow, and when our summers were warm and our winters very severe. Other geologists, however, as Mr. Prestwich, have contended that we have no evidence of warm summers having obtained during palæolithic times; while, yet others have, on the contrary, thought the evidence pointed to a considerably warmer climate than we now enjoy. Sir C. Lyell gives two explanations of the facts, and thinks the commingling of arctic and southern forms of animal life, may point either to a period of strongly-contrasted summers and winters, or to fluctuations of climate; but he is clearly of opinion that all the palæolithic gravels belong to post-glacial times. The Rev. O. Fisher has described a deposit, called by him "trail," which he believes to be the product of land-ice; as it overlies in places palæolithic deposits, he considers that a glacial period has intervened since the disappearance of palæolithic man. Mr. J. Croll has referred to the apparently conflicting evidence of the mammaliferous deposits, as an indication of former changes of climate, and this is the view which Sir J. Lubbock inclines to support, and which is advocated in these pages. I have, however, ventured an explanation of the peculiar distribution of palæolithic gravels, differing from any previously advanced. None of these gravels in my opinion are post-glacial, but all must be relegated to pre-glacial and inter-glacial times. Their absence from the northern districts I account for by showing that they have been swept out of these regions by confluent glaciers, and by the sea during the period of great submergence.

These views I first broached during the discussion that followed upon the reading of an interesting paper by Mr. W. Boyd Dawkins, at the Edinburgh meeting of the British Association in 1871. But as the time allowed for discussion at these meetings is necessarily short, I was unable to do justice to my views; believing them, however, to be worth attention, I gave an outline of them in a series of monthly articles in the *Geological Magazine*, the first of which appeared in the number for December, 1871. These articles I subsequently collected, and re-issued with some additions and rearrangements in the summer of the year following.

As far as I know, these papers were the first attempt to prove, by correlating glacial deposits, that the palæolithic gravels of Southern England could not be of post-glacial age, but ought to be referred to inter-glacial and pre-glacial times. They also bring forward, for the first time, reason to show that a wide land-surface existed in the British area after the disappearance of the ice-sheets, and before the period of great submergence had commenced.

The recent discovery of human remains in the Victoria Cave, where, as Mr. Tiddeman has shown, the deposits are overlaid by a glacial accumulation, is direct evidence in corroboration of the conclusions which I had previously arrived at and published.

In the preparation of this volume I have availed myself of a number of papers, communicated from time to time to various scientific societies and publications, and in fact these chapters may be considered as only an expansion of the articles "On Changes of Climate during the Glacial Epoch," which first appeared in the pages of the *Geological Magazine.*

It would be unbecoming in me, if I did not here acknowledge the obligations I am under to the writings of those who have preceded me in describing the glacial deposits of Scotland, more especially to the successive publications of Messrs. Maclaren, Chambers, Jamieson, and A. Geikie. I had hoped to be able to append a list of works by these and other authors, relating to the glacial and post-glacial geology of the British Islands, together with such foreign papers as I had come across in the course of my reading; but after making some progress, I was reluctantly compelled to abandon my design, as the list threatened to form quite a volume of itself.

To my colleagues on the Geological Survey of Scotland, I am indebted for much valuable assistance, willingly rendered. Of these, my old friend Mr. B. N. Peach, supplied me with the highly characteristic sketches of striated stones, and the beautiful views of Loch Doon, and the Carse of Stirling; while from the facile pencil of Mr. H. M. Skae, I received the fine drawing of the Coolin Mountains, and the illustrations of Alpine, Arctic, and Antarctic scenery which adorn these pages. Mr. R. L. Jack also kindly furnished me with the map and sections of Loch Lomond, and with the chart showing the physiography of western Scotland that would appear upon an elevation of 600 feet. To the same obliging friend I am likewise indebted for many practical hints and suggestions during the progress of my work through the press. Mr. R. Etheridge, Jun., has enhanced the value of these pages by the long list of organic remains which is given in the Appendix; and has, moreover, laid me under many obligations by the trouble he has taken to verify references and consult libraries, of which my absence from town on official duty would otherwise, in many cases, have prevented me availing myself. Mr. J. Croll, Mr. J. Horne, and Mr. D. R. Irvine, have also obliged me with a number of glacial notes.

To several of my colleagues on the English Geological Survey, I have likewise to render my acknowledgments. Mr. A. H. Green furnished me with a long series of notes relating to the glacial deposits of England, which, however, the limits I had set for myself have not allowed me to make use of. But I hope my friend may himself be induced to give some general account of the post-tertiary deposits of England—a work which would be of essential service to the students of glacial geology. Mr. Green was also good enough to read over the proofs of Chapter IX., and to suggest a number of improvements and additions which I have gladly incorporated with the text. Mr. Etheridge examined for me several lists of shells from the "pliocene sands" of northern Italy, and his opinions and advice have been of great assistance. I am also indebted to Mr. Whitaker for a diagram-section and notes, relating to the gravels of the Thames valley.

I take this opportunity of thanking those foreign geologists who have sent me copies of their publications. Among these papers one by Mr. Nathorst has just come to hand, of which, had it arrived sooner, I would gladly have availed myself in the text. Mr. Nathorst describes certain inter-glacial deposits met with in Sweden and Denmark, and gives the following table to show their position, which I make no apology to glacialists for transcribing in this place:—

Post-glacial formations.	Peat with	Quercus sessiliflora, Q. robur.
		Pinus sylvestris.
		Populus tremula, Betula nana.
	Clay with	Betula nana, Salix herbacea, S. reticulata, Dryas octopetala, Cytheridea torosa, Limnea limosa, Pisidium Anodonta.
		Salix polaris.
Glacial Period		Glacial Deposits.
Inter-glacial formation	Clay with	Salix polaris, Dryas octopetala, Limnea limosa, Pisidium, Anodonta, Cytheridea torosa.
Glacial Period.		Glacial Deposits.

CONTENTS.

CHAPTER I.

PAGE

INTRODUCTORY 1

CHAPTER II.

SUPERFICIAL FORMATIONS OF SCOTLAND.—THE TILL.

General distribution of Superficial formations.—Till, the oldest member of the series.—How this is proved.—Character of the till.—Stones in the till.—Unfossiliferous character of the till.—Till developed chiefly upon the low-grounds.—Its aspect in upland valleys.—"Sowbacks" or "drums" of till.—"Crag and tail"—Smoothed and broken rocks below till.—Configuration of mountains and hills.—Lines of stones in till.—Subjacent and intercalated beds.—Résumé 6

CHAPTER III.

EARLY THEORIES.—CAUSE OF GLACIER MOTION.

Early theories accounting for origin of the till.—Debacles and waves of translation.—Glacier theory of Agassiz.—Iceberg theory.—Accumulation of snow above snow-line, how relieved.—Evaporation, avalanches, and glaciers.—Origin of glaciers.—Theories of Forbes and Tyndall.—Canon Moseley's experiments. . . 27

CHAPTER IV.

MR. CROLL'S THEORY OF GLACIER MOTION.—ASPECT OF AN ALPINE GLACIER.

Heat capable of being transmitted through ice.—Temperature of glaciers.—Motion of ice molecules in a glacier.—Regelation.—Alpine glaciers.—River-like character of glaciers.—Crevasses.—Size of glaciers.—Character of the rock-surface and stones below a glacier.—Glacial rivers.—Moraines.—Waste of mountains . 40

CHAPTER V.

GREENLAND: ITS GLACIAL ASPECT.

PAGE

Extent of Greenland.—Character of coast and interior.—The great *mer de glace.*—Size of glaciers.—Phenomena of arctic glaciers, and origin of icebergs.—Submarine moraines.—Scarcity of surface moraines in Greenland.—Glacial rivers.—Circulation of water underneath the ice.—The habitable strip of coast-land.—Formation of ice upon the sea.—The ice-foot.—Waste of cliffs.—Transportation of rock débris by ice-rafts and icebergs.—Action of stranded icebergs upon the sea-bottom 54

CHAPTER VI.

ORIGIN OF THE TILL AND ROCK-STRIATIONS AND GROOVINGS OF SCOTLAND.

Stones of the till are glaciated.—Till not like terminal moraine-matter. —Mud of glacial rivers.—Till not an iceberg deposit.—Rock-striæ produced by glaciers.—Scotland once covered with ice.—Direction of the ice-flow in the Highlands.—In the Southern Uplands.—In the Lowlands of the great Central Valley.—Absence of superficial moraines.—Stones in the till derived from the subjacent rocks, not from precipices overhanging the ice.—Stones and mud below ice forming a *moraine profonde* or ground moraine.—Unequal distribution of this deposit explained 75

CHAPTER VII.

ORIGIN OF THE TILL, AND ROCK-STRIATIONS AND GROOVINGS OF SCOTLAND—*Continued.*

Direction of ice-flow indicated by stones in the till.—Cross-hatching of rock-striæ accounted for.—Intermingling in the till of stones derived from separate districts.—"Debatable land" between rival ice-flows. —Local colouring of till an indication of direction followed by ice-flow.—The Ochils, Pentlands, and other hills completely overflowed by ice.—Deflections of the ice-flow.—Till in upland valleys, why terraced.—Origin of lowland "drums."—Crag and tail, &c.—Islets lying off the coast glaciated from the mainland.—Ice filled up all our shallow seas.—General ice-sheet like that of Antarctic continent 90

CHAPTER VIII.

CAUSE OF COSMICAL CHANGES OF CLIMATE.

Evidences of former great changes of climate during past ages.—Theories accounting for these.—Translation of the solar system through space. Change in the position of the earth's axis of rotation.—Sliding of the earth's crust round the nucleus.—Changes in the distribution of land and sea.—Elevation and depression of the

PAGE

land.—Sir Charles Lyell's speculations upon the effect of such changes.—Climatal effect of winds and ocean currents.—Effect of accumulation of land within the tropics, and of same round the poles. —Peculiar distribution of land and sea inadequate to account for great cosmical changes of climate 103

CHAPTER IX.

CAUSE OF COSMICAL CHANGES OF CLIMATE—*Continued.*

Failure of geologists to furnish an adequate theory.—Astronomical phenomena may perhaps afford a solution.—Movement of the earth on its own axis and round the sun.—Eccentricity of the orbit.—Precession of the equinoxes.—Nutation.—Obliquity of the ecliptic . . 121

CHAPTER X.

CAUSE OF COSMICAL CHANGES OF CLIMATE—*Continued.*

Effect of variation of the eccentricity of the earth's orbit.—Sir J. Herschel's opinion.—Arago's.—Purely astronomical causes insufficient to afford a solution of the problem.—Mr. Croll's theory.—Changes of climate result indirectly from astronomical causes.—Physical effects of a high eccentricity of the orbit.—Extremes of climate at opposite poles.—Modifications of the course followed by ocean-currents.—Perpetual summer and perennial winter.—Physical effects of obliquity of the ecliptic.—Succession of climatal changes during a period of high eccentricity 133

CHAPTER XI.

BEDS SUBJACENT TO AND INTERCALATED WITH THE SCOTTISH TILL.

Beds in and below till.—Seldom seen except in deep sections.—Examples of superficial deposits passed through in borings, &c.—Sections exposed in natural and artificial cuttings.—Examples.—Beds contorted and denuded in and below till.—Examples.—Fossiliferous beds in the till.—Examples.—Striated pavements of boulders in till 150

CHAPTER XII.

BEDS SUBJACENT TO AND INTERCALATED WITH THE SCOTTISH TILL—*Continued.*

Beds below and in the till indicate pauses in the formation of that deposit.—How the aqueous beds have been preserved.—Their crumpled and denuded appearance.—Their distribution.—Character of the valleys in which they occur.—The present stream-courses, partly of pre-glacial, inter-glacial, and post-glacial age.—Old course of the River Avon, Lanarkshire.—Pre-glacial courses of the Calder Water and Tillon Burn.—Buried river channel between Kilsyth and Grangemouth 165

CHAPTER XIII.

BEDS SUBJACENT TO AND INTERCALATED WITH THE SCOTTISH TILL—*Continued.*

PAGE

Stratified deposits passed through in borings.—Probably in most cases of fresh-water origin.—Old lake at Neidpath, Peebles.—Lakes of inter-glacial periods.—Borings near New Kilpatrick.—Pre-glacial valley of the river Kelvin.—Origin of the deposits occupying that buried river-course.—Valley between Johnstone and Dalry.—Flat lands between Kilpatrick Hills and Paisley.—Condition of the country in pre-glacial times.—Loch Lomond glacier.—Ancient glacial lake.—Succession of events during recurrent cold and warm periods 179

CHAPTER XIV.

BEDS SUBJACENT TO AND INTERCALATED WITH THE SCOTTISH TILL—*Continued.*

Inter-glacial deposits of the Leithen Valley, Peeblesshire.—Other examples of similar deposits.—Crofthead inter-glacial beds.—Climatal conditions of Scotland during inter-glacial ages.—Fragmentary nature of the evidence not conclusive as to the climate never having been positively mild.—Duration of glacial and inter-glacial periods.—Inter-glacial shelly clay at Airdrie.—Arctic shells at Woodhill Quarry, Kilmaurs.—Recapitulation of general results . 193

CHAPTER XV

BOULDER-CLAY BEDS OF SCOTLAND.

Stony clays of maritime districts.—Character of boulder-clay as distinguished from till.—Traces of bedding, &c.—Water-worn shells.—Crushed, broken, and striated shells.—Shell-beds.—Boulder-clay passes into till.—Boulder-clay of Stinchar valley; of Berwick cliffs.—Boulder-clay and associated deposits of Lewis; of Caithness and Aberdeenshire.—Boulder-clay of Dornoch Frith; of Wigtonshire and Ayrshire.—Origin of boulder-clay.—Scotland resembled Greenland during the formation of this deposit . . . 205

CHAPTER XVI.

UPPER DRIFT DEPOSITS OF SCOTLAND.

Morainic débris and perched blocks of Loch Doon district; of Stinchar valley; of northern slopes of Solway basin; of Rinns of Galloway; of the Clyde and Tweed valleys; of the Northern Highlands.—Erratics; carried chiefly by glaciers; have travelled in directions corresponding with the trend of the rock-striæ.—Condition of Scotland during the carriage of the erratics that occur at high levels.—Sand and gravel series.—Kames.—Terraces . . . 218

CHAPTER XVII.

UPPER DRIFT DEPOSITS OF SCOTLAND—*Continued.*

PAGE

Distribution of the kames.—Kames of Carstairs and Douglas Water; of the Kale and the Teviot; of the Whiteadder valley; of Kinross-shire; of Leslie and Markinch, in Fifeshire; of the Carron Water; of Perthshire and Forfarshire; of the valleys of the Don, Dee, Ythan, Deveron, Spey, and Findhorn; of the vale of the Ness; of the Beauly Frith.—Sand and gravel along flanks of hilly ground between rivers Clyde and Irvine.—Absence of kames, &c., in Lewis and Caithness. —Evidence of transport down the valleys.—Passage of well-rounded gravel into angular gravel and débris at high levels.—Sand and gravel overlying morainic débris.—Denudation of older drifts in sand and gravel districts.—Evidence of river action.—Fossils in ancient river-gravels, &c.—Isolated cones and ridges of gravel and sand, and high-level terraces.—Evidence of action of sea.—Sand and gravel drift of Peeblesshire.—Limits of submergence of Scotland.—Summary of conclusions 234

CHAPTER XVIII.

UPPER DRIFT DEPOSITS OF SCOTLAND—*Continued.*

Shelly clays of maritime districts.—Position of these deposits with respect to older drift accumulations.—General character of brick-clay sections.—Organic remains.—Ice-floated stones and boulders. Crumpled and contorted beds.—General inferences . . 258

CHAPTER XIX.

UPPER DRIFT DEPOSITS OF SCOTLAND—*Continued.*

Morainic débris.—Its position in Highlands and Southern Uplands.—Essentially a local deposit.—Of more recent date than the other drift deposits.—Denudation of the moraines.—These moraines not to be confounded with the older morainic débris of earlier periods . 268

CHAPTER XX.

UPPER DRIFT DEPOSITS OF SCOTLAND—*Continued.*

Parallel Roads of Glen Roy, old lake-terraces and not raised sea-beaches. —Valley dammed up by local glacier.—General summary of conclusions regarding succession and origin of the upper drift deposits 273

CHAPTER XXI.

LAKES AND SEA-LOCHS OF SCOTLAND.

Different kinds of lakes.—Lakes occupying depressions in drift.—Lakes dammed by moraines and older drift deposits.—Lakes lying in rock-bound basins.—Origin of rock-basins discussed.—Ramsay's theory.—Sea-lochs.—Submarine rock-basins.—Their glacial origin. —Silted-up rock-basins 282

CHAPTER XXII.

POST-GLACIAL AND RECENT DEPOSITS OF SCOTLAND.

PAGE

Raised beaches.—Faintly-marked terraces, &c., at high levels.—Platforms of gravel, &c.—Platforms and notches cut in rock.—Raised beaches at mouth of Stinchar, at Newport, and Tayport.—Highest-level beaches belong to glacial series.—Evidence of cold climatal conditions during formation of some post-glacial beaches.—Oscillations of level.—Raised beaches at the lower levels.—Human relics in raised beaches.—Proofs of oscillations of level.—Submarine peat-mosses, &c.—Blown sand 304

CHAPTER XXIII.

POST-GLACIAL AND RECENT DEPOSITS OF SCOTLAND—*Continued.*

Peat-mosses, their composition.—Trees under peat, their distribution.—Submerged peat-mosses of Scotland, England, Ireland, Channel Islands, France, Holland.—Loss of land.—Continental Britain.—Climatal conditions.—Causes of decay and overthrow of the ancient forests.—Growth of the peat-mosses.—Climatal change.—Decay of peat-mosses 317

CHAPTER XXIV.

POST-GLACIAL AND RECENT DEPOSITS OF SCOTLAND—*Continued.*

Action of the weather on rocks.—Erosion by running water.—Post-glacial erosion insignificant in amount as compared with denudation during last inter-glacial period.—Recent river-terraces.—Silted-up lakes.—Marl-beds.—Organic remains in fresh-water alluvia.—Human relics.—Conclusions regarding succession of events in post-glacial times.—General summary of glacial, inter-glacial, and post-glacial changes 340

CHAPTER XXV.

GLACIAL DEPOSITS OF ENGLAND AND IRELAND.

Necessity of comparing deposits of different countries.—Glaciation of mountain districts of England and Wales.—Till, its character.—Direction of ice-flow in the north-west districts.—Lower boulder-clay of Lancashire, Cheshire, &c.—Middle sand and gravel.—Upper boulder-clay.—River-gravels.—Morainic débris.—Succession of changes.—Lower, middle, and upper glacial series of East Anglia.—Irish glacial deposits.—Till.—Morainic débris.—Lower boulder-clay, middle gravels, &c.—Upper boulder-clay.—Eskers.—Erratics.—Sand-hills.—Marine shells at high levels . . . 355

CHAPTER XXVI.

SUPERFICIAL DEPOSITS OF SCANDINAVIA.

PAGE

Extensive glaciation of mountainous and northern regions throughout the northern hemisphere.—Glaciation of Norway, Sweden, and Finland.—Lower and upper till of Sweden.—Inter-glacial fresh-water deposits.—Till of Norway.—Åsar.—Erratics on åsar.—Clays with marine arctic shells.—Contorted bedding, &c.—Post-glacial deposits with Baltic shells.—Moraines.—Succession of changes.—Southern limits of glaciation in northern Europe.—The great northern drift.—Theories of the origin of the åsar.—Summary 377

CHAPTER XXVII.

SUPERFICIAL DEPOSITS OF SWITZERLAND.

No trace of sea-action or of floating ice.—Erratics of the Jura.—Glaciers of the Rhine and its tributaries.—Glacier of the Rhone.—Moraine-profonde in Dauphiny.—Moraine-profonde of Swiss low-grounds.—Ancient alluvium or diluvium overlying moraine-profonde.—Inter-glacial deposits of Dürnten, &c.—Morainic débris, &c., overlaying inter-glacial deposits.—Reason why the moraines of the second great advance of glaciers are large and well-preserved.—Post-glacial deposits . . . 398

CHAPTER XXVIII.

SUPERFICIAL DEPOSITS OF NORTH AMERICA.

Glaciation of the northern regions.—The Barren Grounds.—Profusion of lakes.—Pre-glacial beds.—Oldest glacial deposits.—Unmodified drift, or till.—Scratched pavements in till.—Marine arctic shells in boulder-clay of maritime districts.—Inter-glacial deposits.—Iceberg drift.—Later glaciation.—Morainic débris.—Gravel and sand mounds, ridges, &c.—Erratics on gravel and sand ridges. Limits of submergence uncertain.—Glacial lake-terraces in White Mountains.—Laminated clays with marine arctic shells.—Local moraines.—Tables showing succession of glacial deposits in Europe and America 410

CHAPTER XXIX.

CAVE-DEPOSITS AND ANCIENT RIVER-GRAVELS OF ENGLAND.

Prehistoric deposits.—Stone, bronze, and iron ages.—Palæolithic or Old Stone period.—Neolithic or New Stone period.—Universal distribution throughout the British Islands of neolithic implements.—Animal remains associated with neolithic relics.—Palæolithic implements.—Absence of intermediate types.—Break between palæolithic and neolithic periods.—Caves and cave-deposits.—Kent's cavern, Torquay.—Succession of deposits.—River-gravels, &c., with mammalian remains and palæolithic implements.—Geographical changes during palæolithic period.—Gap between neolithic and palæolithic deposits 431

CHAPTER XXX.

CLIMATE OF THE PALÆOLITHIC PERIOD.

PAGE

Groups of mammalian remains associated together in palæolithic deposits.—Southern, arctic, and temperate species.—Theory of annual migrations.—Condition of this country during palæolithic period.—Cause of extreme climate of Siberia.—Supposed elevation of Mediterranean area in palæolithic times, untenable.—Probable climatal effects of such an elevation.—Climate of western Europe during palæolithic period not "continental," but "insular."—Conclusions 450

CHAPTER XXXI.

GEOLOGICAL AGE OF THE PALÆOLITHIC DEPOSITS.

Oscillations of climate.—Cold and warm periods.—Evidence of the river-deposits.—Character of the evidence cumulative in favour of former climatal changes.—No evidence of warm post-glacial climate.—Southern mammalia not of post-glacial age.—Age of cave-deposits.—No proof that they are post-glacial.—Relation of river-gravels to glacial deposits.—Age of the boulder-clay at Hoxne, &c.—Boulder-clay upon which palæolithic deposits sometimes rest belongs to the older glacial series.—Distribution of palæolithic gravels.—Comparison between palæolithic gravels of South England and river-deposits of the north.—Palæolithic deposits not met with in districts that are covered with accumulations belonging to the later glacial series, but confined to regions which we cannot prove to have been submerged during the latest period of glacial cold.—Palæolithic deposits of inter-glacial, not post-glacial age.—Bulk of them probably belong to last inter-glacial period.—Recapitulation 467

CHAPTER XXXII.

GEOLOGICAL POSITION OF NEOLITHIC, PALÆOLITHIC, AND MAMMALIFEROUS DEPOSITS OF FOREIGN COUNTRIES.

Palæolithic deposits wanting in Switzerland.—Mammalia of inter-glacial beds.—Swiss post-glacial deposits belong to neolithic, bronze, and more recent periods.—Post-glacial deposits of northern Italy of neolithic or more recent age.—Mammalian remains of Piedmont.—Palæolithic tools and remains of southern mammalia nowhere found in superficial deposits overlying the great northern drift.—Wide distribution of neolithic implements over northern Europe.—Inferences.—Palæolithic man and the southern mammalia not post-glacial.—Distribution of the old mammalia over Siberia and North America.—Proofs of mild climates within Arctic Circle.—Trees in Greenland.—Mammalia absent from districts covered with the later glacial deposits, but abound in the districts beyond.—General conclusions 486

CHAPTER XXXIII.

PAGE

CONCLUSION 504

APPENDIX.

Note A. Table of Sedimentary Strata 511

Note B. Table of Quaternary Deposits of the British Isles, with some of their Equivalents in other Countries . . . 516

Note C. Traces of a Glacial Period in the Southern Hemisphere 518

Note D. Map and Sections of Loch Lomond . 518

Note E. Map showing the Physiography of West of Scotland that would appear upon an elevation of 600 feet 519

Note F. Glacial and Inter-glacial Deposits of Northern Italy 525

Note G. List of Fossil Organic Remains occurring in Scottish Glacial Deposits 533

Note H. Map showing the Principal Directions of Ice-flow in Scotland . 563

LIST OF MAPS, CHARTS, AND FULL-PAGE ILLUSTRATIONS.

I. Loch Doon (lower reach) *to face page* 22

II. Greenland Glacier 56

III. Antarctic Ice-sheet . . 101

IV. Map of part of the Tweed valley, near Peebles . 181

V. Map of Basin of the Clyde, near Glasgow 185

VI. Coolin Mountains, Skye . . 220

VII. Loch Doon (upper reach) . 295

VIII. Loch Broom, from the Admiralty chart 297

IX. The Inner Sound, from the Admiralty chart 298

X. Loch Etive, from the Admiralty chart . . 300

XI. The Carse of Stirling . . 311

XII. Map of Norway, Sweden, and Finland, showing principal lines of glacial erosion . . . 382

XIII. Map and Sections of Loch Lomond . 518

XIV. Map showing the Physiography of West of Scotland that would appear upon an elevation of 600 feet . 519

XV. Map of Scotland, showing direction of ice-flow in pocket.

XVI. Chart of January Isothermal lines . at end of volume.

XVII. Chart of July Isothermal lines . at end of volume.

CHAPTER I.

INTRODUCTORY.

THE earlier students of the physical history of our earth considered that a great gap or strongly-defined boundary-line separated the Present from the Past. Some mighty convulsion of nature was believed to have marked the close of the geological ages, and to have preceded the advent of man and the introduction of the plants and animals with which he is associated. It was hardly doubted that the present distribution of land and water over the earth's surface dated back to a time anterior to the coming of our race, and that when man first entered Britain he had to cross seas that still roll between us and the Continent. In short, it was held that within the human period only a few minor changes had been effected in the physical aspect of our country. It was admitted, indeed, that large areas of forest-land had been displenished, that considerable tracts of peat-moss had grown, and that here and there, where the coasts were formed of incoherent materials, the sea had made some inroads; but no one supposed that greater changes than these had transpired since the first occupation of Britain.

Subsequent research, however, has overturned many of these opinions, and widened our views in regard to the magnitude of the physical changes of which man has been a witness. Not only have great oscillations of climate happened within the human period, but the distribution of land and sea also has undergone very considerable modifications. Seas have vanished and returned, wide areas of land have appeared and disappeared—broad valleys have been hollowed out of solid rocks by running water. It is from a knowledge of these and similar facts that geologists arrive at their estimate of the antiquity of man, and have assured themselves that no mighty convulsion of nature separates the human period from the earlier ages—the deposits which were at one time looked upon as the sure evidence of such a "break in the succession" being now recognised as only so many links in the chain that binds the present to the past.

The study of these deposits has unfolded a deeply interesting and almost romantic history. We are introduced to scenes that are in strangest contrast to what now meets the eye in these latitudes: geographical and physical changes of the most stupendous character pass before us; we see our islands and northern Europe at one time enveloped in snow and ice; at another time well wooded, and inhabited by rude tribes of men and savage animals; now the British Islands are united to the Continent—again, the sea prevails, and a large part of Britain, together with all the low grounds in the north of Europe, are overwhelmed beneath the waters of an arctic ocean, across which float rafts and bergs of ice. Yet again, we behold the land rising slowly out of the water, and Britain once more becoming continental, and re-peopled. Finally, we follow the working of those physical influences by which at last the present order of things is brought about.

Those who hear of this history for the first time may well be excused if they listen with some incredulity. It seems difficult to understand how the records of such extraordinary events could be preserved; or how, having been preserved, geologists are able to interpret them. Yet there is really no mystery about the matter. Difficulties undoubtedly do arise, and sometimes problems suggest themselves, for which it is hard or even impossible at present to find a solution. Nevertheless the whole matter resolves itself into a question of circumstantial evidence. The facts are patent to every one who will take the trouble to examine them, and the interpretations adduced by geologists are capable of being attested by an appeal to what is actually taking place in the world around us. In the following pages, therefore, I propose to give an outline-sketch of the evidence, mentioning only such details as may serve to bring the salient points clearly before the mind, and endeavouring at the same time to put the reader, who may chance to be not specially skilled in geology, into a position to judge for himself as to the reasonableness of the explanations advanced.

The earlier pages will be occupied with an account

of the later chapters in the geological history of Scotland. We shall trace out the succession of events that marked the origin of certain loose and incoherent materials which overlie the solid rocks of that country, and are represented by similar accumulations covering vast areas throughout the northern regions of our hemisphere. The consideration of the Scottish deposits will naturally lead us to inquire into the principles upon which they and their equivalents in other lands must be interpreted. We shall then describe in succession the superficial accumulations of other portions of the British Islands, of Scandinavia, of Central Europe, and of North America, for the purpose of ascertaining how far the conclusions arrived at by geologists in different countries harmonize with each other. Having traversed this wide field of inquiry, and become aware that the deposits, which were at one time slumped together and vaguely believed to represent a period of wild cataclysms and convulsions, are really the records of a long series of changes, each of which flowed as it were into the other, we shall finally take up the subject of the antiquity of man.

In considering this difficult but important and interesting question, it will be necessary to treat first of the special evidences which have been adduced by archæologists and geologists, to prove the great age of our race. Thereafter we shall endeavour to determine what is the exact position in the geological history of those deposits which contain the very oldest traces of man. Our aim, in short, will be to discover, if possible, at what stage during those great climatal and geographical revolutions, which shall have previously engaged our attention, man certainly occupied Britain. If we are able to determine this point, we shall have paved the way for eventually arriving at some approximately definite estimate of the antiquity of man in Western Europe.

CHAPTER II.

SUPERFICIAL FORMATIONS OF SCOTLAND.—THE TILL.

General distribution of Superficial formations.—Till, the oldest member of the series.—How this is proved.—Character of the till.—Stones in the till.—Unfossiliferous character of the till.—Till developed chiefly upon the low-grounds.—Its aspect in upland valleys.—"Sowbacks" or "drums" of till.—"Crag and tail."—Smoothed and broken rocks below till.—Configuration of m untains and hills.—Lines of stones in till.—Subjacent and intercalated beds.—Résumé.

THROUGHOUT the length and breadth of Scotland occur numerous scattered heaps and ragged sheets of sand, gravel, and coarse débris, and widespread deposits of clay, beneath which in many places, especially in the lowland districts, the solid rocks that form the framework of the country are in great measure concealed. The general character of these superficial heaps and gatherings must be familiar to every one. They appear in the scaurs and bluffs that overhang our streams and rivers, and are often well-exposed by the wash of the waves along certain sections of the sea-coasts. The traveller by rail can hardly fail to notice them as he is swept along—here capping the rocks with a few feet of sand and gravel, there thickening out so as to form the whole face of the cutting from top to bottom. In the numerous quarries with which the country is pitted the rock is commonly crowned with a more or less thick covering of similar materials; while in sinking for coal and ironstone, and in digging foundations for houses and bridges, superficial accumulations of such débris no less frequently occur. Bricks and tiles are manufactured in large quantities from the beds of clay, and the heaps of sand and gravel, occurring as they often do at a great distance from the sea, are much in request by builders, farmers, and others.

So widely are the superficial deposits distributed that they may be said to be common to every part of the country, for they are met with from Zetland to the Cheviots, and from the Outer Hebrides to the east coast. But while they occur over so wide an area they are at the same time very unequally aggregated. In the highland and upland districts they appear to be for the most part restricted to valleys—the craggy broken mountains of the north and the rounded swelling hills of the south of Scotland, showing but little trace of them at the higher elevations. Over the intervening Lowlands, however, they spread in broad but somewhat ragged sheets, which are often continuous across wide tracts.

The materials of which these deposits are made up consist principally of stony clay, fine brick-clay, silt, sand, gravel, and a kind of loose débris of earthy clay and stones. At first sight these various beds appear to be confusedly intermingled, and to show little order in the mode of their occurrence. Sometimes stony clay, at other times sand or gravel, overlies the solid rocks. Again, these deposits may be absent and a fine brick-clay, or a coarse débris of stones and large blocks may cumber the ground instead. But this confusion is only apparent—a regular succession does really exist. It frequently happens that in deep artificial excavations, or natural sections, several varieties of these loose materials occur together. And when this is the case we invariably find that the lowest-lying member of the series consists of a tough stony clay. Above this stony clay, or *till*, as it is called, come beds of sand and gravel, or, as the case may be, a loose earthy débris of stones and large blocks and boulders. But in the neighbourhood of the sea it often happens that the first deposit resting immediately upon the till is fine brick-clay. Thus, whenever the till or stony clay appears in the same section with any of the other superficial deposits, it invariably lies at the bottom. Hence we conclude that of all these deposits the till is the oldest, since it must have been laid down in its present position before the other heaps of material could have gathered over its surface.

It is only now and then, however, that the lowest-lying or oldest superficial accumulations are overlain by later formed deposits. Throughout wide districts stony clay alone occurs, just as in other regions heaps of sand and gravel form the only covering of the solid rocks. Yet we can have no difficulty in deciding as to the relative age of the beds; for having already satisfied ourselves that the till constantly underlies the other deposits, when all occur together in one section, we can have no doubt that the former must be the older accumulation, and that the latter, even when they rest directly upon rock (the stony clay being altogether absent from the district) must have been formed at a later date. A reference to the accompanying diagram (Fig. 1), which represents an ideal cutting or *section*, will help to render these remarks a little clearer. The figure is intended to give a general view of the relation of the underlying till to the overlying sand and gravel series. In this section, *t S* are the superficial deposits resting upon *r r*, the solid rocks. It will be observed that the stony clay or till, *t*, is distinctly covered by the sand and gravel *S*. At $S^{\times}$ the sand and gravel repose directly upon the rocks *r*, the till being absent at that point; while at $t^{\times}$ till alone occurs. When the superficial formations are viewed upon the large scale they are invariably found to follow the order indicated.

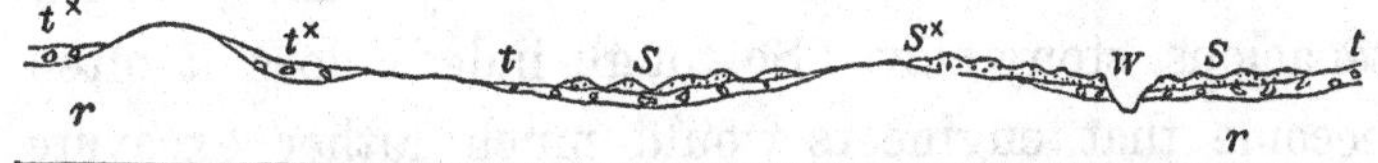

Fig. 1.—Diagrammatic section, showing relative position of till t $t\times$ and overlying sand and gravel series S $S\times$. *W.* = river valley.

It has already been mentioned that besides sand and gravel various other kinds of materials sometimes overlie the till. To determine the relative position of these accumulations the same kind of reasoning applies. There are other methods, however, by which this is ascertained, but a consideration of these must be deferred to a subsequent page. At present our attention is confined to the stony clay or till. This deposit and the overlying beds have received one general name—the Drift or Glacial formation; and it is usual to speak of Lower and Upper Drift, according as we refer to the stony clays or the deposits above them. In this and the following chapter I shall consider the character and phenomena of the Lower Drift, for which purpose it will be necessary to enter into some little detail. But these details are absolutely necessary if the reader would understand clearly the nature of the problem which a survey of the drift-phenomena suggests. It would, however, lead me far beyond due limits were I to attempt to give anything like an exhaustive account of the till. All I shall try to do will be to gather together into a short space what appear to be the more salient points in the evidence, from an attentive consideration of which the reader will be able to judge for himself how far the inferences set forth in the sequel are justified.

The deposit known as *till* is usually a firm, tough, tenacious, stony clay. So tough indeed does it often become that engineers would much rather excavate the most obdurate rocks. Hard rocks are more or less easily assailable with gunpowder, and the numerous joints and fissures by which they are traversed enable the navvies to wedge them out often in considerable lumps. But till has neither crack nor joint—it will not blast, and to pick it to pieces is a very slow and laborious process. Occasionally, however, the clay becomes coarser and sandier, and when this is the case water soaks through it. It then loses consistency, and is ready to "run" or collapse as soon as an excavation is made.

Sometimes the stones in the till are so numerous that hardly any matrix of clay is visible. This, however, does not often happen. On the other hand, they occasionally appear more sparsely scattered through the clay, which may then be dug for brick-making; but this occurs still less frequently.* As a rule, it is hard to say whether the stones or the clay form the larger percentage of the deposit in a mass of typical till. Generally speaking, however, the stones are most numerous in the till of hilly districts; while at the lower levels of the country the clayey character

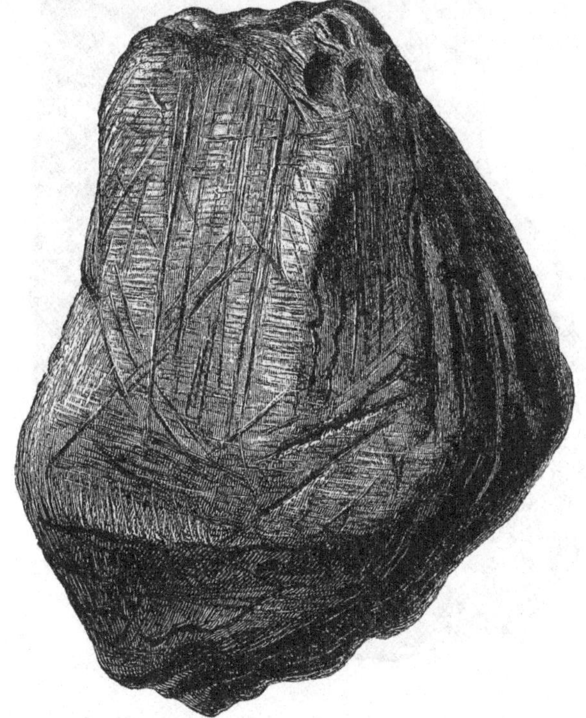

Fig. 2.—Scratched stone (black shale), from the till. (B. N. Peach.)

* My friend Mr. R. L. Jack informs me that at Port Dundas bricks are made out of a typical stony till, the stones being crunched up by a machine.

Fig. 3.—Scratched stone, from the till (limestone). (B. N. Peach.)

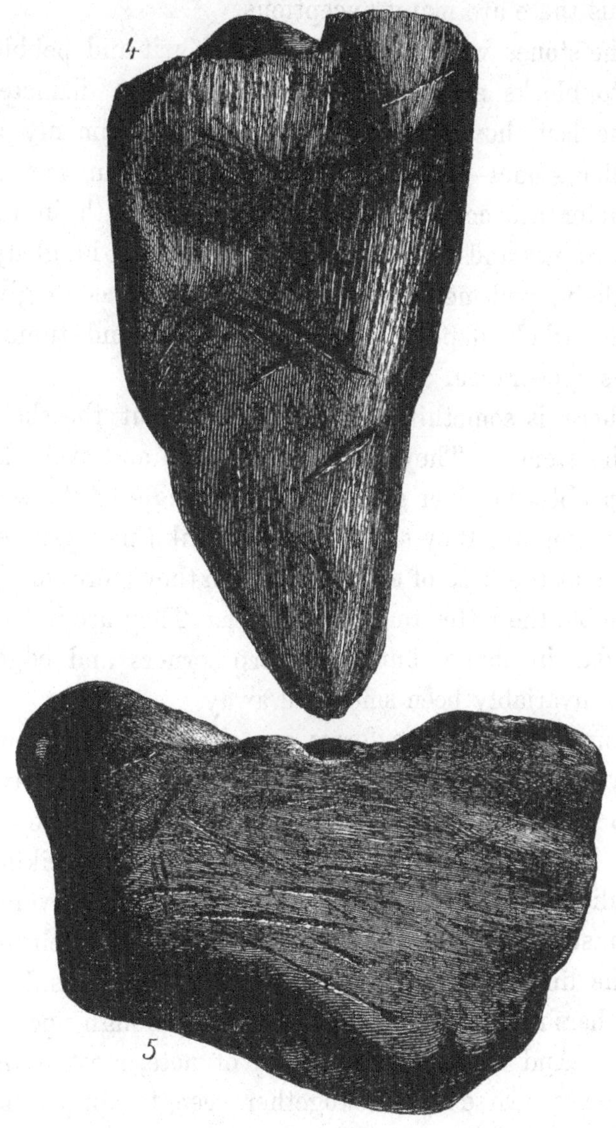

Figs. 4 and 5.—Scratched stones from the till. 4 Clay-ironstone, 5 Limestone. (B. N. Peach.)

of the mass is upon the whole more pronounced. But to this there are many exceptions.

The stones vary in size from mere grit and pebbles up to blocks several feet or even yards in diameter. These last, however, do not occur so commonly as smaller stones—indeed boulders above four feet in diameter are comparatively seldom met with in the till. Stones and boulders alike are scattered higgledy-piggledy, pell-mell through the clay, so as to give to the whole deposit a highly confused and tumultuous appearance.

There is something very peculiar about the shape of the stones. They are neither round and oval like the pebbles in river gravel or the shingle of the sea-shore, nor are they sharply angular like newly-fallen débris at the base of a cliff, although they more closely resemble the latter than the former. They are indeed angular in shape, but the sharp corners and edges have invariably been smoothed away.

Some characteristic forms are shown in the accompanying illustrations (Figs. 2, 3, 4, 5) which have been drawn from actual specimens. Their shape, as will be seen, is by no means their most striking peculiarity. Each is smoothed, polished, and covered with striæ or scratches, some of which are delicate as the lines traced by an etching needle, others deep and harsh as the scores made by the plough upon a rock. And what is also worthy of note, most of the scratches, coarse and fine together, seem to run parallel to the longer diameter of the stones, which, however, are scratched in many other directions as well. These appearances are displayed in Figs. 3, 4, and 5. The distinctness of the markings depends very much upon the nature of the stones. Hard fine-grained rocks like limestone and clay ironstone have often received a high polish, and retain the finest striæ. Figs. 3, 4, 5 represent fragments of these. Soft and coarse-grained rocks, like grit and sandstone, are not usually so well polished and scratched, and the same may be said of other easily-decomposed masses, such as granite, and various igneous rocks. Fig. 2 shows a fragment of dark carboniferous shale; it will be noticed that this specimen is not so distinctly oblong as the others, and that the direction of the striæ does not coincide in so marked a manner with the longer axis of the stone —some of the more strongly pronounced scratches, however, do. No other appearances connected with the till are more striking than these scratchings and smoothings. They become to the geologist what hieroglyphics are to the Ægyptologist—the silent but impressive records of an age long passed away, enabling him to realise the former existence in these islands of a state of things very different indeed to that which now obtains. We must travel to other and distant lands before we can hope to appreciate the full significance of the "scratches," or realise all that they suggest.

The stones which, as we have seen, occur so abundantly in the till consist of fragments of the various kinds of rock to be met with in Scotland. It must not be supposed, however, that any single section of till will yield specimens of all these varieties.

On the contrary, if we desired to collect from the clay a complete set of the Scottish rocks, we should have to traverse an area as wide as that occupied by those rocks themselves. The till, in short, is quite local in character, for in districts where sandstone occurs most abundantly, the stones in the clay likewise consist almost exclusively of sandstone. And, similarly, in regions where hard volcanic rocks prevail, the overlying till is invariably crowded with fragments of the same. Not only the stones, however, but also the colour and texture of the clay itself are influenced by the character of the rocks in whose neighbourhood the till occurs. Thus, in a district where the rocks consist chiefly of dark shales, clays, and thin sandstones, with occasional seams of coal, the overlying till is usually hard and tough, and its prevailing colour a dark dingy grey or dirty blue; while in a region of red sandstone it is tinted red or brown, and commonly shows a more open or sandier texture.

Remains of the great woolly elephant, or mammoth, and the reindeer, and fragments of various kinds of trees, such as pine, birch, and oak, have occasionally been found in the till, in which they appear precisely in the same manner as stones or boulders; but not a single trace of any marine organism has yet been detected in true till.

It is in the lower-lying districts of the country where the till appears in greatest force. Wide areas of the central counties are covered with it continuously to a depth varying from two or three feet up to one hundred feet and more. But as we follow it towards the mountain regions it becomes thinner and more interrupted—the naked rock ever and anon peering through, until at last we find only a few shreds and patches lying here and there in sheltered hollows of the hills. Throughout the Northern Highlands it occurs but rarely, and only in little isolated patches. It is not until we get away from the steep rocky declivities and narrow glens and gorges, and enter upon the broader valleys that open out from the base of the highland mountains to the low-lying districts beyond, that we meet with any considerable deposits of stony clay. The higher districts of the Southern Uplands are almost equally free from any covering of till. Occasionally, however, this deposit puts in a bold appearance in certain hilly regions, as, for instance, in many of the valleys of Peeblesshire, Galloway, and the Border counties, where its aspect is often highly suggestive of its origin. In the localities referred to it frequently forms a more or less broad terrace, sloping gradually with the inclination of the valley in which it occurs. Through this terrace a stream usually cuts a course for itself, and by winding from bank to bank gradually undermines the till, and in some cases has nearly succeeded in

Fig. 6.—Diagrammatic section across two upland valleys: t = till; W^1 W^2 = stream courses; r solid rock.

clearing it away altogether. A transverse section across two such valleys is given in the diagram annexed (Fig. 6). The cuttings made by the streams are seen at W^1 W^2; t represents the till, and r the underlying rocks. Fig. 7 gives a pictorial representation of the same appearances.

Fig. 7.—Greskin Burn, Dumfriesshire: stream cutting through terrace of till. (H. M. Skae.)

In the lowland districts the till never shows this terrace-like appearance, but rolls itself out in a series of broad undulations. It is especially worthy of note, too, that the long parallel ridges, or "sowbacks" and "drums," as they are termed, which are the characteristic forms assumed by the till in broad valleys like those of the Tweed and the Nith, invariably coincide in direction with the valleys or straths in which they lie. A section, therefore, drawn across a till-covered area in the Lowlands would give the general aspect represented in the diagram (Fig. 8), where the ridges of stony clay are seen at *t d*.

Throughout the midland districts a number of

prominent crags and bosses of rock project beyond the general surface of the ground in such a way as to

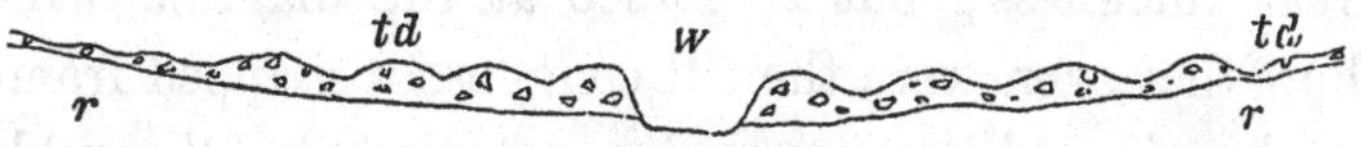

Fig. 8.—Diagrammatic section across a lowland valley: *td* = "sowbacks" or "drums;" *W* = river course; *r* solid rock.

form conspicuous features in the landscape. Behind these crags the till often accumulates to a considerable depth. Edinburgh Castle Rock affords a good example of this appearance, and Arthur's Seat is another instance on a larger scale. From each of these hills a ridge or tail proceeds with a long gentle slope, and similar appearances characterize the numerous isolated boss-like hills which are scattered up and down the Lowlands. The leading phenomena associated with this "crag and tail" arrangement are shown in the woodcut (Fig. 9). The "crag," *c*, itself is usually

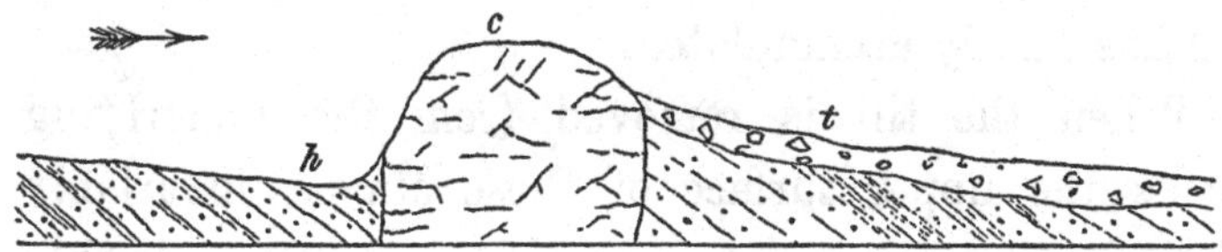

Fig. 9.—"Crag and tail."

composed of some hard volcanic material, like basalt or dolerite,* which intersects softer and more yielding rocks, such as sandstone and shale. At the base of the "crag" there is not infrequently a hollow (seen at *h*) scooped out of the solid rocks. Here also the

* The geological reader will understand that the "crag" is not necessarily formed of intruded igneous rock. The outcrop of any hard bed, that dips with accompanying softer beds at a low angle, has a tendency to produce "crag and tail" when the dip happens to coincide with the direction in which the till has travelled.

till is either very thin or does not occur at all. On the lee-side of the "crag," however, it often attains a great thickness; but a glance at the diagram will show that even were the till completely stripped from the lee-side of the "crag" a notable "tail" would still remain, for the rock on the lee-side is considerably higher than the rock at *h*. The phenomena of "crag and tail" therefore are not entirely dependent upon the presence of the stony clay. They may, and in point of fact do, exist where no till is visible. All that we can affirm in regard to the part played by the latter in the formation of "crag and tail" is simply this—that in districts where till abounds it is usually heaped up on one particular side of projecting crags or bosses. In Edinburghshire, for example, a greater depth occurs on the east than on the west side of prominent isolated hills. In other districts again it may be on the north, west, or south side where the till has chiefly accumulated.

When the till is removed from the underlying rocks the upper surface of these almost invariably shows a smoothed (see Fig. 10) and often highly

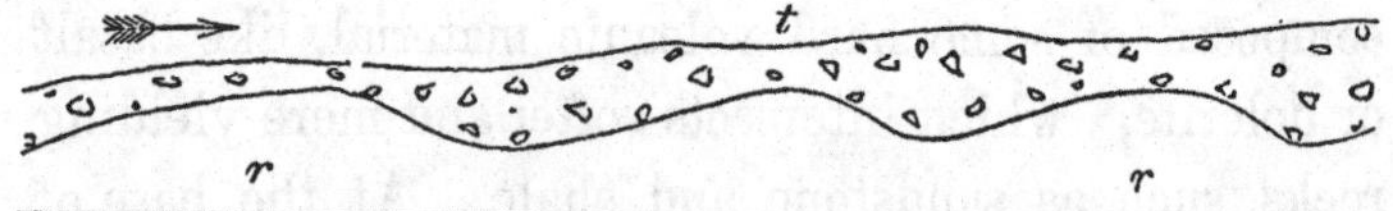

Fig. 10.—Smoothed rock surface below till.

polished appearance, and the whole pavement is marked with those peculiar scratches or striæ that form so characteristic a feature of the stones included in the till. But the extent to which this polishing is carried depends very much upon the nature of the rocks. We have seen that the most perfectly smoothed and striated stones of the till consist of such close-grained rocks as limestone and clay ironstone—the striæ upon fragments of sandstone being often faint and ill-defined. The same rule holds good in regard to the rocky pavement upon which the till reposes. Limestone invariably yields a beautifully smoothed and polished surface, and some other rocks, such as serpentine, receive nearly as fine a dressing as a lapidary's wheel could give them. But soft sandstones and highly-jointed rocks are much less finely marked, and often show a broken and shattered surface below till (Figs. 11, 12).

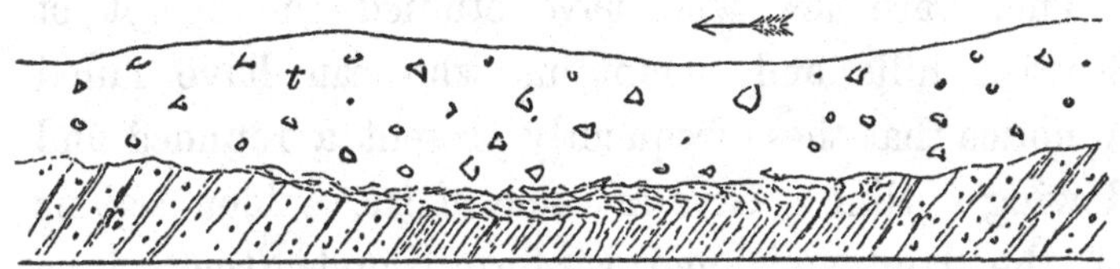

Fig. 11.—Broken rocks below till: near Peebles, river Tweed.

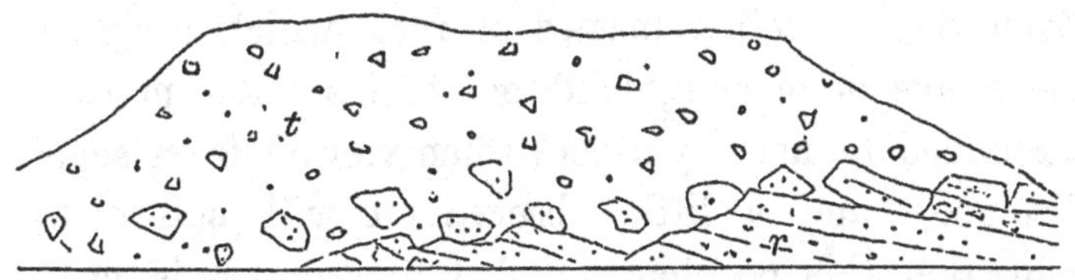

Fig. 12.—Broken sandstone below till: near Union Bridge, river Tweed.

The direction of the scratches, ruts, and grooves upon the rock-head, usually coincides with the trend of the valley in which the till occurs. If, for instance, we took the compass-bearing of a considerable valley running from east to west, we should almost certainly find that the striæ pointed in the same direction. But this rule does not hold true for many of the smaller valleys of the Lowlands. In the Northern Highlands, however, and in the Southern Uplands, the striæ constantly follow the same direction as the valleys, and thus radiate from the high-grounds to almost every point in the compass. In the intervening Lowlands, their prevailing course is from east and north-east to west and south-west, and from west to east, according as they occur in the western or midland and eastern districts. But a fuller account of these wonderful markings will be better appreciated when we come to discuss the origin of the till.

There are few who have studied the aspect of Scottish hills and mountains who can have failed to notice that these frequently present a rounded and flowing outline. Save in some of the wilder regions of the Highlands and western islands, there is a general absence of abrupt sharp peaks and ridges. Even the projecting masses of rock which roughen the flanks of our highland mountains often present a rounded hummocky aspect when viewed from some distance: and a little observation will suffice to show that this peculiar rounded appearance is most pronounced when the slopes of the rugged glens are scanned in a direction coinciding with the inclination of the valleys. Let us take as examples of what is meant, the well-known Glen Rosa in the island of Arran, or Glen Messan in Argyllshire. As the observer advances up these glens he sees nothing

Loch Doon (lower reach): illustrating rounded outline of hills and hill-slopes. (By B. N. Peach.)

To face p. 22.

of the rounded outline referred to. The slopes of the mountains bristle with irregular crags and projecting rock-masses, amongst which we look in vain for any such appearance as I have tried to describe. And so the broken and confusedly tumbled rocks continue until we reach the corries at the heads of the glens. But no sooner do we turn to retrace our steps than the whole aspect of the rocky slopes appears to change, and even the most rugged projecting bosses show as if they had been rounded off by some force pressing over them in a direction down the glens. And so it is with all the other deep valleys and sea-lochs of the Highlands—the smoothed and rounded rocks look up the glens, the broken and jagged masses face down.

In describing the till I remarked that the irregular manner in which the stones were scattered through that deposit imparted to it a confused and tumultuous appearance. The clay does not arrange itself in layers or beds but is distinctly unstratified. We cannot dig it with greater ease in any one direction. It shows no lines of division, but is a homogeneous mass from top to bottom. Occasionally, however, a layer of stones or boulders stretches horizontally across the clay, and not infrequently nests of sand and gravel and irregular lenticular beds of similar deposits make their appearance in some places.

A very good example of the phenomena in question is seen in the accompanying sketch section (Fig. 13). At *l* a line of stones may be observed running along the face of the till; *d* represents earthy sand with stones; *c*, well rounded gravel and shingle. Several little nests of sand are scattered here and there through the body of the clay, which, despite the presence of these and the gravel and layer of

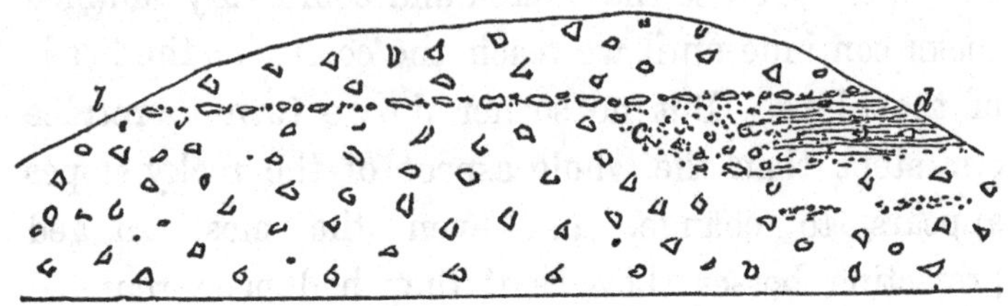

Fig. 13.—Line of stones in till: Glen Water, Ayrshire.

boulders, is a perfectly unstratified and amorphous mass of clay, and angular stones, scratched and polished. But although this is the invariable character of the till itself, it nevertheless contains here and there more or less regular beds of gravel, sand, silt, mud, and brick-clay, and occasionally similar deposits underlie the till and separate that deposit from the subjacent solid rocks. To the appearances presented by some of these intercalated and underlying beds I shall direct attention farther on, when I hope to show that they aid us greatly in our attempts to unravel the geological history of the Drift.

Not wishing to oppress the reader with too many minutiæ on the very threshold of his investigations, I have omitted from the preceding brief account of the stony clay many interesting details, which, however, will come before our attention after we have gained some insight into the mystery of the till itself. But before proceeding to consider some of the theories which have been advanced to explain the origin of this deposit, it may be well to recapitulate the principal facts brought forward in the preceding pages. These may be briefly stated as follows:—

1. The loose accumulations of gravel, sand, clay, and other materials that cover so large a portion of the country, belong, for the most part, to what is termed the *Drift* or *Glacial* formation.

2. This formation is divisible into at least two series, namely, *Upper Drift* and *Lower Drift*—the latter being the older series of the two.

3. The most characteristic member of the lower drift is a more or less tough unstratified stony clay, called *till.*

4. The stones in this till are scattered confusedly through the clay, and are not arranged with any reference to their relative specific gravity.

5. They are invariably of a blunted angular and subangular shape, and exhibit smoothed, polished, and striated surfaces.

6. They are generally local in character, that is to say, the till of a sandstone district is charged chiefly with fragments of sandstone; and with boulders of volcanic rocks in a region composed for the most part of rocks of igneous origin.

7. A similar connection obtains between the colour and texture of the clay and the prevalent character of the rocks:—in a red sandstone district the till is red and sandy; in a coal and black shale district

it is of a dark dingy grey or blue, and usually excessively tough.

8. Till lies thickest in the Lowlands, and thins away as it approaches the hills.

9. In upland valleys it has a tendency to assume a terrace-like aspect, while in lowland tracts it shows a broad undulating surface, and is frequently arranged in long round-backed ridges, which are parallel to one another and the direction of the principal valley of the district.

10. Till is often heaped up on one particular side of prominent hills and rocks, especially in the Lowlands.

11. The pavement of rock below till is frequently smoothed, polished, and striated, more especially if the rock be hard and fine grained; softer rocks, like sandstone and shale, and highly-jointed rocks like greywacké, and many igneous masses, often show a broken surface below till.

12. The direction of the rock-striations coincides generally with the trend of the principal valleys.

13. The hills and mountains commonly exhibit a more or less rounded and flowing outline—the rounded appearance of the crags upon a mountain-slope being most conspicuous to an observer looking down the glens. Viewed in the opposite direction, the rounded and smooth outline disappears.

14. Till, while itself quite unstratified, yet shows an occasional line of stones or boulders, and a few irregular nests and patches of sand and gravel.

CHAPTER III.

EARLY THEORIES.—CAUSE OF GLACIER MOTION.

Early theories accounting for origin of the till.—Debacles and waves of translation.—Glacier theory of Agassiz.—Iceberg theory.—Accumulation of snow above snow-line, how relieved.—Evaporation, avalanches, and glaciers.—Origin of glaciers.—Theories of Forbes and Tyndall.—Canon Moseley's experiments.

IN the infancy of the science, or, perhaps, some little time before that, a number of wonderful "theories of the earth" and "histories of creation" were ushered into the world, some of which continued to be popular long after their absurdities had been exposed. These curious productions gave full rein to the imagination; they abounded with striking and often terrible pictures, and may still be read with a certain degree of interest. But even after geology had made some progress, and an appeal to nature was recognised as the only safe guide to the interpretation of the facts that began to accumulate, ardent and poetical minds were occasionally too prone to supply the missing links in a chain of evidence by drawing upon the fancy, and when no sufficient "natural cause" could be discerned, imagination, of course, was bound to supply one. The theories which from time to time have been advanced in explanation of the phenomena of the drift, afford copious illustrations of the truth of these remarks. It would be a great mistake, however, to suppose that an erroneous theory is always or even often wholly pernicious; for, not infrequently, it does good service by warning the student into closer and more continuous observation, and thus opens up new channels of discovery which otherwise might have remained for a longer time unsuspected and unknown.

To describe the different theories which have appeared and disappeared, ever since the superficial deposits began to attract the attention of geologists, would be a thankless and perhaps a useless task. One of these theories conceived that somehow and somewhere in the far north a series of gigantic waves was mysteriously propagated. These waves were supposed to have precipitated themselves upon the land, and then swept madly on over mountain and valley alike, carrying along with them a mighty burden of rocks and stones and rubbish. Such deluges were styled "waves of translation," and the till was believed to represent the materials which they hurried along with them in their wild course across the country. The stones and boulders as they rattled down one mountain's side and up another, were smoothed and scratched, and the solid rocks over which they careered received a similar kind of treatment. After the water disappeared the stones were found confusedly huddled together in a paste of clay, but occasional big blocks, often far-travelled, might be seen perched in lonely positions on hill-tops, or scattered over hill-slopes like flocks of sheep; while mounds of sand and gravel rolled themselves out here and there in the Lowlands. Such in a few words was the theory of great debacles or waves of translation. It was unfortunate for this theory that it violated at the very outset the first principles of the science, by assuming the former existence of a cause which there was little in nature to warrant. Large waves it was known had certainly been raised by sudden movements of the earth's crust, and had several times caused great damage to seaport towns; but gigantic rushes of water across a whole country had fortunately never been experienced within the memory of man. The theory, indeed, had only been advanced in a kind of desperation, for geologists were quite at their wits' end to discover any natural cause that would account for the peculiar phenomena of the drift—they could see neither rocks being rounded and scratched nor till being formed anywhere—the causes to which these were due had apparently ceased to operate—and so the perplexed philosophers were in a manner compelled to fall back upon waves of translation. They did not, however, attempt to uphold this theory very strenuously. It was felt that, even granting the probability of great waves of translation having swept across the land, still a number of facts remained which could neither be accounted for nor yet explained away. There was the scratching and polishing of the stones for instance. It seemed impossible that the stones could have been so dressed by running water, no matter how rapidly it flowed. The stones in the bed of a mountain-torrent show nothing like the scratches so characteristic of those in the till. They are water-worn, truly, but the mysterious markings that streak the till-stones from end to end are nowhere visible. There was also great difficulty in conceiving how large fragments of rock could be carried from the mountains of the Highlands to the south of Scotland by rushes of water. The water might roll them down from one hill into a valley, but it could hardly push them up another hill, and so repeat the process often in a distance of many miles. Then, again, how did the stones come to be intermingled with fine clay—surely the current or wave that was sufficiently powerful to force along blocks of stone several feet in diameter, must have swept the finer matter—sand and clay—to infinitely greater distances. Thus there were a great many loose screws in the theory, and every new fact discovered threatened to make it collapse altogether. Day by day the great waves of translation became more and more apocryphal, until at last they ceased even to be hinted at, and sunk quietly to rest for ever.

Next succeeded the glacier theory of Agassiz, of which more anon, and this again was followed by the iceberg hypothesis. The latter accounts in a natural way for the transport of large blocks and boulders from one part of the country to another. It supposes the land to have been submerged to a certain depth during the accumulation of the drift, at which time icebergs setting sail from the tops of the mountains, which then existed as frozen islets, carried with them loads of earth, sand, and rock, which they scattered over the bed of the sea as they floated on their way to the south. This theory is still upheld by some geologists as a satisfactory explanation of the origin of the till, but, as I shall endeavour by-and-by to show, it does not discriminate between deposits which have been accumulated at different times and under different conditions. For the present, however, we must leave it; but the phenomena of icebergs and what these floating masses are capable of doing, will come before our attention in another place.

The older observers had clearly shown that the agents of geological change as represented in this country failed to supply an adequate solution to the enigma of the mysterious till, with its scratched stones. But the phenomena of glacier regions threw a new light on the subject. Geologists, following the lead of Agassiz, gradually became convinced that in some way or other ice had to do with the origin of the till; and this conviction deepened as the effects of glacial action became better known. I do not suppose that there is any one nowadays who has given this matter the slightest consideration without coming to a firm belief in the ice-origin of the till. But as regards the precise way in which that stony clay owes its origin to ice, there is still some dispute; and what I propose now to do is to go through the evidence bit by bit for the purpose of ascertaining whether it is not possible to come to a definite conclusion. It is quite clear that no theory can be considered satisfactory which does not explain and account for the various appearances detailed in the preceding chapter. No half-explanation will suffice; the key which we obtain must open a way into every obscure hole and corner; each and every fact must have full recognition in the theory which may be ultimately adopted. Our first duty, then, must be to find out whether the process of dressing rock-surfaces and smoothing and polishing rock-fragments is now in operation, and whether at present any deposits like our till are being formed and accumulated. I have just said that geologists agree in ascribing the origin of these and other phenomena of the drift to glacial action. What then, let us inquire, is the nature of that action, and how do the appearances presented by ice-covered regions serve to explain the origin of the stony clay or till?

Every one is aware that the watery vapour which the heat of the sun sucks up from lake and stream and sea, by-and-by condenses and returns to earth. At certain elevations it almost invariably comes down in the form of *snow;* at lower levels it falls as *rain.* It is customary to speak of an imaginary line, called the *snow-line*, above which more snow falls than is melted or evaporated. This line rises to a great elevation in tropical countries; but as we follow it to north and south it begins gradually to descend, until in the icy regions about the poles it drops to the level of the sea.* The

* This is not exactly true for arctic regions. Even in the highest north latitudes hitherto reached, it would seem that the short summer is sufficiently warm to melt the snow away from the land immediately bordering on the sea.

line, however, is very far from forming an equal curve from the equator to the poles. Many circumstances conspire to render it most irregular and often rapidly undulating. If the winds that blow across a mountainous country convey much moisture, the snow will descend to a lower level than it would were these winds less charged with moisture. Hence it frequently happens that the snow-line actually rises as it recedes from the equator. The Cordillera of Mendoza (22° S. lat.), for example, is snow-clad at a height of 13,200 ft., while the Sierra Famatina, which is four degrees further to the north, shows no snow at a height of 14,764 ft. This difference is no doubt due to extensive radiation and the relative dryness of the atmosphere. It is for the same causes that the limit of permanent snow reaches a lower level on the southern than it does on the northern slopes of the Himalaya. The winds that sweep up from India are laden with moisture, while those that blow from the north have been sucked dry by the heated plains of Central Asia. All the moisture precipitated above the ideal line takes the form of snow; but there are limits to its vertical increase—it does not grow in thickness to an indefinite extent. The hill-tops are being constantly relieved of a portion of their wintry mantle by evaporation, and occasionally great shreds of this mantle suddenly drop away, and disappear down the steeper hill-slopes, carrying ruin and desolation with them when they chance to reach cultivated ground. But neither evaporation nor the constant discharge of avalanches suffices to drain the mountains of their vast reservoirs of snow. This is effected by a regular system of ice-rivers or *glaciers*. Enthusiastic physicists have described these wonders of nature in endless volumes, memoirs, papers, and pamphlets, and each observer has vied with his predecessor in attempting some novel or fuller explanation of their phenomena. One might think it a cold subject enough to discuss, but the records of science tell another story. Many a warm word, many a hot fight, have these ice-rivers occasioned; and the battle still goes on, though less briskly now than it did some years ago. Even the briefest account of the various theories which have been advanced from time to time in explanation of the phenomena of glaciers would occupy more space than I can afford. I must therefore content myself by merely recapitulating what appear to be the general results of the controversy.

Pressure, as every schoolboy knows, will convert a handful of new-fallen snow into a hard lump; and if the pressure be sufficiently severe, the hardened snow will become ice. Over seemingly frail bridges of snow the adventurous mountaineer will traverse yawning clefts and chasms—for the snow, trodden firmly down, becomes stiff and rigid. It is this property of snow which makes a glacier or ice-river possible. When snow has accumulated to a great depth, its own weight squeezes down its lower strata; and should the pressure of the overlying mass be sufficiently great, the under portions of the snow will eventually be compacted together into true ice. The passage of snow into ice is simultaneously carried on, but to a very limited extent, by alternate thawings and freezings. When the sun shines upon the snow, or a warm wind passes over its surface, the upper layer partially melts, and the disengaged water trickles down drop by drop into the layers below, where it solidifies. Every spring and summer this process is repeated with the last winter's snow, and thus partly, by thawing and freezing, but chiefly by pressure, the whole mass tends to harden into ice.

Thus solidified and apparently rigid, one would at first suppose that the hardened snow or ice would be as immovable as the rock of the mountain upon which it reclined. We know that a bed of tough clay will rest upon a very considerable slope without sliding downward, and even the loose stones and débris which cover so many hill-sides in a highland country find repose upon an incline of 30°. At Fourneaux the débris shot out from the mouth of the great Cenis tunnel forms a still steeper slope. Mr. Whymper tells us that "its faces have as nearly as possible an angle of 45°."* But ice, which is a much more rigid body than even the hardest clay, will move upon a slope that is inappreciable by the eye. And thus it happens that wherever snow attains a sufficient thickness to compress itself into ice, this ice, as the overlying mass continues to receive the tribute of each winter, begins to steal down every slope, however gentle the incline may be. Thus it is that in alpine countries the valleys become more or less filled with streams and rivers of ice, which are constantly fed from the snow-

* *Scrambles amongst the Alps*, p. 62.

fields above. The snow, packed and pressed into ice, gradually moves off the mountains, and the frozen streams thus formed collect at the head of every valley, down which they pour in those great solid masses which are known as glaciers.

Why, then, does this hard body behave so differently from other solid substances? What property does ice possess which enables it to creep upon slopes adown which only fluids and semi-fluids can move?

James David Forbes, whose works on glacial phenomena rank among the classics of physical science, came to the conclusion that the motion of a glacier was due to the viscous or plastic nature of the ice, which moved upon a slope much in the same way as a substance of the consistency of treacle or tar, its own weight being sufficient to urge it forward. When the ice was exposed in the glacier to a peculiarly violent strain, its limited plasticity necessitated the formation of an infinity of minute rents, and the internally bruised surfaces were forced to slide over one another, still producing a quasi-fluid character in the motion of the whole. The same eminent physicist further held that reconsolidation of the bruised glacial substance into a coherent whole might be effected by pressure alone acting upon granular snow, or upon ice softened by imminent thaw into a condition more plastic than ice of low temperature.*

At a later date Professor Tyndall proposed a different explanation of the phenomena. The lamented Faraday had previously pointed out that when two

* *Occasional Papers on the Theory of Glaciers*, p. xvi.

pieces of ice with moistened surfaces are brought into contact they immediately freeze together; and the same phenomenon took place, according to Dr. Tyndall, when a piece of ice (at 32°) was subjected to pressure and compelled to change its shape. Under such conditions the ice is broken by the pressure into fragments; but these fragments being in close contact, immediately re-unite to form a solid mass. If this be true of small pieces of ice, it must be equally true of large masses; and Dr. Tyndall therefore concluded that the motion of a glacier was due to "fracture and regelation." Under the pressure of the superincumbent snow and ice, fissures were everywhere forming and closing again; and this rupturing and healing process extended to the smallest particles of the frozen mass. Its own weight crushed, squeezed, and pushed forward the ice, which seemed to flow with a viscous motion, owing simply to the fracture and regelation of all its particles.*

Quite recently, however, certain interesting experiments made by the late Canon Moseley suggest some difficulties which are hard to overcome. He proved conclusively that the mere weight of a mass of ice is not sufficient to cause that ice to move upon a moderate slope. Forbes' theory of viscosity, and Tyndall's

* A different explanation of the phenomenon of regelation under pressure has been given by Professor James Thomson, and his brother, Sir William Thomson. These physicists ascribe the consolidation of the ice crushed in a mould under Bramah's press to simultaneous liquefaction, which commences at every point of the interior of the ice to which the pressure extends, and to subsequent solidification when the pressure is removed. *Proceedings of the Royal Society of London*, 1857, 1858. See also Professor Tait's remarks in *Life and Letters of Principal Forbes.*

theory of fracture and regelation, alike imply that ice descends a slope by virtue of its weight alone. But if ice always remains throughout its entire mass a solid body, it is evident, as Canon Moseley has shown, that the resistance offered by the ice cannot possibly be overcome by the mere weight of the mass. Besides the weight there must be some other force pushing the ice forward. A substance of the consistence of clay, which is a much less rigid body than ice, will nevertheless remain perfectly inert upon a slope adown which ice would move with the greatest ease. The clay cannot descend by gravitation, and if ice were the solid mass which it is commonly supposed to be, it would be just as impossible for it as for a mass of clay to descend a slight incline by gravitation. Yet we know that ice, hard and brittle though it be, does nevertheless move upon such a slope. This movement, however, is not effected by the sliding of the ice over its bed; a glacier does not slip down its valley as a disengaged slate or slab of stone would slide off the roof of a house. "All the parts of a glacier," as Canon Moseley has remarked, "do not descend with a common motion; it moves faster at its surface than deeper down, and at the centre of its surface than at its edges. It does not only come down bodily, but with different motions of its different parts; so that, if a transverse section were made through it, the ice would be found to be moving differently at every point of that section. . . . There is a constant displacement of the particles of the ice over one another and alongside one another, to which is opposed that force of resistance which is known in mechanics as *shearing-force*." Now the question comes to be, what power is it that compels the ice-particles so to move? The motion of fluids and semi-fluids upon a slope is undoubtedly due to gravitation; can the movement of a glacier be owing to the same cause? If so, then how and in what manner can the glacier assume the properties of a fluid?

Mr. J. Croll has suggested an explanation of the difficulty, which is so ingenious, and appears to me so suggestive, that I cannot deny myself the pleasure of attempting to describe it, which I shall proceed to do in the following chapter.

CHAPTER IV.

MR. J. CROLL'S THEORY OF GLACIER MOTION.—ASPECT OF AN ALPINE GLACIER.

Heat capable of being transmitted through ice.—Temperature of glaciers.—Motion of ice molecules in a glacier.—Regelation.—Alpine glaciers.—River-like character of glaciers.—Crevasses.—Size of glaciers.—Character of the rock-surface and stones below a glacier.—Glacial rivers.—Moraines.—Waste of mountains.

WATER in the act of freezing gives out heat, and while passing into the solid state expands; so that ice occupies, weight for weight, a greater space than water. On the other hand, when ice melts it absorbs heat, and upon becoming water takes up less room than it did in the crystallized or solid state. Bearing this in mind, let us suppose that the sun shines upon a glacier, and imparts to its surface a certain amount of heat. The application of this heat will not and cannot possibly raise the temperature of the ice (taking that temperature at 32°), but its immediate consequence will be to melt or turn a certain portion of the ice into water. When this is done, a fresh surface of ice will of course be exposed to the same action, and were the process to be continued long enough it is obvious that the glacier would by-and-by disappear altogether. But as soon as the sun ceases to shine upon a glacier, the water which has been set free will freeze again. Meanwhile, however, it will have shifted its position. The force of gravity compelled it while in the liquid state to flow from a higher to a lower level; and now, when it becomes ice once more, it necessarily rests at some distance below its former position. But what is true of the surface of a glacier must also to some extent be true of every portion of the frozen mass.

We know that heat can be transmitted through a slab of ice at 32°, and yet the ice as a mass will still continue solid. This curious result is thus explained. When heat is applied to the surface of the slab it cannot by any possibility raise the temperature of the ice. The molecules at the surface merely pass from the solid or crystalline state into the fluid condition. The heat is therefore converted into a certain amount of energy which enables the molecules to overcome their tendency to assume the crystalline form. The moment the molecules are allowed to crystallize they give back in the form of heat the energy which had forced them asunder; or, to put it more simply, heat is given out in the act of freezing. The same process is repeated with the next adjoining molecules. No sooner have their neighbours resumed the crystalline condition, than the heat given out instantaneously attacks the others, and forces them to melt. "That peculiar form of motion or energy called heat disappears in forcing the particles of the crystalline molecule separate, and for the time being exists in the form of a tendency in the separated particles to come together again. But it must be observed that although the crystalline molecule when it is acting as a conductor takes on energy under this form from the heated body, yet it only exists in the molecule under such a form during the moment of transmission—that is to say, the molecule is melted, but only for the moment. When B accepts of the energy from A, the molecule A instantly assumes the crystalline form. B is now melted, and when C accepts of the energy from B, then B also in turn assumes the solid state. This process goes on from molecule to molecule, till the energy is transmitted through to the opposite side, and the ice is left in its original solid state."*

Such are the phenomena that take place during the transmission of heat through ice. Let us now see what light they throw on the origin of the motion of glaciers.

The numerous experiments made by Agassiz lead to the conclusion that the ice of glaciers has a comparatively high temperature. He found, for example, that in a hole sunk to a depth of 200 ft. in solid ice, the thermometer gave an average temperature of 31° 24′ Fahr. In winter-time, however, the thermometer marked 28° 24′ Fahr. This, however, appears to have been a somewhat exceptional temperature. At all events the experiments clearly proved that in winter the ice at the heart of a glacier is not nearly so cold as the external atmosphere, while in summer

* "On the Physical Cause of the Motion of Glaciers," *Philosophical Magazine*, March, 1869, and Sept., 1870.

the mass of the glacier possesses as near as may be a temperature of 32° Fahr.

The heat that acts upon a glacier comes from various sources. There is first the heat received directly from the sun; then, in spring and summer warm winds will sometimes blow across it, and mild rains will fall upon its surface. Again, there is the heat of the rocky cradle in which it lies, and, not least, that which is derived from the friction of the ice and stones upon its bed as the glacier slowly creeps away. In summer-time the numerous small cracks and fissures that penetrate the ice in all directions are filled with water which has trickled down from the surface. In this manner heat is rapidly conveyed from one portion of the glacier to another. And just as we found that heat when applied to a slab of ice at 32° will be transmitted through that slab without the ice as a mass changing its condition; so must heat in the same way pass from the various sources mentioned right through the solid ice of the glacier. But in doing so it will compel the icy mass to change its position. When a molecule of glacier ice, A, has momentarily melted owing to the reception of energy in the form of heat, it will by virtue of its fluidity tend to move downwards, but the instant that it resumes the solid state the heat evolved will immediately attack the molecule behind it, B. This molecule will then melt and contract in bulk, and will descend by its own gravity until it stops against A. Here then it will freeze again. But as A now occupies a lower level than it did before the heat dissolved it, so B in like manner solidifies a little below its former position. The heat emitted by B at the moment of its passing into ice is next transmitted to C, whereupon the same changes ensue as before, and so on throughout the entire mass. Thus, while acting as conductors of heat, each individual molecule of ice is compelled by the force of gravity to move from a higher to a lower level. The result of this general molecular motion is that the whole body of a glacier moves gradually downwards, impelled by gravitation, in the same way as fluids and semifluids are urged forwards upon a slope. As a hard and brittle body, ice could not possibly flow—its own weight would be quite unable to compel the particles of ice to lose their cohesion. It is the transmission of heat through a glacier which by momentarily converting each atom of its substance into water renders motion possible. Could heat be entirely withdrawn the motion of a glacier would be at once arrested, and the ice would remain as inert and immovable a body as any other substance equally hard and unyielding.

Heat is thus the great lever which forces the hard masses of compacted snow and ice from higher to lower levels, and relieves the mountains of their loads of frozen water. If it were not for that peculiar property of ice which enables it to behave in many respects like a viscous or semifluid body, all the waters of the earth, the myriad rivers, and lakes, and seas, would gradually be lifted up by the heat of the sun, and carried on the wings of the wind to the mountains, there to accumulate in vast and constantly growing masses until ocean and all its feeders had been exhausted. But the heat of the sun which, falling upon clay, sand, or solid rock, merely raises the temperature without changing the condition of these masses, pulses through the great piles of ice that cumber the higher elevations of Alpine countries. The temperature of the ice itself cannot rise, but every atom of its bulk is set in motion, and slowly and gradually the solid heaps creep down hill-slope and valley, their progress being accelerated or retarded according to the degree of heat acting upon and passing through them. It is thus that during day the downward motion of the ice is less sluggish than at night; and for the same reason a glacier in summer-time moves more quickly on its way than in winter, when its motion is exceedingly slow, sometimes not reaching to half the summer rate.

The motion of ice, then, being due to the transmission of heat, which, by momentarily converting into water each molecule or atom of the frozen mass in succession, allows these to descend by gravitation, it is quite evident that a body of ice will move down any slope, however gentle the inclination may be. Its motion is precisely that of running water, and hence we need not be surprised to find it stealing slowly down inclinations which are so slight as to appear to the eye like level plains.

But it may be asked how it happens that a body built up of atoms which are constantly passing from the solid to the liquid state, and *vice versâ*, nevertheless retains an unvarying hard and brittle condition. This is due to the property of regelation. When two fragments of ice at 32° are brought into contact, they instantly unite, as we have already seen, so as to form one solid piece. What is thus true of ice in the mass is equally true of its most minute atoms. Whenever these impinge upon one another they immediately freeze firmly together. No sooner does a molecule, which has momentarily melted, resume the solid state than it immediately unites to the other solid particles by which it is surrounded; hence the continuity of the whole is preserved, and throughout its entire bulk the ice remains a solid body. Such is the ingenious theory advanced by Mr. Croll.

We have now learned what are the means which Nature adopts to prevent the increase of snow to an indefinite extent upon the mountains. The under portions of the frozen heaps are gradually squeezed by the superincumbent masses into ice, and this ice thereupon begins to creep outwards and downwards. At the head of an alpine valley streams of ice collect from the contiguous slopes, and, becoming welded together into one mass, creep down the bottom of the valley, forming a glacier or ice river. It depends upon the size of a glacier, whether its journey is to be a long or short one. Often the ice reaches many thousand feet below the limits of permanent snow, amongst which it takes its birth. But sooner or later its progress is at last arrested by the increasing temperature, which melts the ice away as fast as it comes down.

In its course from the regions of perennial snows

down to where it is cut off by the increasing warmth the glacier has many analogies with a stream or river of water. The velocity of a river varies according to the inclination of the valley down which it flows, and this it would seem is equally true of glaciers. When the course of a river is rocky and falls rapidly, its waters are broken, and hurry on tumultuously. In like manner, a glacier that makes its way along rough and rocky valleys shows a broken and tumbled surface. There are waterfalls and ice-cataracts. Again it is known that rivers flow more swiftly at the surface and the middle than they do at the bottom and sides, where the water is retarded by friction: the same is the case with glaciers. The thread of a river's current moves from right to left of a medial line according as the river winds from one side of the valley to the other; the icy current of a glacier follows the same direction. Nay, we may carry the parallel yet further, for the twigs and branches and trunks of trees which drop into a river and float upon its bosom are represented in alpine valleys by the blocks and stones and débris which fall upon a glacier from crag and cliff, and are borne upon its surface down the valley. From its river-like character and the wonderful manner in which it accommodates itself to the form of a valley, narrowing when that contracts to a gorge, and expanding when the valley again widens out, it is no wonder that many eminent physicists have maintained that ice is a viscous body, and not the hard and brittle substance that it seems. When, owing to the nature of its bed, the glacier becomes subjected to strain or tension, however, it is always more or less seamed with gaping cracks or clefts which descend to a great depth in the ice. These *crevasses*, as they are called, have undoubtedly been produced by the snapping of the ice, owing to its inability to resist strain or tension by stretching out. The beginning of a crevasse is often notified by a loud report—the rupturing of the ice. At first the crack may be only wide enough to admit the edge of a knife, but it gradually opens, until frequently it yawns into a wide impassable chasm. The origin of these crevasses is due to the unequal or differential motion of the ice—the glacier moving faster at the surface and the centre than it does at the bottom and the sides, and thus bringing tension into play. Many fissures and crevasses also are caused by the inequalities of the bed over which the glacier flows, for the bottom of a valley does not always, or even often, slope at the same angle throughout: some portions incline more rapidly than others. Now the ice when it reaches in its downward progress the edge of one of these steeper inclines will naturally tend to move more quickly, and to drag forward the ice behind it. At this point then a state of tension is brought about, to which the ice can only yield by snapping asunder. When, afterwards, the ice arrives at a gentler and more equal slope, its motion is at once checked, the gaping crevasses begin to close up, and by-and-by disappear altogether, until finally, under the influence of regelation, the ice becomes solid as before. Thus in its course from the snow-fields to its termination the ice is being continually broken by mechanical strain, and just as constantly these breakages are being repaired by the regelation or freezing together again of the broken faces.

Glaciers, like rivers, are of all sizes. Many have a depth of several hundred feet, and some in polar regions are probably not less than 3,000 or even 5,000 ft. in thickness. It may easily be conceived that the pressure of such enormous masses of ice must have a prodigious effect. When a glacier advances beyond its usual limits everything goes down before it: loose soils and débris are pushed forward, and the strongest and thickest trees are overborne just as if they were so many straws. But striking as these examples of the irresistible force of a glacier may be, the destructive and overwhelming power of ice in motion becomes still more noteworthy when the rocks over which a glacier passes are examined. This may be done in summertime, when the glaciers shrink from the sides of their valleys. Creeping in below the ice, which it is often possible to do for some little distance, we find the rocks finely smoothed and polished, and showing long striæ and ruts, that run parallel to the course followed by the glacier. If we pick out some of the stones that are sure to be scattered about below the ice we shall find that many are smoothed, polished, and striated in the same manner as the surface of the rock itself. All this is the work of the glacier. The rocky precipices and mountains that hem in a glacier are always splitting up under the influence of frost, and tons of rock and rubbish are continually rolling down and gathering upon the surface of the ice below. Much of this débris falls into crevasses, and must no doubt frequently reach the bottom of the ice. The stones then get jammed into the frozen mass, and are pressed against the underlying rock-head with all the weight of the superincumbent glacier. Graving tools must also be supplied to the glacier by the wrenching of fragments from its own rocky bed. These and the stones received from the crevasses, thus firmly held in the grasp of the ice, become potent agents in grinding and scratching the pavement over which they are forced, while the smaller stones and sand and mud that result from the grinding process, complete the smoothing and polishing of the glacier's bed. Could the glacier be removed, we should find the whole bottom of the valley smoothed and polished, and streaked with long parallel ruts. Every high projecting boss would be rounded and dressed on the side that looked up the valley; while the rock on the lee side, sheltered from the attack of ice-plough, and chisel, and graver, would retain all its roughnesses. Smaller and less abrupt knobs of rock would be rounded and polished all over, and every dimple and hollow would be similarly smoothed and dressed.

The finer-grained materials employed by the ice in polishing its bed, the impalpable mud and silt, are carried out from beneath by the stream that issues at the foot of the glacier. In this manner almost all glacier rivers have imparted to them a turbid appearance, the colour of the water depending upon that of the sediment which it holds in suspension. The water may thus be bluish or milky, yellow or dark grey. The stones employed by the glacier in grinding and graving its rocky bed are themselves ground and engraved, and numbers of such smoothed, polished, and striated fragments are pushed out at the foot of the glacier. But they here become intermingled with the vast heaps of débris which have been discharged from the mountains and brought down upon the surface of the slowly moving ice.

These heaps are called *moraines*. They are composed for the most part of rough, unpolished, angular fragments of all sizes, from mere sand and grit up to blocks many tons in weight. As might be expected, they show no trace of any arrangement into beds or layers—large and small stones, huge blocks and angular gravel, grit and boulders, are all confusedly mixed together. The moraines are not scattered irregularly over the whole surface of the ice, save when the glacier is very narrow, but gather chiefly upon the sides, at the base of the mountain-slopes. The glacier in this manner becomes fringed from its origin to its termination with long mounds or bands of débris, which are constantly dropping over the end of the ice, and adding to the immense piles of rubbish collected there. These latter are the *terminal moraines*, those fringing the sides of the glacier being its *lateral moraines*. Some terminal moraines attain a great size, forming mounds of débris several hundred feet in height.

One may see from the size of the moraines how

great must be the waste of the mountains. In all mountainous regions, indeed, the action of frost upon

Fig. 14.—Alpine Glacier. (H. M. Skae.)

the rocks becomes abundantly evident. Even in our own country the tops and slopes of our higher hills are often buried to a great depth with the débris which alternate freezings and thawings have wrenched from the solid rocks. In high latitudes the exposed rocks are almost everywhere broken up in this way. Von Baer found the hills on the west coast of Nova Zembla literally covered with their own wreck—no rock, however hard or fine-grained, being able to withstand the summer moisture and intense winter frost of that desolate country. Mr. Kennan also, in his lively-written "Tent Life in Siberia," tells us how, upon crossing the mountains of Kamchatka, he found the table-land and hill-tops crowded with great square and angular blocks and slabs of rock, which looked for all the world as if they had rained from the skies! They were undoubtedly the ruins of the solid rock which they covered and concealed, and had been detached by frost acting along the natural joints and fissures. In alpine countries this wrecking of the mountains goes on chiefly by day, and in summer-time. During night, and at early morn, dead silence reigns among the snowy peaks: no streams are heard, no water trickles over the surface of the ice; but when the power of the sun begins to be felt, then the noise of water running, leaping, and falling grows upon the ear; soon the glaciers are washed by numberless little streams; great avalanches, wreathed in snow-smoke, rush downwards with a roar like thunder; masses of rock wedged out by the frost of the previous night are now loosened by the sun, and dash headlong down the precipices, while long trains of débris hurry after them, and are scattered far and wide in wild confusion along the flanks of the glaciers.

It is not necessary at the present stage of our inquiry to dwell longer upon the geological phenomena of alpine glaciers. Some additional details will be referred to and described when we come to consider the history of the Upper Drifts.

CHAPTER V.

GREENLAND: ITS GLACIAL ASPECT.

Extent of Greenland.—Character of coast and interior.—The great *mer de glace*.—Size of glaciers.—Phenomena of Arctic glaciers, and origin of icebergs. — Submarine moraines. — Scarcity of surface-moraines in Greenland.—Glacier rivers.—Circulation of water underneath the ice.—The habitable strip of coastland.—Formation of ice upon the sea.—The ice-foot.—Waste of cliffs.—Transportation of rock-débris by icerafts and icebergs.—Action of stranded icebergs upon the sea-bottom.

WE have now acquired some knowledge of facts that bear upon the origin of the Scottish till, but we shall gather yet further aid in our attempts to decipher the history of that deposit by taking a peep at some arctic country. For this purpose we cannot do better than select ice-covered Greenland. That desolate region of the far north, despite the bleak and barren aspect of its coasts, and the horrors of the ice-choked seas that must be traversed to reach its more northern shores, has nevertheless been frequently visited by daring navigators, who have pushed their investigations many hundred miles north of the Danish settlements. The accounts which they give are chiefly taken up with descriptions of the wild ice-bound coast of Greenland, few attempts having been made to penetrate into the interior.

But that cannot be said to be altogether a terra incognita, for, although it has never been, and probably never will be, traversed, yet enough is known to leave us in little doubt as to the general character of these unvisited desolations.

The western shores of Greenland have been traced northwards from Cape Farewell in the latitude of the Shetland Islands, to beyond the 80th parallel. The eastern and north-eastern coasts have not been so continuously followed, but our knowledge of these has been considerably increased during recent years, thanks to the exertions of German and Swedish geographers. The superficial area of Greenland cannot be less than 750,000 square miles, so that the country is almost continental in its dimensions. Of this great region, only a little strip extending to 74° north lat. along the western shore is sparsely colonised—all the rest is a bleak wilderness of snow and ice and rock.

The coasts are deeply indented with numerous bays and fiords or firths, which, when traced inland, are almost invariably found to terminate against glaciers. Thick ice frequently appears, too, crowning the exposed sea-cliffs, from the edges of which it droops in thick tongue-like and stalactitic projections, until its own weight forces it to break away and topple down the precipices into the sea. The whole interior of the country, indeed, would appear to be buried underneath a great depth of snow and ice, which levels up the valleys and sweeps over the hills. The few daring men who have tried to penetrate a little way inland from the coast, describe the scene as desolate in the extreme—far as eye can reach nothing save one dead dreary expanse of white. No living creature frequents this wilderness—neither bird, nor beast, nor insect—not even a solitary moss or lichen can be seen. Over everything broods a silence deep as death, broken only when the roaring storm arises to sweep before it the pitiless blinding snow.

But even in the silent and pathless desolations of central Greenland the forces of nature are continuously at work. The vast masses of snow and ice that seem to wrap the hills and valleys as with an everlasting garment, are nevertheless constantly wearing away, and being just as continuously repaired. The peculiar properties of ice that prevent it accumulating upon the land to an indefinite degree, are just as characteristic of the snow-fields of Greenland as of those of alpine countries. Fast as the snows deepen and harden into ice upon the bleak wilds of Greenland, that ice creeps away to the coast, and thus from the frozen reservoirs of the interior, innumerable glaciers pour themselves down every fiord and opening to the sea. Only a narrow strip of land along the coast-line is left uncovered by the permanent snow-field or mer de glace—all else is snow and ice.

Some of the glaciers attain a vast size. The great Humboldt glacier is said by its discoverer, Dr. Kane, to have a breadth of 60 miles at its termination. Its seaward face rises abruptly from the level of the

PLATE II.

Greenland Glacier. (By H. M. Skae.)

To face p. 56.

water to a height of 300 feet, but to what depth it descends is not known. Other glaciers of large size occur frequently along the whole extent of the north-western shores of Greenland. Among these is that of Eisblink, south of Goodhaab, which projects seaward so as to form a promontory some thirteen miles in length. This immense glacier flows from an unknown distance in the interior, and buries its face to a great depth in the sea. A submarine bank of débris forms a kind of semicircle some little way in front of it, and may owe its origin, in part, to the stream that issues from underneath the glacier, but, as we shall see presently, a bank would necessarily gather in the same place, even although no water whatever circulated below the ice.

I have already remarked that the Greenland glaciers discharge into the sea by fiords and indentations of the coast. If the ice-filled fiords could be cleared out, we should find that these arms of the sea would occupy deep hollows, continuous with long valleys stretching into the heart of the country. The west coasts of Scotland and Norway afford excellent examples of the kind of scenery that Greenland would present were its fiords and valleys to be freed of ice. In Scotland the fiord valleys are watered by streams shed from the hills of the interior, but in not a few of the Norwegian valleys, the streams that enter the fiords, when followed up, are found to issue from glaciers. In North Greenland, however, the ice generally fills up the whole valley, and pushes forward into the sea. Only in a few cases do the glaciers terminate inland and thus give rise to rivers. Yet even when they enter the sea, fresh water continues to escape from underneath the ice, discolouring the sea with the sediment which it sweeps out along with it, and even to some extent diminishing its saltness.

Many arctic glaciers are so thick and massive that they glide boldly on over the bed of the sea, and thus displace the water often for many miles. Instead of the deep fiords being filled with sea-water, as is the case in Scotland and Norway, they are occupied entirely by ice. When the glacier in its downward progress first entered the sea at the head of a fiord, it must have towered for many hundred feet above the level of the water. But as it continued on its course, and crept onward over the deepening bed of the fiord, it gradually buried its lofty face in the waves, until, when it reached the lower end of the fiord and entered the open sea, its front rose only a little height above the reach of the tides. Thus, the sloping platform of ice that faces the sea, however lofty it may be, must bear only a small proportion to the much greater thickness of ice concealed below.

It is well known that ice is not by any means so heavy as water, but readily floats upon its surface. Consequently, whenever a glacier enters the sea, the dense salt water tends to buoy it up. But the great tenacity of the frozen mass enables it to resist the pressure for a time. By-and-by, however, as the glacier reaches deeper water, its cohesion is overcome, and large segments are forced from its terminal front and floated up from the bed of the sea to sail away as *icebergs*. As there is considerable misapprehension about the formation of icebergs, it is necessary to look a little more closely at the facts, because, as we shall afterwards see, they have a strong bearing on the origin of the Scottish till. Some have supposed that the whole mass of the glacier, after it has entered the sea for some little distance, becomes buoyed up, retaining at the same time its continuity with the landward portion of the ice or mer de glace. But from what has already been explained regarding the total inability of ice to yield to mechanical strain in any other way save by breaking, it becomes obvious that the seaward portion of an arctic glacier cannot by any possibility be floated up without sundering its connection with the frozen mass behind. So long as the bulk of the glacier much exceeds the depth of the sea, the ice will of course rest upon the bed of the fiord or bay without being subjected to any strain or tension. But when the glacier creeps outwards to greater depths, then the superior specific gravity of the sea-water will tend to press the ice upward. That ice, however, is a hard, continuous mass, with sufficient cohesion to oppose for a time this pressure, and hence the glacier crawls on to a depth far below the point at which, had it been free, it would have risen to the surface and floated. If at this great depth the whole mass of the glacier could be buoyed up without breaking off, it would certainly go to prove that the ice of arctic regions, unlike ice anywhere else, had the property of yielding to mechanical strain without rupturing. But the great tension to which it is subjected takes effect in the usual way, and the ice yields, not by bending and stretching, but by breaking. The diagram (Fig. 15) will give a clearer notion of the relation

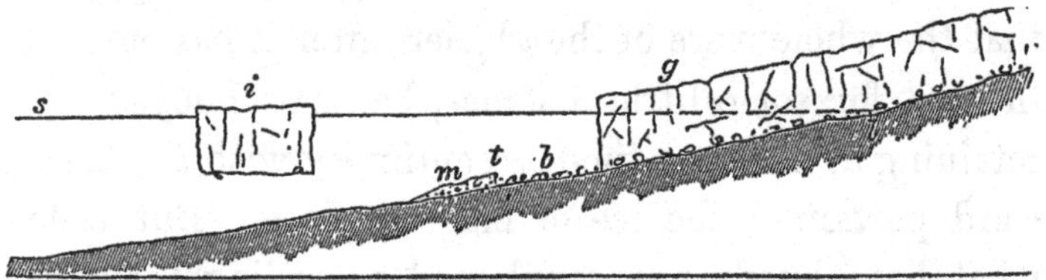

Fig. 15.—Greenland glacier shedding an iceberg.

which an arctic glacier bears to its rocky bed and the sea which it displaces. Let *s* represent the level of the sea. The glacier *g* enters the sea and creeps on until it reaches a point *t*, at which the pressure of the water overcomes the cohesion of the ice, and forces off a large segment from the front of the glacier. This segment then floats away as an iceberg *i*, with probably some stones frozen into its bottom. The débris underlying the ice will now be partially exposed at *b*—the place formerly occupied by the iceberg. After a time the glacier will again advance to *t*, pushing before it some portion of the débris seen at *b*. By the successive shedding of icebergs and the continuous advance of the glacier, a mound of rubbish will tend to collect at *m*—the materials of which will probably be partially arranged by tides and any streams of fresh water that may happen to issue from the bottom of the glacier.

A few stones may occasionally remain frozen into the bottom of the detached icebergs, but it is evident that the greater portion of the sub-glacial deposit must remain upon the bottom of the sea. The existence of such frontal or terminal submarine moraines is not merely hypothetical. They are well known to occur—the Tallert Bank in front of the glacier of Eisblink being a very good example.

Thus, from its origin in the "central silence" of Greenland, to its termination in the sea, the glacier clings pertinaciously to its bed. It nowhere floats so as to allow the sea to get in below, but when the pressure of the water becomes too much for it, immense fragments break away, and rise to the surface, causing the sea sometimes to "boil like a pot."

But before attempting to describe the phenomena connected with floating ice, I have still to glance at some of the appearances exhibited by the ice upon the land.

In general appearance the glaciers of Greenland do not differ, save in size, from those of other countries. When the bed of an arctic glacier is rough and irregular, the ice becomes intensely broken and crevassed, just as we saw was the case with the smaller ice-rivers of the Alps. The arctic glaciers are also in summer-time washed by innumerable streamlets due to the melting of the ice; it is only occasionally, however, that any scattered stones and débris appear upon their surface. This is owing to the fact that the whole interior of the country is so effectually concealed beneath its coat of snow and ice that no bare rocky slopes from which fragments might be detached are left exposed to the action of the frost. All the inland valleys appear to be filled up and levelled to the tops of the hills, only the extreme tips of which appear here and there above the bleak wastes of the mer de glace. Hence there is well-nigh a total absence of those long trains of débris* that thunder down the steeps of the alpine mountains, and gather in heaps along the sides of the glaciers. It is not until the glaciers of Greenland descend to the sea-coast, where the cliffs and mountains that overlook them are more exposed to the action of the weather, that they begin to receive a goodly tribute of blocks and boulders. But the fiord valleys in which the glaciers lie, are in general so broad, that vast expanses of ice show no speck of stone or dust—it is only here and there along their flanks that some bare cliff is able to shower down upon them a heap of frost-riven débris. The greatest apparent waste of rocks takes place upon the exposed sea-coasts, where the frost has full freedom to split up the crags and hurl them downwards. But when we think of the immense extent of the glacier system of Greenland, and how in the interior every hill is covered and every valley filled to overflowing with a moving sea of snow and ice, we

* Dr. Rink obtained from the top of a mountain at Upernavik, a good view into the interior of Greenland, and saw lines of stones dotting the surface of the ice as far as eye could reach; from which he inferred that still further to the east there must be bare cliffs or precipices rising above the mer de glace.

can hardly overestimate the tremendous tear and wear to which the buried country must be subjected. We have seen what effect the small glaciers of the Alps have in smoothing and scoring the rocks of Switzerland, and underneath the ice of Greenland similar grinding, and scratching, and polishing must be taking place. Rough crags and sharp projecting bosses of rock will have all their asperities removed; the tops and sides of mountains will be smoothed and dressed, nor can we doubt that valleys will be gradually deepened, and heaps of striated and polished stones will accumulate and be dragged on underneath those mighty ice-rivers which are ever slowly making their way from the interior to the sea.

But all the glaciers of Greenland do not reach the sea. Some even terminate at a distance of many miles from the coast. From the foot of such glaciers streams of water issue and flow all the year round. In some cases these streams unite so as to form considerable rivers, one of which, after a course of forty miles, enters the sea with a mouth nearly three-fourths of a mile in breadth. Dr. Kane, who discovered and named this river the Mary Minturn River, seems to have been much impressed with the appearance of such a body of fresh water flowing freely at a time when the outside sea was thickly covered with ice. It is highly probable, however, that water circulates to some extent below every glacier. The intense cold of an arctic winter penetrates to only a comparatively little distance from the surface of the ground. If it were otherwise—if the winter temperature of North Greenland could penetrate to any depth, it is clear that not a drop of water in any of its valleys would be permitted to remain in the liquid state, and the short summer would be unable to set free any considerable portion of the frost-bound water. It is well known, however, that upon digging down through the snow to the underlying soil, the temperature of the latter is found to be considerably higher than that of the external atmosphere. Snow and ice are bad conductors of heat, and thus the warmth imparted to the ground in summer is never entirely dissipated, but imprisoned, as it were, by the investing sheet of snow. In like manner, the rocks that are permanently concealed under the mer de glace and the great glaciers, must retain pretty nearly the same temperature all the year round. It is quite impossible that the intense cold of winter can pierce entirely through the thick ice; or, on the other hand, that the higher temperature of the rocks can make its escape upwards. Hence any natural springs that may rise below the glaciers will continue to flow on beneath the ice, while the temperature of the rocks themselves, and the heat derived from the intense friction of the glacier grinding upon its bed must tend to melt the under portions of the gelid mass, and thus materially add to the volume of water in circulation betwixt the ice and its pavement.

Between the edge of the mer de glace, or snow-field, of Greenland and the sea, there intervenes a narrow strip of country, from which in summer-time the snow almost entirely disappears. In the sheltered nooks of this narrow tract of land the short summer suffices to waken from their long winter sleep numerous feeble flowerets that gleam and twinkle timidly among thick tufted grasses. Here the purple lichnis and white-starred chickweed, with many other sweet little plants, blossom and bloom under the fleeting sunshine. Dwarf heaths, willows, and alders are also plentiful, and the barren rocks put on a gay livery of orange-coloured lichens. The musk ox, the reindeer, the arctic fox, and hare frequent these solitudes, and numerous flocks of sea-birds enliven the coasts; only to disappear, however, as soon as the fading day warns them of the approach of the long night of winter. Of course, the Esquimaux are entirely confined to this narrow belt of ground adjoining the sea; of the interior of the country they know nothing.

A glance at a map of the western hemisphere will show that Greenland is separated from Labrador, and the bleak islands that flank the northern coasts of America by a broad belt of water, the wider portions of which are known as Davis Strait and Baffin's Bay. Towards the far north this water-belt suddenly contracts to a comparatively narrow strait at Smith's Sound, but afterwards expands again into Kennedy's Channel, beyond which nothing until recently was definitely known. Throughout the greater part of the year this wide belt of sea is always more or less clogged with ice. In winter-time it is nearly all frozen over, but in summer the ice breaks up into a tumbled archipelago of floating islands, through which an adventurous voyager may make his way, with great difficulty and danger, up to where the belt begins to contract at Smith's Sound. There can be little doubt that the comparatively sheltered character of much of this region aids the formation of the ice upon the sea; for when the expeditions of Dr. Kane and Dr. Hayes traced the coast-lines to the far north, they found, where these suddenly retired, that the ice-choked water of Kennedy's Channel expanded into a wide open sea that rolled its great billows against long lines of black cliffs, and stretched away far as eye could reach towards the pole.* Again, upon the eastern shores of Greenland, which are exposed to the full swell of the ocean, ice never accumulates to such an extent as it does in Baffin's Bay. The high cliffs, therefore, that overlook the water-belt of Davis Strait and Baffin's Bay must protect the sea to some extent from those fierce storms which in open ocean throw the water into violent commotion, and prevent ice forming. Dr. Hayes, with a small party, climbed from the shore near Port Foulke, and ventured upwards of sixty miles upon the snow-covered table-land, but was overtaken by a storm of such violence that it was with great difficulty he succeeded in making good his return. The travellers found as they descended towards the coast that they gradually escaped the fury of the wind; and when at last they reached the sea-level, all there was peaceful

* The recent voyage of the "Polaris" seems to have verified the inference of Kane and Hayes as to the existence of an "Open Polar Sea.'

and quiet, although they could see by the great clouds of snow-dust, which continued to stream out from the crest of the towering cliffs, that the storm still raged with undiminished fury on the bleak table-lands above them.

The ice formed upon the surface of the sea by direct freezing rarely attains a greater thickness than 18 or 20 ft., and where the water is liable to more or less agitation it is usually much less. There is a limit to the influence of frost upon the sea just as there is upon the solid ground. The cake of ice protects the underlying water even as snow shelters the earth. As the ice rapidly thickens, the warmer temperature of the sea finds increasing difficulty in soaking upwards through its crystalline roof, until at last it becomes practically imprisoned. Thus, with the warmth of the sea shut in, and the cold of the external air shut out, the ice-cake comes to assume its maximum thickness soon after the winter sets in—the intense cold of the later winter months adding little, if anything, to its depth.

The sea-ice where it abuts upon the land reaches a much greater thickness than that which is formed off shore. Along the coast from near the Arctic Circle, up to Kennedy's Channel, a narrow shelf or platform, varying from 60 to 150 ft. or so in breadth, adheres to the rocks, accommodating itself to every sweep and indentation of the coast-line. In the higher latitudes this shelf never entirely disappears, but further south it breaks up and vanishes towards the end of summer. It owes its origin to the action of the tides. The first frost of the late summer covers the sea with a coat of ice which, carried upwards along the face of the cliffs by the tide, eventually becomes glued to the rocks. In this position it remains, and gradually grows in thickness with every successive tide until it may reach a height of 30 ft., and sometimes even more, presenting to the sea a bold wall of

Fig. 16.—Greenland Ice foot. (H. M. Skae.)

ice, against which the floes grind and crush, and are pounded into fragments. Its growth only stops with the advent of summer, when it begins to yield to the kindly influence of the sun, and to the action of the numerous streams that issue from the melting glaciers, and lick out for themselves deep hollows in the shelf as they rush outwards to the sea.

During summer vast piles of rock and rubbish crowd the surface of the ice-foot. These are of course derived from the cliffs, to the base of which the ice-foot clings.

To such an extent does this rock-rubbish accumulate, that the whole surface of the shelf is sometimes buried beneath it and entirely hidden from view. In the far north, where the ice-foot is perennial, it becomes thickly charged with successive deep layers and irregular masses of rock and débris—the spoil of the summer thaws. And when, as frequently happens, portions of this ice-belt get forced away from the land by the violent impact of massive floes, the current carries southward the loaded ice, which ere long will drop its burden of rock and rubbish as it journeys on, and warmer temperatures begin to tell upon it. Along that part of the coast of Greenland where the ice-foot is shed at the end of every summer, the quantities of rock débris thus borne seawards must be something prodigious.

But the rafts detached from the ice-foot must occasionally float from the coast other records of the land besides fragments of its bleak cliffs. Dr. Kane describes the skeleton of a musk-ox which he saw firmly embedded in the ice of the ice-foot, along with the usual stones and débris. We cannot suppose that this is an isolated and solitary case. On the contrary, when we consider the position of the ice-foot, stretching as it does along the whole coast-line, and constantly receiving the waste of the land, it does not appear at all improbable that the remains of the arctic mammalia may not infrequently get frozen into the ice-foot, and eventually be carried out to sea. It is quite true that these animals do not abound through out the maritime regions of Greenland, yet here and there in favoured spots they collect in considerable herds.

The ice-foot is not the only carrier of stones from Greenland. Glaciers, as we have seen, enter the sea at many places along the arctic coasts—often filling up those long deep sea-valleys or fiords which in lower latitudes form commodious natural harbours, and frequently penetrate for many miles into the interior of a country: of such a character are the friths and fiords of Scotland and Norway. A glance at a good chart of Greenland will show that similar inlets of the sea occur very numerously along the west coast of that country as far north as Upernavik. But as we follow the coast-line to still higher latitudes the sea no longer invades the land in the same way as to the south of Upernavik. The deep fiord-valleys still continue, but they are choked up with glaciers, which have pushed out the sea and occupied its place. As these glaciers slowly creep on to profounder depths a point is reached at which, as already described, the pressure of the dense sea-water becomes too strong for the tenacity of the glacier to resist; and thereupon the ice ruptures, and great masses surge upwards and float off as icebergs. Some of these bergs attain a prodigious size. Dr. Hayes measured one which had stranded off the harbour of Tessuissak to the north of Melville Bay, and estimated it to contain about 27,000,000,000 of cubical feet. This berg could not have weighed less than 2,000,000,000 of tons; it was aground in water nearly half-a-mile in depth. What, then, must have been the thickness of the glacier from which it had been detached? Captain Ross, in his first voyage, describes another iceberg of gigantic proportions. This mass of congelation had stranded in sixty-one fathoms of water, and its weight was estimated at 1,292,397,063 tons.

It is highly possible, as I have shown at page 60, that icebergs carry away with them stones which were frozen into their bottoms at the time when they formed part of the glaciers. But the proportion of stones thus transported must be very small; only a few stones at most can adhere in this way to the ice. In places, however, where the glaciers are overhung by rocky precipices, as is frequently the case just before the ice-rivers pour themselves into the sea, the glaciers become sprinkled along their sides with rocks and débris detached by frost from the cliffs above. But owing to the great breadth of the glaciers, it can be only an infinitesimal portion of their surface that is so sprinkled. The great Humboldt glacier has a breadth of upwards of sixty miles, and is continually shedding icebergs along its whole vast extent of frontage. But with the exception, perhaps, of those icebergs that break away from the extreme corners at the north and south, none of the others carry seaward any stones whatever, save what fragments may have become jammed and frozen into their bottoms. By far the larger number of the arctic icebergs therefore contain no extraneous matter, and melt away in mid-ocean without leaving behind them any record of their voyage. Now and then, however, icebergs which have at one time formed portions of the side of a glacier are heavily laden with débris, and as bergs float much deeper than detached masses of ice-foot, they come more under the influence of oceanic currents, and thus, despite winds and tides, are frequently carried immense distances before they finally melt away. Sailors have met with them as far south as the Azores, so that memorials of the arctic lands must be widely scattered over the bed of the Atlantic Ocean. It is curious to speculate upon the manner in which these memorials will be distributed across the floor of the sea. Many deep-sea soundings have made us aware that the ocean is of very irregular depth; there are submarine plateaux and hills and valleys just as there are subaërial table-lands, and mountains and dales. And we can easily imagine how as the melting icebergs drift southwards and drop their burdens as they go, fragments of rock, chipped by the frost or torn by the glacier from the bleak cliffs and mountains of Greenland, will come to rest sometimes upon submarine hill-tops, sometimes in submarine valleys.

Occasionally icebergs run aground, and in this position are rocked to and fro, and sometimes wheeled about by the force of the currents. This oscillatory movement is usually accompanied by loud noises, and the sea becomes turbid often for more than a mile with the mud which the rocking berg stirs up from the bottom. It frequently happens, too, that when a strong swell is running in upon a stranded berg, the ponderous mass, after for some time swinging fearfully from side to side, will heel right over, and split up into smaller fragments, which thereupon float away.

Icebergs do not grate continuously along the bottom of the sea. When they once run aground their progress is stopped, until by gradual melting or by splitting up into several smaller pieces, they are again floated off and swept on by the currents. Now and then, however, a berg propelled by the tide may work its way for a short distance over a shoal or up a gently sloping beach; but it is evident that it will do so in a most irregular manner, and will very soon cease to advance. When a berg has stranded, all that currents can do is to drive it forward into any soft mud and sand that may happen to be lying upon the sea-bottom; but the motion of the ice will soon be arrested by the accumulation of débris pushed on in front. A mass of ice 3,000 ft. thick would certainly make havoc of any loose incoherent beds of silt and sand into which it might plough; or, should it run aground upon a reef, it would doubtless pound and crush the hard rocks that formed the pivot upon which it oscillated. But although the rocky coasts of North America have often been examined with a view to discover striated surfaces that could be shown to be the work of icebergs, yet nothing has been observed to lead us to believe that striations and markings, like those produced by glaciers, are ever the result of iceberg action.

In this rapid sketch of certain phenomena of the arctic regions attention has been confined to such facts as have a geological bearing. Nor have all these been exhausted; there still remain some interestin questions to be discussed in connection with the marir life of the arctic regions. But this part of my subje(must be deferred to a subsequent page, when I com to consider the history of the beds that overlie th till.

CHAPTER VI.

ORIGIN OF THE TILL AND ROCK-STRIATIONS AND GROOVINGS OF SCOTLAND.

Stones of the till are glaciated.—Till not like terminal moraine-matter.—Mud of glacial rivers.—Till not an iceberg deposit.—Rock-striæ produced by glaciers.—Scotland once covered with ice.—Direction of the ice-flow in the Highlands.—In the Southern Uplands.—In the Lowlands of the great Central Valley.—Absence of superficial moraines.—Stones in the till derived from the subjacent rocks, not from precipices overhanging the ice.—Stones and mud below ice forming a *moraine profonde* or ground moraine.—Unequal distribution of this deposit explained.

IN Chapter II. some account was given of the till, the lowest-lying, and therefore the oldest, of the superficial deposits of Scotland. It will be remembered that this deposit was described as a more or less tough tenacious clay, crammed with a pell-mell assemblage of stones—these stones being of all shapes and sizes, and almost invariably showing smoothed, polished, or scratched faces. Now, from what we know of glaciers and glacial action we can have no difficulty in coming to the conclusion that in some way or other ice has been concerned in the production of till. We look in vain for striated stones in the gravel which the surf drives backwards and forwards on a beach, and we may search the detritus that brooks and rivers push along their beds, but we shall not find any stones at all resembling those of the till. Running water is powerless to produce anything of the kind; it will round and smooth rock-fragments, no matter how hard they be, but it cannot cover them with striations. The boulders and stones of the till undoubtedly owe their shaping and scratching to the action of ice. Just such stones, as we have seen, are exposed beneath the overhanging sides of a glacier when the sun has caused the ice to shrink back and disclose a portion of its rocky bed, and numbers might be picked out of its terminal moraines—those heaps of rubbish which a glacier brings down or pushes before it. But we cannot fail to remark that although scratched and polished stones occur not infrequently in the frontal moraines of alpine glaciers, yet at the same time these moraines do not at all resemble the till. The moraine consists for the most part of a confused heap of rough angular stones and blocks, and loose sand and débris; scratched stones are decidedly in a minority, and indeed a close search will often fail to show them. Clearly, then, the till is not of the nature of a terminal moraine. Each stone in the till gives evidence of having been subjected to a grinding process. Almost every fragment has been jammed into the bottom of a glacier, and, held firmly in that position, has been grated along the rocky surface underneath, or over a pavement of the tough stony clay itself. In such a position the stones would naturally arrange themselves in the line of least resistance; hence it is that the most distinct ruts and striæ coincide with the longer diameter of the stones. But when the stones and boulders which are dragged on underneath a glacier approach a round or oval shape they can have no tendency to lie in any particular way, and so will come to be scratched equally well in all directions. For obvious reasons, soft rocks, like sandstone, will not attain so good a polish as hard limestone or close-grained shale; nor shall we expect to find the stones rounded like gravel or shingle, for they cannot move so freely under ice as pebbles do under water. Occasionally, however, they will be rolled over and compelled to shift their position; but this process will only result in smoothing off their sharper edges and in marking them with irregular striæ.

Now all these appearances, as we have seen, are actually found to characterize the stones in the till; and such being the case, we can hardly resist the conclusion that the whole deposit—clay and stones alike—has in some way or other been formed below ice. We look in vain, however, amongst the glaciers of the Alps for such a deposit. The scratched stones we may occasionally find, but where is the clay? We take our stand at the foot of a glacier and watch the river as it leaps forth from its cave of blue ice. Not a few visitors, I suppose, have been surprised at the turbid appearance of the ice-born river. Why should the melting glacier give rise to such a milky-white, or, as is sometimes the case, yellow-brown stream. If we lift some of the water in a glass and examine it, we shall find that its colour is due to the presence of a very fine impalpable mud. In the more sheltered reaches of a glacier river this mud will occasionally accumulate to some depth. It is an unctuous, sticky deposit, and only requires pressure to knead it into a tenacious clay. There can be no doubt whatever that it owes its origin to the grinding power of the glacier. The stones and sand which the ice forces along are crushed and pulverised upon the rock below, and the finer material resulting from this action is what renders the glacial rivers turbid and milky. If there were no water to wash out the mud formed in this way below a glacier, it is evident that not only scratched stones but clay also would gather underneath the ice, and be pushed out at its termination; and this clay, owing to excessive pressure and to the finely-divided nature of its ingredients, would be hard and tough. The Scottish till, when it has been exposed to the influence of the weather, sooner or later crumbles down, and, when water washes over it, then that which was once a hard tough clay becomes a soft, sticky, unctuous mud, that clings persistently to everything it touches. No one who compares this mud with that derived from the glacial waters of the Alps will fail to notice their similarity. Thus, whether we consider the character of the stones in the till, or the nature of the clay, we are almost equally convinced that both have had a glacial origin.

It is clear, however, that the conditions for the gathering of a stony clay like the till do not obtain (as far as we know) among the alpine glaciers. There is too much water circulating below the ice there to allow any considerable thickness of such a deposit as till to accumulate. Neither can till owe its origin to icebergs. If it had been distributed over the sea-bottom it would assuredly have shown some kind of arrangement. When an iceberg drops its rubbish, it stands to reason that the heavier blocks will reach the bottom first, then the smaller stones, and lastly the finer ingredients. There is no such assortment visible, however, in the normal till; but large and small stones are scattered pretty equally through the clay, which moreover is quite unstratified.* Again, putting aside the unstratified character of the till, we cannot fail to remark that the great mass of stones and débris which icebergs carry seawards, consists almosts exclusively of rough, unpolished angular fragments that have tumbled upon the surface of the ice from cliffs and precipices. The only polished and striated stones that an iceberg can possibly steal away with are the few that may have got jammed into its base before it was shed from its parent glacier. Such being the character of the débris borne seawards upon glaciers, it is evident that the till, with its pell-mell accumulation of finely-polished and striated stones, cannot be of the nature of iceberg droppings. These are strong reasons for rejecting the iceberg theory of the origin of till, yet they are not by any means the most cogent, as will be seen by-and-by.

Since till, then, cannot be formed in and deposited by water in the same way as gravel and sand—since

* To certain appearances of stratification presented by some stony clays, I shall refer in the sequel, see Chapter XV.

no such deposit accumulates as a terminal moraine in alpine valleys, nor can possibly be the result of iceberg droppings—what other explanation of its origin can be given? To answer this question, we must for a little recall certain other phenomena associated with the till.

When that deposit is removed from the underlying rocks these almost invariably show either a well-smoothed, polished, and striated surface, or else a highly confused, broken, and smashed appearance. But scratched and polished rock-surfaces are by no means confined to till-covered districts. They are met with everywhere and at all levels throughout the country, from the sea-coast up to near the tops of some of our higher mountains. The lower hill-ranges, such as the Sidlaws, the Ochils, the Pentlands, the Kilbarchan and Paisley Hills, and others, exhibit polished and smoothed rock-faces on their very crests. Similar markings streak and score the rocks up to a great height in the deep valleys of the Highlands and Southern Uplands, and throughout the inner and outer Hebrides and Orkney and Shetland the same phenomena constantly occur.

The direction of these parallel ruts and striations coincides, as a rule, with the line of the principal valleys. In the Northern Highlands, for example, they keep parallel to the trend of the great glens; and in the Southern Uplands, likewise, they follow all the windings of the chief dales and "hopes." In the Lowlands, however, their direction does not appear to be influenced so much by the configuration of the ground; for they often cross low valleys at right angles or nearly so, and sweep up and over intervening hills, even when these happen to have an elevation of upwards of 1,800 ft. above the sea.

The scratches upon the rocks have exactly the same appearance as those that crowd the surface of the stones in the till; but whereas the striations on the stones may cross and recross, those upon a surface of rock usually run in one and the same direction. Sometimes, however, we meet with exceptions to this rule, when two or even three sets of striæ may be observed upon the same surface of rock. But such cross-hatchings do not occur very often, and seem to be confined to the lowland districts—at all events, I have never seen them in any of the valleys of the Highlands or Southern Uplands. No one who shall compare the dressed rocks with the scratched stones can have any doubt that both owe their origin to the same cause. If glacier ice scratched the stones, then the rocks must have been smoothed and dressed by the same agency. The work cannot possibly have been done by icebergs, for floating ice has no power to grate along the sea-bottom, so as to polish and dress submarine hills and valleys. The agent that performed the work has actually clung to the ground, and accommodated itself to every inequality of surface—here rounding and smoothing knobs and bosses of rock, there sliding into and polishing dimples and depressions. In short, the appearances tally precisely with what has been observed in the valleys of the Alps and elsewhere. When we have the opportunity of examining the deserted bed of a glacier, we find it smoothed and dressed in every part—wherever the ice has been able to get at the rock it has ground, scratched, and polished it. Nor can any reasonable person resist the conclusion that the dressed rocks of Scotland have been worked upon by ice in the same way. We must believe that all the hills and valleys were once swathed in snow and ice—that the whole of Scotland was at some distant date buried underneath one immense mer de glace, through which peered only the higher mountain-tops. This is no vague hypothesis or speculation founded on uncertain data, no mere conjecture which the light of future discoveries may explode. The evidence is so clear and so overwhelmingly convincing that we cannot resist the inevitable conclusion. Suppose some visitor who had only newly arrived in our country were to stumble in the course of his wanderings upon a deserted line of railway, where the old, battered rails gave evidence of having been well used, he surely would require no reflection to conclude that cars and waggons must frequently have passed along the line. What would be thought of our visitor's sanity if he were to reason in this way:—"Although this looks very like a railway, with its embankments and rails and sleepers, yet I cannot think it is so, for no trains run upon it, and I have been here several months, but in all that time have never seen it used." Now old embankments and worn-out rails are no more convincing proofs of the former passage of wheeled carriages, than the smoothed, scratched, and rounded rocks are of the grinding action of old glaciers; and the incredulity that would reject the evidence of the latter might well be expected to treat the former in a similar way.

Since, then, we must believe that the dressed and rounded rocks could only have been so dressed and rounded by land-ice, it follows that wherever such rock-surfaces occur there at one time a glacier must have been. Now the scratches may be traced from the islands and the coast-line up to an elevation of at least 3,500 ft.; so that ice must have covered the country to that height at least. In the Highlands the tide of ice streamed out from the central elevations down all the main straths and glens, and by measuring the height attained by the smoothed and rounded rocks, we are enabled to estimate roughly the probable thickness of the old ice-sheet. But it can be only a rough estimate, for so long a time has elapsed since the ice disappeared, and rain and frost together have so split up and worn down the rocks of these highland mountains, that much of the smoothing and polishing has vanished. But although the finer marks of the ice-chisel have thus frequently been obliterated, yet the broader effects remain conspicuous enough. From an examination of these, we gather that the ice could not have been less, and was probably more than 3,000 ft. thick in its deeper parts. What wonder, then, that a mass of this bulk, gliding from the mountains down to the sea during a long course of ages, should have left such an impress of its grinding-power upon the rocks that the lapse of thousands of years has not succeeded in removing it. For even when the fine smoothing and polishing have disappeared, the hills yet show in their rounded and flowing outlines that peculiar configuration which is so characteristic of ground over which a glacier has passed.

It is well known that the glaciers of Switzerland are mere pigmies compared to what they have been formerly. The slopes of the alpine valleys are all smoothed, scratched, and scored up to a considerable height above the present surface of the glaciers; and these smoothed rocks are often separated from the rough, broken, and craggy rocks above by a well-marked line, indicating the height reached by the glaciers in days gone by. But in Scotland such a distinct line of division rarely marks out the upper limits of the glaciation. Frost and rain have made havoc of the ice-work at the higher elevations of the country, and roughened the exposed rocks into crags and peaks.

In the Southern Uplands the ice moved, as in the Highlands, from the central high-grounds down all the main valleys—its track being well marked out by an abundant series of finely preserved striæ.* From the mountains of Galloway, and the uplands of the south-east, vast glaciers descended in every direction. The valleys of the Annan, the Nith, and the Dee were filled to overflowing with great confluent glaciers that poured their united volume into the Solway Frith and the Irish Sea. In like manner a vast stream of ice that flowed north-east and then south-east, buried the

* See the general map showing the direction of glaciation.

great vale of the Tweed between the Cheviots and the Lammermuirs.

When the ice-markings are followed into the lowlands of the central valley, we find that in the vale of the Forth their general tendency is towards the east; while in the lower reaches of the Clyde valley their trend is east, south-east, south, and south-west. The meaning of this apparent confusion is perceived when we trace out the track of the glaciers that issued from the Highlands, and follow the spoor of those that crept down from the Southern Uplands. It then becomes apparent that a great current of ice from the high-grounds of Lanarkshire set down the valley of the Clyde, and was met above Hamilton by a vast glacier coming in the opposite direction. Hence the two opposing streams were deflected to east and south-west—on the one hand sweeping across the Lothians into the Frith of Forth and the German Ocean, on the other overflowing the uplands of Renfrewshire, and passing south-west into Ayrshire, so as to unite with the glacier masses descending from the Galloway mountains.

Underneath these great streams of ice the whole surface of the country would be subjected to excessive erosion. Hill-slopes would be ground and polished, valleys deepened and smoothed; here the rocks would be finely dressed and striated, there crushed and broken. And what would be the character of the débris that resulted from all this grinding and graving-work? In the glaciers of the Alps we have every reason to believe that a considerable proportion of the stones used as chisels and stylets by the ice are introduced from above. They tumble from the crags upon the surface of the ice, and drop into those deep crevasses which must sometimes cut a glacier to its bottom. But when ice buried Scotland to a depth of several thousand feet, only a few hill-tops would rise above the general level of the mer de glace. Consequently, little débris would be showered upon the ice; and, even supposing considerable heaps of blocks and rock-rubbish did acccumulate here and there at the base of some isolated hill, it is nevertheless very unlikely that any portion would ever work a way to the bottom of the thick ice-sheet. The gravers employed by the ice in dressing the Scottish hills and valleys could not have been derived from above; they must have been obtained from the rocks lying below.

It is quite certain, however, that the ice when it first overflowed the land would find a plentiful supply of loose stones lying upon the ground ready for use. For long ages before the country became locked in ice the climate must have been getting colder and colder. The result of this intense frost would be to split up the rocks everywhere: nor would this be a difficult matter. We must remember that the present deep subsoils, that bury the solid rocks to such a depth, had no existence before the advent of the ice-sheet. The rocks would not be covered with a deeper soil than is the case in countries where no drift deposits exist. I have already referred to the heaps of broken rock that cumber the exposed ground in northern regions, where whole hills are well-nigh buried in their own ruins. In all arctic and alpine countries, and, indeed, wherever a rock is exposed to the action of frost, it is sure to split up sooner or later. The moraines of the Alps, and the prodigious piles of débris that collect upon the ice-foot of the arctic regions, are sufficient evidence of what frost can do. It is certain, therefore, that when the ice began to creep over Scotland, it would have to make its way through piles of broken fragments and over shattered rock-surfaces. Ice-chisels would thus be prepared for it beforehand, which would aid in the work of dislodging others from the rocks. As the crushing and grinding continued, few stones would escape being smoothed and striated, while the fine mud resulting from all this work would get mixed up with the stones, and form a stony clay. It is true that water would circulate below the ice to some extent, as we know it does underneath the glaciers of Greenland, and no doubt much glacial mud would be carried away by this means; nevertheless, all that could possibly escape would bear but a very small proportion to what remained behind. Thus both mud and stones would tend to collect under the ice; and as that great mass moved onwards, pressing with prodigious weight, the mud and the stones would be squeezed and dragged forward so as to become a confused and pell-mell mixture of clay and stones, with here and there traces of water-action in the form of irregular patches and interrupted bands of stones, gravel, and earthy sand—in short, till. Such, then, would appear to be the origin of that remarkable deposit: it is the ground moraine, or *moraine profonde*, of the old ice-sheet.

We must not suppose, however, that till gathered equally underneath every portion of the confluent glaciers. On the contrary, wherever the inclination of the ground was such as to cause the ice to quicken its flow—on steep hill-slopes, for example—clay and stones would not readily collect; but in places where the motion was slow, there the till would have a tendency to accumulate: in short, the distribution of the till would be regulated by the varying pressure of the ice above. Some have objected that the moment a layer of till was formed between the ice and the subjacent rock, all wear and tear of the latter would cease, and therefore that the formation of till itself would suddenly come to an end. It would just be as reasonable, however, to infer that all wearing-away of a river-channel must stop the moment that the channel becomes filled with gravel and sand. But who does not know that the materials in the bed of a stream are continually travelling onward, no matter how slowly. It is quite true that so long as a bank of sand and gravel shall lie in one place, the rock on which it rests will escape the rasp of the river. But the river that piles up such banks will by-and-by sweep them away again, and employ the sand and gravel as agents for wearing down, scouring, and filing the rocks which they formerly protected. And so, no doubt, it must have been with the ice-sheet and its débris. Over many portions of its bed there would be a continual travelling onwards of clay, sand, and stones; while in other areas masses of débris which had collected here and there would be ever and anon ploughed up again, and pushed and dragged forward from one position to another. In this way the underlying rocks would be alternately protected and exposed.

CHAPTER VII.

ORIGIN OF THE TILL, AND ROCK-STRIATIONS AND GROOVINGS OF SCOTLAND—*Continued.*

Direction of ice-flow indicated by stones in the till.—Cross-hatching of rock-striæ accounted for.—Intermingling in the till of stones derived from separate districts.—"Debatable land" between rival ice-flows.—Local colouring of till an indication of direction followed by ice-flow.—The Ochils, Pentlands, and other hills completely overflowed by ice.—Deflections of the ice-flow.—Till in upland valleys, why terraced.—Origin of lowland "drums."—Crag and tail, &c.—Islets lying off the coast glaciated from the mainland.—Ice filled up all our shallow seas.—General ice-sheet like that of Antarctic continent.

THE course followed by the ice-sheet in its downward progress from the high-grounds to the coast is indicated, as described in last chapter, by the direction of scratches and furrows and flutings, and by that peculiar rounded outline which the grinding of the heavy mass has imparted to our mountain-slopes and hill-tops.* But even when these markings do not appear, either on account of the obliterating effect of weathering, or else because they lie concealed below a superficial covering of drift, yet the till itself often furnishes evidence as to the direction of ice-flow. If, for example, we know from what part of the country

* Rocks which are so rounded, whether striated or not, are known as *roches moutonnées.* The name was probably given to them on account of a fanciful resemblance to the rounded shape of a sheep's back.

the scratched stones in the till have been derived, it is obvious that we ascertain at the same time the course followed by the ice that brought them. Hence we are enabled to track out the trail of the mer de glace over all the country. And it is worthy of note that the evidence supplied by the stones always corroborates that afforded by the *roches moutonnées* and striated rocks. If these last owe their origin to a current of ice that came from the north, then the stones also will be found to have travelled in the same direction. And, curiously enough, in those districts where the rocks exhibit a "cross-hatching" of striæ, or where the striations on two contiguous rocks do not agree in direction, there also the till shows an intermingling of stones derived from separate districts. Now what does this prove? Clearly this, that the ice-currents were occasionally deflected and forced to go another way. The great stream that crept across the central valley of Scotland was certainly at times turned out of its normal course—now towards the south by the pressure of that powerful current of ice that poured down from the Highlands, and again towards the north when the ice-stream coming from the Southern Uplands overpowered and forced back the other; in short, there was a "debatable ground" between the northern and the southern currents, over which sometimes the one and sometimes the other prevailed.* The right of possession to the hilly tract that lies between Paisley and Kilmarnock seems frequently to

* Reference will be made in a succeeding chapter to certain instances of cross-hatching which are to be explained in a different way, see Chapter XV.

have been disputed by the rival ice-streams—the rocks of that area being sometimes striated from north to south, and sometimes in the opposite direction. We find also an intermingling of stones—fragments of mica-schist and gneiss from the highland mountains occurring now and again in the till of the valley of the Irvine; while stones derived from the high-grounds to the south of that river appear here and there in the till that sweeps up to the crests of the hills overlooking the basin of the Clyde. A similar intermingling of stones from the north and south is seen in the till of the valley of the Esk near the Moorfoot Hills, in Edinburghshire.

But beyond this "debatable land" striated rocks and scratched stones alike point to a persistent ice-flow in determinate directions. In the near neighbourhood of the Highlands all the stones without exception tell of a move outwards from the mountains, and it is the same with rock-striations. The till in the valleys of the Southern Uplands has in like manner invariably been derived from the contiguous high-grounds. Following the till from the base of the Grampians, where it is crammed with fragments of mica-schist, granite, gneiss, quartz, &c., &c., down into the basin of the Forth, we find the number of these highland stones gradually decreasing, until by-and-by they disappear altogether. And the same is the case with the till that stretches northwards from the Southern Uplands. At first the fragments brought from these uplands are in the majority, but they gradually fall off northwards, until finally we cease to meet with them. It is curious also to notice how the stones lose in size as the parent rock is left further and further behind—the longer the distance travelled, the greater having been the degree of crushing and grinding undergone.

The local colouring assumed by the till is another strong proof of transportation by land-ice. As described in a previous chapter, this deposit varies both in colour and texture, according to the nature of the rocks near which it lies. Thus it becomes red, and shows a sandy texture in districts where red sandstone is the prevailing rock; but in a region where coal and black shales abound, there we encounter a hard, tough, tenacious deposit, having a dark greyish blue colour. The reason for this difference is obvious. The clay is derived from the grinding and crushing of the underlying rocks, and consequently changes its character as the rocks change theirs. But just as the included stones not infrequently have been dragged for a long distance from their parent rocks, so in like manner has the clay formed in one place travelled onward to another. Hence it often happens that the till of a given district—a red sandstone region, for example—will be found to have invaded and covered adjoining ground where the rocks are neither red nor composed of sandstone.

If space permitted, some special proof might be offered in support of a statement already made—namely, that the ice overflowed the hill-ranges and isolated hills of central Scotland. The Ochils, for example, that separate the basin of the Forth from Strathearn are ground off and smoothed in such a way as to indicate that the mass of ice must first have crossed the valley of the Earn from the Grampian mountains, and thereafter overflowed the Ochils and passed on south-eastwards across the Lomonds and the Cleish Hills into the valley of the Forth. The evidence afforded by the till that covers the southern slopes of the Ochils points precisely to the same conclusion, for that deposit is abundantly charged with fragments of gneiss, granite, and other rocks that could only have come from the Highlands. Mr. B. N. Peach found a considerable mass of till at a height of 2,200 ft. on the shoulder of Ben Cleugh, and numerous scratched stones occurred on the very top of the hill (2,300 ft.) Similar proof of the passage of land-ice over considerable eminences in the lowland districts might easily be given. For instance, on the very top of Allermuir Hill, one of the highest points in the Pentlands, Mr. Croll got a patch of till containing, amongst other local stones, certain fragments which have been brought from the north or north-west—thus clearly showing that these hills also were overtopped by the mer de glace. I have already mentioned the fact that both rock-scratches and till indicate that the high-grounds between Paisley and Kilmarnock have been surmounted by land-ice. According to my brother, Professor Geikie, relics of the till are found near the top of Tinto Hill (in Lanarkshire), which rises to a height of upwards of 2,300 ft. above the sea.

That the ice should have overflowed the land up to

such heights will not surprise one; for, by a stream of ice some 3,000 ft. or so in thickness, hills like the Ochils, the Pentlands, and Tinto would be as easily surmounted, as stones and boulders are overflowed in the bed of a river. And yet, just as these boulders will deflect that portion of the river's current that strikes upon them, so the heights to which I refer appear to have partially turned aside the stream of ice that beat against them. This is shown by the manner in which the flutings and groovings bend round the sides of a hill before they finally cross it and resume their normal direction.

The flutings and groovings in the valleys of the Southern Uplands show distinctly that the ice to which they owe their origin not only filled the valleys, but swept across the intervening hills. The markings referred to run in nearly a horizontal direction along the steep slopes of the hills, so that they appear to rise as the valley descends; and thus, while we follow the stream, they gradually mount higher and higher until the crest of the hill is reached, over which they eventually disappear. The beautiful valley of the Yarrow below Gordon Arms inn affords a fine example of the phenomena in question.

Reference has already been made to the unequal distribution of the till. It lies thickest in the valleys, and thins away towards the hills, being found for the most part in patches when we get above a height of 1,000 ft. In the hilly districts of the south of Scotland it occurs chiefly on the bottoms of the valleys, but it may sometimes be met with nestling in hollows even up to a height of 1,800 or 1,900 ft. The whole appearance of the deposit, however, shows that it never did attain any thickness at these heights—the force of the ice-stream on steep slopes and exposed places having prevented its accumulation, just as a river will not allow sediment to accumulate upon the tops and exposed sides of the large boulders in its bed. But in the lower reaches of the upland valleys, notably in Peeblesshire, the till often attains some depth. It gradually lessens, however, as we trace it up towards the heads of the valleys, where it eventually disappears.

The general aspect presented by the deposit in these valleys is that of a flat-topped terrace inclining gently from the hill-sides and sloping gradually down in the direction followed by the stream, at about the same angle as the bottom of the valley itself. I believe that this terrace-like appearance of the till was most probably assumed underneath the ice-sheet. In narrow and deep hollows, like the upland valleys, the ice was not liable to such deflections as took place over the "debatable grounds;" and the till forming below it consequently escaped being squeezed to and fro; the valleys were filled with streams of ice flowing constantly in one and the same direction, and the probabilities are therefore strong that the débris which accumulated below would be spread out smoothly.

In the Lowlands the effect produced by the varying direction and unequal pressure of the ice-sheet is visible in the peculiar outline assumed by the till. Sometimes it forms a confused aggregate of softly swelling mounds and hummocks; in other places it gives rise to a series of long smoothly-rounded banks or "drums" and "sowbacks," which run parallel to the direction taken by the ice. This peculiar configuration of the till, although doubtless modified to some extent by rain and streams, yet was no doubt assumed under the ice-sheet—the "sowbacks" being the glacial counterparts of those broad banks of silt and sand that form here and there upon the beds of rivers. Perhaps the most admirable example in Scotland of this peculiar arrangement or configuration of the till occurs in the valley of the Tweed, between the Cheviot Hills and the Lammermuirs. In this wide district all the ridges of till run parallel to each other, and in a direction approximately east and west. This, too, is the prevailing trend of the rock-striations and roches moutonnées in the same neighbourhood.

The phenomena of "crag and tail" afford yet another indication of the path followed by the ice. A familiar illustration of the mode in which "crag and tail" have been formed may be obtained by placing a large stone in the current of a stream, and watching the effect produced upon the carriage of sediment by the water. The current sweeps against the stone, and is deflected to right and left—there being of course considerable commotion in front and quiet water behind. The current thus stemmed is forced downwards with a stronger pressure upon the bed of the stream by the water continually advancing from behind, and the result is that a hollow is gradually scooped out in front of the stone, and for some way along its sides. In the rear, where there is comparatively little stir in the water, silt and sand speedily accumulate, until a long sloping "tail" is formed, stretching away from the stone for as great a distance as the quiet water extends. If for a stone we substitute a big crag, standing up in a broad valley, and for the little stream of our illustration a deep current of land-ice, we shall have no difficulty in explaining the origin of "crag and tail." In the valley of the Forth, where isolated hills and bosses of rock are not uncommon, the till is invariably heaped up on the east side of the crags, showing that the set of the ice-stream was from west to east; the direction in which the till has travelled, and the course followed by the rock-striations, both lead us to the same conclusion.

Thus on every hand we are furnished with abundant proof of the former existence of a great mer de glace in Scotland. From the tops of some of our higher mountains down to the edge of the sea, no part of the country has escaped abrasion. The hills are worn and rounded off, and the valleys are cumbered with the wreck and ruin of the rocks. Nay, most of the islands which lie off the coasts plainly indicate by striations and other glacial markings that ice has swept over them also. They are smoothed not from the centre to the circumference, as would have been the case had they supported separate glaciers of their own, but the striations go right across them from side to side. It cannot be doubted therefore that the ice, to the grinding action of which these striations are due, actually crossed from the mainland over what now forms the bed of the sea. Perhaps the most striking example of this is furnished by Lewis, the northern portion of the Long Island, which I found to be glaciated across its whole breadth from south-east to north-west. The land-ice that swept over this tract must have come from the mountains of Ross-shire—a distance of not less than thirty miles. Leaving the mainland, it must have filled up the whole of the North Minch (sixty fathoms in depth), and overflowed Lewis to a height of 1,250 ft. at least.

In like manner the Island of Bute has been scored and smoothed from end to end by a mass of ice which, streaming out from the highlands of Argyleshire, filled up the Kyles, and then passed southwards over the whole island to occupy the bed of the Frith of Clyde between Bute and Arran. Many of the islets scattered along the western coasts tell the same tale. Again, it may be mentioned as a very striking fact that the lofty cliffs along the south-west coast of Ayrshire are striated along their tops in a direction parallel to the trend of the coast—that is, from north-east to south-west. An examination of the general map will show how these striations have been produced by a mass of ice that filled up the bed of the adjacent sea, and streamed south-west towards the northern coast of Ireland. From these and similar facts geologists have been inclined to infer that at the time the mer de glace covered Scotland the whole of our country stood at a higher level relative to the sea than now; in other words, that a large part of what in these days forms the floor of the sea was at that time in the condition of dry land. This being so, the ice from the central parts of the country would creep outwards and overflow what are now islands, in the same way as it surmounted the Ochils, the Pentlands, and other hill ranges of central Scotland. But, as Mr. Croll remarks, it is quite unnecessary to suppose that the land during the great extension of the ice-fields, stood any higher above the sea. A mass of ice, upwards of 2,000 ft., and in parts attaining 3,000 ft. in thickness, would fill up every fiord-valley, and dispossess the sea in all the sounds, straits, and channels that separate the islands from themselves and the mainland. A glance at the Admiralty charts will show how this could be. From them it will be learned, that, between the mainland and the islands, the sea seldom attains a greater depth than 70 fathoms or 420 ft., and even this depth is quite exceptional. The German Ocean between England and the coasts of France and the Netherlands, does not average more than some 150 or 160 ft. in depth: and the soundings show that the water deepens very gradually northwards. To reach the 100-fathom line, we approach quite close to the coast of Norway, and the same line lies considerably north of the Shetland Islands, from which it sweeps west by south, keeping outside of the Hebrides and Ireland. In no part of our seas, then, could the water have been of sufficient depth to float those prodigious masses of ice which we can prove were generated in Scotland during the

PLATE III.

ANTARCTIC ICE-SHEET. (Reduced by H. M. Skae from the drawing by Sir J. C. Ross.) To face p. 101.

glacial epoch. Before ice will float, it requires water deep enough to accommodate some seven or eight parts of its bulk below the surface of the sea; and therefore the great mer de glace, being unable to float in 600 ft. of water, must have pushed back the shallow seas that flow around our coasts, so as even to coalesce with the ice-sheet that crept out at the same time from Norway. In short, the ice would flow along the bottom of the sea with as much ease as it poured across the land, and every island would be surmounted and crushed and scored and polished just as readily as the hills of the mainland were. But the ice-sheet would not only enfold the western islands and join them to the mainland, it would also extend still further seawards, and terminate at last in precipitous or vertical cliffs, resembling that great wall of ice which Commodore Wilkes and Sir J. C. Ross encountered in the Antarctic Seas. How far westward the ice would extend into the Atlantic would, of course, depend entirely upon its thickness and the depth of the sea. If it retained at its outskirts only one-third of the great depth under which it buried central Scotland, it must have gone out as far at least as the 100-fathom line. Sir J. C. Ross's striking account of the mighty ice-sheet under which the Antarctic continent lies buried, gives one a very good notion of the kind of appearance which the skirts of our own ice-sheet presented. After reaching the highest southern latitude which has yet been attained, all his attempts to penetrate further were frustrated by a precipitous wall of ice that rose out of the water to a height of 180 ft. in places, and effectually barred all progress towards the pole. For 450 miles he sailed in front of this cliff, and found it unbroken by a single inlet. While thus coasting along, his ships (the *Erebus* and *Terror*) were often in danger from stupendous icebergs and thick pack-ice, that frequently extended in masses too close and serried to be bored through. Only at one point did the ice-wall sink low enough to allow of its upper surface being seen from the masthead. Ross approached this point, which was only some fifty feet above the level of the sea, and obtained a good view. He describes the upper surface of the ice as a smooth plain shining like frosted silver, and stretching away far as eye could reach into the illimitable distance. The ice-cliff described by Ross is the terminal front of a gigantic mer de glace, which, nurtured on the circumpolar continent, creeps outward over the floor of the sea until it reaches depths where the pressure of the water stops its farther advance by continually breaking off large segments and shreds from its terminal front, and floating these away as icebergs. And such must have been the aspect presented by the margin of the old ice-sheet, which, in the early stages of the glacial epoch, mantled Scotland and its numerous islets, filling up the intervening straits and channels of the sea, and terminating far out in the Atlantic Ocean in a flat-topped vertical cliff of blue ice.*

* For further details respecting Scotch ice-sheet, see Appendix, Note E.

CHAPTER VIII.

CAUSE OF COSMICAL CHANGES OF CLIMATE.

Evidences of former great changes of climate during past ages.—Theories accounting for these.—Translation of the solar system through space.—Change in the position of the earth's axis of rotation.—Sliding of the earth's crust round the nucleus.—Changes in the distribution of land and sea.—Elevation and depression of the land.—Sir Charles Lyell's speculations upon the effect of such changes.—Climatal effect of winds and ocean currents.—Effect of accumulation of land within the tropics, and of same round the poles.—Peculiar distribution of land and sea inadequate to account for great cosmical changes of climate.

THE stony record everywhere assures us that from the earliest times of which geologists can take cognizance down to the present, our globe has experienced many changes of climate. The plants of which our coal-seams are composed speak to us of lands covered with luxuriant growths of tree-ferns and auracarians, and the fossils in our limestones tell us of warm seas where corals luxuriated in the genial waters. Nor is it only in our own latitudes that scenes like these are conjured up by a study of the rocks. Even in high arctic regions, where the lands are well-nigh entirely concealed beneath the snow, and where the seas are often choked with ice all the year round, we often meet with remarkable proofs of genial and even warm climates having formerly prevailed at several widely separated periods. Limestones containing fossil corals, and numerous remains of extinct chambered shells, such as are now represented by the nautilus of the Pacific Ocean, occur frequently in the highest latitudes yet reached by man. Dr. Hayes brought from the bleak shores of Grinell Land certain fossils, the nature of which clearly indicates that at some distant date a genial ocean capable of nourishing corals and chambered shells, must have overspread that region. Similar results have been obtained by many of our most distinguished arctic voyagers, and from their observations it is now well ascertained that over all the regions within the Arctic Circle which have yet been visited, genial climates have prevailed at different times during past geologic ages —climates that not only nourished corals and southern molluscs in the seas, but clothed the lands with a rich and luxuriant greenery.

A close and careful scrutiny of the rock formations of our own country shows us, moreover, that, in the distant past, those warm and genial conditions that extended from our own latitudes up to polar regions, ever and anon disappeared, and were replaced by cold and even arctic climates. Coral seas and swampy jungles pass away, and are succeeded in time by snow-covered mountains, by glaciers creeping down the valleys, and icebergs sailing drearily away from frost-bound coasts. Again a sunny picture rises up before us, to be replaced, as the ages roll on, by yet other scenes of arctic sterility. It was long, however, before geologists began to recognise the evidence for this remarkable succession, or, as we may call it, rotation of climates. The old belief used to be, that the climates of the globe, owing in great measure to the escape of the earth's internal heat into space, had gradually and regularly cooled down: so that the older formations were thought to represent ultra-tropical conditions, while the later deposits contained the records of less tropical and temperate climates. The astronomer and cosmogonist assure us that there was a time when this earth existed as a mass of gaseous matter, and that this matter, by parting with some portion of its heat, passed ere long into a fluid condition. After some time a hard crust formed over the surface of the molten liquid, and when long ages had passed away, and the outer shell or skin had sufficiently cooled down to allow of all the varied phenomena of evaporation and rain and rivers, then by-and-by life appeared, and those wonderful organic forces began to act, which, under the guidance of infinite intelligence, have culminated in the beautiful creation of which we form a part. And the geologist, taking up this strange history where the astronomer and cosmogonist left off, fancied that he could trace in the stony record the continuation of the same great world-change. In the earlier pages of the record he found evidence for the former existence in the British area, of tropical conditions that seemed gradually to fade away as he continued his researches into the later chapters of geological history. So that he found, or fancied he found, a slow transition from an age of tropical forests and warm oceans to the temperate climate which we at present enjoy. But the rapid accumulation of facts proved fatal to this as to many other theories. It is no longer questioned that the climates of the past were due to the very same causes by which the climates of the present are brought about. The earth, no doubt, still radiates heat into space, but in considering the history of the past, so far back as it is revealed by geology, this cooling of the earth may safely be disregarded. The climates of the world in our times vary according to the proportion of heat received either directly or indirectly from the sun, and so it must have been during all the ages of which any records have come down to us. At the very earliest time of which the geologist can speak with confidence, the climates of the world were probably as well-marked as they are now. We must look elsewhere than to the secular cooling of our globe for the causes which have at several periods induced a mild and even genial temperature within the Arctic Circle—periods during which the whole northern hemisphere enjoyed a kind of perpetual summer. For we know now that such genial conditions had been preceded and were eventually succeeded by climates of more than arctic rigour, when our hemisphere, which had luxuriated in one long-continued summer, became the scene of great snow-fields and glaciers and floating ice. This alternation of genial climates with arctic conditions, obviously cannot be accounted for by the cooling of the earth, due to the radiation of its central heat into space. If we find the remains of full-grown trees in Greenland, and ammonites and corals even farther north, we may be quite sure, that, owing to some cause, apparent or obscure, these regions must at one time have received from some external source a greater proportion of heat, either directly or indirectly, than they do now. And so, conversely, if in our own land we discover traces of great snow-fields and massive glaciers, we cannot hesitate to conclude, that in the ages when such frigid conditions prevailed, this area was deprived of much of the heat which now reaches it. But if this be so, we may well ask what the nature of that action is which can alternately visit our hemisphere with long-continued ages of fruitful summer, or render it bleak and barren with perpetual snow and ice.

There have been many attempts to account for the phenomena. Some have speculated upon the possibility of the whole solar system travelling onwards through the boundless realms of space, and passing in its course through warmer and colder tracts than that in which it now moves. When the sun with its attendant orbs swept through those hypothetical warm regions, the whole climate of the world it was supposed would be affected, and tree-ferns and cycads would then flourish within the Arctic Circle, while the northern seas would be tenanted by large chambered shells and corals. But when, on the other hand, the colder abysses of space were traversed, a total change of climate would be experienced; the luxuriant vegetation would fade away from the polar regions, and a bleak ice-cap would cover the poles of the globe and spread outwards as the cold increased, until the snow and ice might reach down to latitudes like our own.

Others, again, have imagined a change in the position of the earth's axis of rotation, due to the elevation of extensive mountain tracts somewhere between the poles and the equator. This, they think, would be sufficient to shift the axis so as to confer upon regions which once were circumpolar, the temperature of lower latitudes. But it has been demonstrated that the protuberance of the earth at the equator, so vastly exceeds that of any possible elevation of mountain masses between the equator and the poles, that any slight change which may have resulted from such geological causes, could have had only an infinitesmal effect upon the general climate of the globe.

Another ingenious writer sought to account for the remains of large trees that are found in Greenland, and for the traces of glacial cold in this country, by considering whether it might not be possible that the external crust or shell of the globe had actually slid round its fluid or semi-fluid nucleus, so as to bring the same areas of the external surface under very different conditions. Thus it was suggested that lands, which at one time basked under a tropical sun, might, in the slow course of ages, be shifted to some more northern region, while countries which had for long years been sealed up in the ice of the

Arctic Circle might eventually slide down to temperate latitudes.

But the theory which has taken firmest hold of the geological mind, is that so strenuously upheld by Sir Charles Lyell. This theory maintains that the climates of the past may be accounted for by that continuous change in the distribution of land and sea which has been going on all through the geologic ages. There is no fact more patent than that sea and land have frequently changed places. What are the rocks, of which the continents are mainly composed, but the hardened sediments of mud and sand that gathered upon the floors of ancient oceans? And what are all our so-called "formations" but just so many fossil sea-bottoms, as it were, piled one on top of the other? In the loneliness of the desert, in the streets of populous cities, in deep valleys, and on the crests of lofty mountains—everywhere we meet with traces of the sea. Along some coast-lines abundant evidence shows that the land is sinking down, and the sea slowly but surely gaining ground. In other regions the reverse takes place—the shores extend and the sea retreats. Such changes are due to those mysterious subterranean forces that give birth to earthquakes and volcanoes, and every bed of rock testifies to their unceasing activity. And, therefore, when this theory asks us to believe that the distribution of sea and land must sometimes have been very different to what it is at present, we are asked nothing which the facts do not fully justify us in admitting.

Lyell conceives, that, if land were massed chiefly in the region of the equator and the tropics, the climate of the globe would be such that tree-ferns might grow luxuriantly on any islands that might happen to lie within the Arctic or Antarctic Circle. For the land, heated to excess under the equatorial sun, would give rise to warm currents of air, which, sweeping north and south, would carry with them the heat of the tropics, and thus temper greatly the climate of higher latitudes. And some such condition of things, he thinks, may have obtained during, for instance, the Carboniferous period,* when tree-ferns and their allies flourished within the Arctic Circle. On the other hand, were the land to be grouped chiefly round the poles, the reverse of all this, he believes, would come about; for with no land under the equator to soak up the heat of the sun, and give it to the winds to carry north and south to polar regions, the climates of the northern and southern hemispheres would be so greatly affected, that snow and ice would then gather upon the ground, and creep gradually outwards down to those low latitudes where we now meet with their traces.

In these interesting speculations it will be observed that Sir Charles considers the atmosphere as the chief medium by which the heat derived from the sun is carried from one latitude to another: one of the principal reasons he gives for the intenser cold of the

* For Table of Geological Formations see Appendix, Note A.

Antarctic as compared with the Arctic regions, being the absence of land in the south temperate zone, where its presence would warm the atmosphere and so give rise to genial winds. It may be questioned, however, whether such would be the result. It seems more likely that the presence of land in the quarter referred to would only serve to increase the cold by affording another gathering-ground for snow and glaciers. Even if we supposed the land to remain uncovered with snow, and to succeed in warming the atmosphere, it is difficult to see how this could have any effect in ameliorating the climate of the higher latitudes about the pole. The heated air would certainly rise and flow towards the pole, while a cold under-current would set from the pole to restore the equilibrium. Long before the north wind, however, arrived at its destination, it would have radiated all its heat into the colder regions of space, and so would eventually reach the surface of the earth in its progress towards the pole as a cold, and not a warm wind.

On the same principle it does not appear likely that the massing of the land in the tropics, and under the equator, would have the effect which is supposed. The air, heated to excess over the equatorial and tropical regions, would, of course, rise and flow towards the poles, but its warmth would be filched from it, and dissipated into space before it could again return to the level of the sea. Thus, even supposing most of the land to be distributed between the tropics, it is evident that, as far as aerial currents themselves are concerned, the temperature of the seas about the poles might never rise above the freezing point, while any islands that might chance to lie within the Arctic and Antarctic Circles would be sealed up in persistent snow and ice.

The south-west winds, to which in the present economy of things we are indebted for the temperate character of our climate, do not derive their heat directly from the equator. The equatorial heat which they carry upwards is taken from them in their lofty flight towards the north, and thus they reach the level of the sea as cold, dry winds. But, blowing for many leagues in the Atlantic athwart the heated waters of the Gulf-stream, they gradually become warmed again and laden with moisture, and this warmth and moisture they yield up when they reach our coasts. Were there no broad currents of warm water setting towards the north, from which the cold dry winds on their descent to the sea-level might receive warmth and moisture, there is good reason to believe that the temperature of the northern hemisphere would be greatly depressed, and the cold of the arctic regions might then equal in intensity that of their antipodes. Were it not for the genial influence of the Gulf-stream, Scotland would experience a climate as severe at least as that of Labrador, while the greater part of Norway would be uninhabitable. As it is, however, the temperature of the winter at Hammerfest, in the north of Norway, is only 9° below the freezing point, while in the same latitude in Greenland the winter temperature is 5° below zero; the difference in summer being 41°

in Greenland and 50° at Hammerfest. Again, the temperature in the month of January in Caithness (58° N. lat.) is about 36,° or 4° above freezing point; but in the same latitude in Labrador the winter temperature falls to 4° below zero; and the winter temperature of Caithness is not attained on the American continent until we descend to 39° N. lat. on the shores of Chesapeake Bay. Lisbon, which is in the same latitude as the last-named place, has a winter temperature of about 47°, or 15° above freezing point.

These wide differences arise solely from the presence of the Gulf-stream. That great current of warm water, coming from the equatorial regions, washes all the western shores of Europe from the coast of Spain to the north of Norway. It laves the shores of Nova Zembla, and can even be traced north of Spitzbergen, where its waters are still appreciably warmer than those of the surrounding seas. The air in contact with this broad ocean-stream is everywhere warmed, and the winds licking up the heated water as they sweep across the Atlantic reach Europe laden with moisture, which is condensed and precipitated as soon as the winds are cooled by coming in contact with high tracts of land or with colder strata of air. For this reason our west coasts, which are the first to receive the winds, enjoy a more genial, but at the same time a more humid climate than the corresponding latitudes on the other side of the island.

The differences of temperature resulting from the

presence or absence of oceanic currents become still more striking when we compare the climates of inland tracts like central Siberia with the corresponding latitudes in western Europe. At Jakutsk, Siberia, (62° N. lat.), the mean winter temperature is 36°·6 below zero, and the thermometer has registered as low as 40° below zero. But Jakutsk lies only some 6° further north than Edinburgh, and is in nearly the same latitude as Bergen, where the temperature of January is not under 32°, or freezing point; while the western shores of Nova Zembla, lying between latitudes 71° and 76° N., have a winter temperature of about 10° above zero. This last example is exceedingly striking.

But perhaps the best way to get an adequate conception of the influence of a warm ocean current upon climate is to trace out a line of equal temperature. If every portion of the globe received all its heat directly from the sun, and were there no such disturbing influences as marine currents, winds, &c., then a line traced through all those places which enjoyed the same degree of warmth would of course run due west and east, and would correspond in direction with a parallel of latitude; in short, a place would be warm or cold according as it approached or receded from the equator. But owing to a variety of causes (chief among which is the influence exerted by marine currents and winds), this is not exactly the case. The lines are always undulating, and often rapidly so. Let us follow, for example, the isochimenal line * of +14°

* Lines of equal winter temperature are termed *isochimenal* lines (ἴσος, equal;

from Asia into Europe, and we shall find it crossing several parallels of latitude instead of conforming in direction to these. Taking it up a little to the east of the Caspian Sea, in lat. 45° N., we follow it as it undulates northward to lat. 55° N., not far from Smolensk. It then strikes due north until it reaches 60° N. lat., a little east of Petersburg, after which it trends north-west and west, crossing the Gulf of Bothnia in 64° N. lat. Its direction is now south-west, to 61° N. lat. in the south of Norway, and after this it suddenly curves round and doubles upon itself, running away to the north-east and keeping parallel to the Norwegian shores, but at some distance from the sea, for it intersects only a few of the longer fiords with which the coast-line of Norway is indented. After this it turns more to the east, still keeping parallel to the shore, crosses and recrosses 70° N. lat., turns to south-east, and finally terminates on the borders of Finland in lat. 68°. All places to the north and north-east of this line have of course a colder winter temperature than places south and south-west.* In short, as we proceed inland from the western shores of Europe, we find the cold of winter becomes more and more intense the further we penetrate into the interior, until central Siberia is reached, where we

χειμὼν, winter); of summer, *isotheral* (ἴσος, equal, θέρος, summer); while those indicating equal annual temperature are known as *isothermal* lines (ἴσος, equal, θερμὸς, heat). See Charts at end of volume.

* The wonderful manner in which this line is deflected to the north is due primarily to the influence of the Gulf-stream. Nor can we estimate too highly that influence, which succeeds so far in overruling the effect of mere latitude, as to be able to confer the same temperature on places situated so far apart as the borders of the Caspian Sea, Petersburg, and the north of Norway.

meet with a winter truly arctic, outside of the Arctic Circle. Before the south-west winds, which temper our winter, can reach this distant region, they are robbed of all their warmth, and only tend to increase the cold.

Thus much, then, is certain, it is the currents of the sea that are the chief carriers of heat from the tropics towards the pole; and such being the case, it seems quite reasonable to conclude that were the area over which the sea is heated and warm currents formed to be materially limited, as must certainly have been the case had the great continental masses ever been congregated within the tropics, then the chief secondary source of heat supplied to the temperate and frigid zones would be diminished, and the climates of those areas would become colder than they are at present.

While it is extremely unlikely, therefore, that a great accumulation of land within the tropics would have any ameliorating influence whatever on the climate of the poles, it may be admitted that were the continents to be grouped immediately round the poles, an excessively cold climate might possibly result. This low temperature, however, would not be due to the chilling influence of cold winds blowing from the south. Were an excessively cold climate to be the consequence of such a distribution of land as that last referred to, the low temperature would doubtless be caused by the exclusion of oceanic currents. It is quite possible that, among the many vast geologic changes which have taken place, the land may at some time in the past have been so closely gathered round the poles as to exclude the sea to comparatively low latitudes. If such had ever been the case, a vast breadth of land would then circle round the north pole, along the borders of which wet winds coming from the sea would deposit their moisture. But long before these winds could reach any great distance towards the interior of the supposed arctic continent, they would necessarily part with all their moisture, and arrive at the circumpolar regions as dry winds. In these regions, therefore, there could not be any accumulation of snow and ice. It is clear, then, that if we exclude the sea from our hypothetical arctic continent, we shall certainly lower the temperature of the polar climate; but we shall not on that account produce a glacial period, for the exclusion of the sea means absence of damp winds, and consequently no deposition of snow and no accumulation of ice.

But although it is possible, yet at the same time it is very improbable, that the land ever was so closely aggregated around the poles as to exclude the sea to comparatively low latitudes. All analogy would lead us to infer that even during the greatest extension of land in arctic and antarctic regions the ocean must have indented this land with broad straits, with fiords, inlets, bays, and seas. Supposing then that, with Sir Charles Lyell, we imagine all the continents, shaped as they are now, to be grouped around the poles and a vast sea to extend over equatorial and tropical regions; what, let us ask, would be the result? Why surely this, that a much wider extent of sea being exposed to the blaze of the tropical sun, the temperature of the ocean in equatorial regions would rise above what it is at present. This warm water, sweeping in broad currents, would enter the polar fiords and seas, and everywhere heating the air, would cause warm moist winds to blow athwart the land to a much greater extent than they do at present; and these winds, thus distributing warmth and moisture, might render even the high latitudes of north Greenland habitable by civilised man.

But on any supposition, it is impossible to conceive of such a distribution of land and sea as would cause an influx of warm water into the frigid zone of sufficient extent to raise the whole temperature of the arctic regions, so as to permit the growth of tree-ferns in the higher latitudes, and to nourish in the depths of the polar seas reef-building polyps and chambered shells of species resembling in general appearance those that are met with only in the genial waters of the sunny south. Were a greater body of warm water than the present Gulf-stream, owing to some revolution of the earth's surface—some readjustment of the relative position of sea and land—to flow into the arctic regions, it would doubtless bring about some ameliorating effect upon the climate; but, as I have said, we can hardly imagine that a mere redistribution of the land would induce currents of warm water to flow from the tropics towards the pole in such immense volumes as to preclude the possibility of ice forming within the Arctic Circle. And we know that there have been times in the past when snow and ice either did not occur within the Arctic Circle, or at all events could have been present only in very small quantities. Not much certainly could have floated in the polar sea that nourished the delicate corals and molluscs, whose fossil remains have been disinterred from the rocks of the bleak shores of the Arctic Ocean. In our day, at all events, reef-building corals do not live in seas the temperature of which is under 66°. Nor is it in any degree likely that while the vegetation of the coal-measures was flourishing in what is now the frigid zone, large glaciers could have descended from the higher grounds of those regions in broad, deep streams, so as to reach the sea as they do in our day. If glaciers did exist at all, they must have been few and local, and confined to the recesses of lofty mountainous tracts, did any such elevated areas happen to exist within the arctic regions at that time.

It would really seem, then, after allowing all possible influence to such geological changes as may have resulted in redistribution of land and sea, that these changes yet fail to solve the problem before us. No doubt climates may be and have been varied repeatedly by local disturbances of level, by submergence here and elevation there. But, however extensive such changes may have been, yet their influence could not so far affect the general climate of the globe as to confer at one and the same time upon the whole northern hemisphere, down to low latitudes of the temperate zone, a severe arctic climate, and at another period a climate warm and genial, with no extremes of heat and cold, but a kind of perpetual summer.*

* The reader will find this interesting question discussed by Mr. Croll in *Philosophical Magazine* for Feb. 1870, Oct. 1870, and Oct. 1871.

CHAPTER IX.

CAUSE OF COSMICAL CHANGES OF CLIMATE—*Continued.*

Failure of geologists to furnish an adequate theory.—Astronomical phenomena may perhaps afford a solution.—Movement of the earth on its own axis and round the sun.—Eccentricity of the orbit.—Precession of the equinoxes.—Nutation.—Obliquity of the ecliptic.

SEEING that the phenomena of elevation and depression fail to account for such great cosmical changes of climate, is it not possible that a solution of the problem may be found in the relations of our planet to the sun? Geologists of the modern school have always been jealous, and justly so, of attempts to explain or account for the facts of their science by reference to causes other than those they see at work in the world around them. And perhaps the frequent failure of physicists and astronomers to frame a satisfactory theory for those great changes of climate to which the rocks bear emphatic testimony may sometimes have been viewed by geologists with a kind of secret satisfaction. Of late years, however, the opinion has been gaining ground among our hammer-bearers that in this matter of cosmical climate they must, after all, be content to follow the guidance of the astronomer and the physicist, seeing that their own principles refuse, in this particular at least, to yield as much assistance as would be desirable. Nor, after he has sufficiently questioned all the natural causes with which his own peculiar studies have made him familiar, ought the geologist to feel surprised that these sometimes fail to explain the phenomena that come under his cognizance. There is no hard and fast line separating the domain of one science from that of another, and as the circle of knowledge widens boundary divisions become more and more difficult to determine. Perhaps of no physical science is this more true than that of geology. At one time the investigator into the past history of our globe had the field almost entirely to himself, and the limits of his study were as sharply defined as if they had been staked off and measured. Now, however, it would be hard to say on which of the territories of his scientific neighbours he must trespass most. He cannot proceed far in any direction without coming in contact with some worker from adjacent fields. His studies are constantly overlapping those of the sister sciences, just as these in turn overlap his. It will, therefore, only be a further proof of the unity of Nature if those intricate problems which have hitherto baffled the geologist should eventually be solved by the researches of astronomers and the conclusions of physicists.

I have already referred briefly to certain astronomical and physical theories which have from time to time been advanced in explanation of the grand climatic changes we have been considering. Of these, as I have indicated, some are too problematical to be relied upon as guides in our investigations into the origin of cosmical climates; while the most probable of them all, that, namely, which is supported by Sir Charles Lyell and many geologists of the modern school, appears to be inadequate to explain the more important phenomena.

Now, if all these theories be rejected, what have we to supplant them? Well, there are certain considerations arising from the fact that, when long ages are taken into account, our earth upon its circuit round the sun and in its own diurnal revolution does not always hold exactly the same position with reference to the sun. But before entering upon these considerations it may be well to remind the reader of certain astronomical data, without a thorough comprehension of which it will be impossible to convey a clear conception of the theory which I shall attempt to describe.

There are, as every one knows, two principal movements proper to the earth. The first of these is its translation through space, during which it circles round the sun from west to east, and gives us the succession of the seasons in a period of one year. The second movement is that of rotation on its own axis, a movement which, as no one needs to be told, results in the phenomena of day and night.

The orbit or path described by the earth round the sun is not exactly circular, but rather an ellipse, and this elliptical path is known to astronomers as the *ecliptic*. Neither is the sun placed quite in the centre of the ecliptic or orbit, but somewhat to one end of it, or, in other words, it occupies one focus of the ellipse, so that the earth, during its annual revolution, is nearer or farther from our great luminary according as it approaches or recedes from that focus of the ellipse in which the light-giver hangs. The point in the ecliptic at which the earth approaches nearest to the sun is called *perihelion*, and the point where it reaches its greatest distance from that luminary becomes in astronomical language its *aphelion*. But it must not be supposed that the earth as it journeys year by year round the sun, pursues always exactly the same elliptical path. The attractions of the other planets are producing day by day a slow change on the *shape* of its orbit: in this way its track approaches at one time nearly to a circle, and at another time has a more oval and flattened outline. But these deviations are confined within certain limits, between which they are constantly oscillating backwards and forwards. The eccentricity during a long lapse of years goes on decreasing till it sinks to a minimum value, and the orbit then approaches most nearly to a circle, without, however, ever becoming actually circular. After passing this point the eccentricity begins to increase, and the orbit becomes more and more flattened, till a maximum eccentricity is reached. Then the cycle of changes comes slowly round again in reversed order; the orbit gradually draws nearer and nearer to a circle till a minimum is arrived at, when it again begins to grow elliptical, and continues to alter in this direction till the next maximum is reached. At present, the path is slowly approaching the more circular route, and in about 24,000 years from this date the ellipticity will reach one of the minimum points. After that the earth will again begin to follow a more and more elliptical course round the sun, until, when thousands of years have elapsed, its orbit shall have attained its maximum eccentricity; and then the ellipticity will again slowly diminish as before. It is important to notice that the intervals between consecutive turning-points are very unequal in length, and the actual maximum and minimum values of the eccentricity are themselves variable. In this way it comes about that some periods of high eccentricity have lasted much longer than others, and that the orbit has been more elliptical at some epochs of high eccentricity than at others.

When the maximum of ellipticity is attained, the earth in aphelion will, of course, be farther from the sun than it is now, at that point in the ecliptic; while on the other hand, the earth in perihelion will be nearer. At present, the earth in aphelion is distant about ninety millions of miles from the sun, but when the eccentricity of the orbit is at its superior limit, in other words, when the planets by the force of their attraction have succeeded in pulling the earth as far from the sun as they can, the earth in aphelion will then be rather less than ninety-eight and a half millions of miles from the sun, or eight and a half millions further than it is at present. Of course the reader will understand that while the earth's orbit varies in its degree of ellipticity, the time taken by the earth to complete a revolution round the sun never does. All that the planets do is to modify the shape of the path traversed by the earth.

Besides its movement of translation through the heavens round the sun, the earth, as we know, revolves or rotates on an axis of its own. Now this axis is inclined to the plane of the ecliptic at a certain angle which we will for the present consider to remain constant, so that the axis is always parallel to itself. This is as much as to say that an imaginary line continued through the poles to the skies will be found to touch nearly the same vanishing point in the heavens all the year round, so that no matter what the season may be—winter, spring, summer, or autumn, nor whether the earth is in aphelion or perihelion, a pole will always point in one and the same direction—the point in the heavens to which our north pole is directed being situate quite near to the Pole star.

If we draw through the centre of the sun a plane parallel to the earth's equator, this will cut the earth's orbit at two points, which are termed the *vernal* and *autumnal equinoxes*. The two points half-way between the equinoxes are called the *summer* and *winter solstices*. On the days when the earth passes through the equinoxes the day and night are of equal length all the world over. As the earth leaves the vernal equinox the days in the northern hemisphere begin to get longer than the nights, and continue to increase in length up to the summer solstice; they then decrease till the autumnal equinox is reached, when day and night are again equal. In other words, the time taken by the earth to pass from the vernal to the autumnal equinox is our summer half of the year. In exactly the same way, the time occupied by the earth in travelling from the autumnal to the vernal equinox is our winter half of the year, when the days are shorter than the nights. A very little consideration will show that in the southern hemisphere just the opposite condition of things exists: while it is our summer they are passing through their winter half of the year, and summer comes to them at the same time as winter is passing over our heads. Further, it happens that our mid-winter now occurs very nearly when the earth is in perihelion, or at that point of its annual circuit which is nearest to the sun, and from this important consequences flow. If we look at the two portions into which a line joining the equinoctial points divides the earth's orbit, we shall see that one is longer than the other. It is the shorter bit of the two which corresponds to the winter of the northern hemisphere, and moreover this is just the part of its path over which the earth, on account of its greater proximity to the sun, moves fastest. Both these causes work together to make our winter at present shorter than our summer. In fact, if we compute from the 20th March, or vernal equinox, to the 22nd September, or autumnal equinox, we shall find that the earth takes 186 days to swing round that portion of its circuit within which the northern hemisphere has its spring and summer—the other half of its orbit, which brings the southern hemisphere its summer and us our winter, being traversed in 179 days: so that the summer period of our antipodes is seven or eight days shorter than our own.

Such is the arrangement of the seasons in the two hemispheres at present—our summer occurs when the earth is in aphelion, and is longer than our winter, while that of the southern hemisphere arrives in perihelion, and the winter is now there longer than the summer. But this arrangement has not always obtained. There was a time when our hemisphere had its winter in aphelion, and a period will again arrive when the present condition of things will be reversed, and the seasons in the two hemispheres will completely change. That is to say that our December solstice, instead of being our winter will become our summer, while our future winter solstice will happen in what are now our summer months. This great change is effected by a movement which is known as the *precession of the equinoxes.*

It is not strictly true that the axis of the earth always remains, as we have been hitherto supposing, parallel to itself: for short periods this is nearly true, but nevertheless there is a slight change going on, which in the course of ages comes to be considerable. The attraction of the sun and moon on the protuberant parts of the earth around the equator are for ever, by their unequal pull, causing the axis slightly to shift its position, and the sum total of these displacements results in the motions known as *precession* and *nutation*.

It is the first of them only with which we are specially concerned: it may be thus illustrated. Take two straight sticks, unite one end of one to one end of the other by a loose joint, and connect the other ends by a bit of string; now while one stick is held steadily at rest, move the other stick round it, *so that the string is always kept tight;* the motion of the second stick will then resemble, in everything except speed, the movement of the earth's axis known as *precession.* The extremity of the movable stick evidently describes a circle in space, and if we in imagination conceive the axis of the earth to be prolonged so as to touch the heavens, we should find that in the great cycle of its revolution the end of the imaginary line we had protracted would trace out a circle among the stars. Now, bear in mind that the equinoxes are the points where a plane through the sun's centre, perpendicular to the earth's axis, cuts the earth's orbit. Since the axis is always in motion, this must also be the case with this plane, and therefore with the equinoxes. As the axis swings round it necessarily carries the last with it, and they travel along the ecliptic in a direction opposite that of the earth's annual revolution; and calculations show that the average distance traversed by them in a year is just what would be passed over by the earth in twenty minutes and twenty seconds, so that equal day and night come to us every year some twenty minutes earlier than they did the year before. Thus the places on the earth's orbit of those points on which the seasons depend are constantly shifting round. The summer solstice, for instance, will not always, as now, nearly coincide with the aphelion, it will slowly draw up to the perihelion, and we shall then have a disposition of the seasons corresponding to that which now obtains in the southern hemisphere: afterwards, a continuation of the motion will as slowly bring round again our present arrangement.

This great oscillatory movement would run through a complete cycle in about 26,000 years, were it not for another complex movement, due to the action of the planets, which succeeds in shortening the great cycle by some 5,000 years; so that in half that time, or 10,500 years, our seasons will have completely changed, and the northern will then assume the arrangement of the seasons which at present characterizes the southern hemisphere. In 10,500 years more the equinox will regain its initial position, and the distribution of the seasons that now obtains will return.

This movement of the axis is effected by the attractions of the sun and moon, which act unequally on the globe, owing to the fact that our earth is not an exact sphere, but somewhat flattened at the poles and swollen at the equator. If the world were a perfect sphere no such inequality of action could take place, but the axis would always remain exactly parallel to itself.

The portion of the motion of the earth's axis called *nutation* consists in small deviations, first to one side and then to the other, from the position it would have at any time if precession alone were taken into account. It is as if the string in our illustration were slightly elastic, and kept alternately lengthening and shortening itself a little. Under these circumstances the path of the end of the movable stick will be like the edge of a circular disc with a very slightly crimped or wavy outline; and this is the character of the path actually described by the pole of the earth's axis in space. The effect of nutation is alternately to increase and diminish the inclination of the equator to the ecliptic, but its amount is so very small that for all the purposes we are now concerned with it may be neglected. Nevertheless astronomers have long been aware that this inclination does vary to a greater degree by reason of a movement due to the joint action of all the planets. This movement is at present gradually bringing the earth's equator to coincide more nearly with the plane of the ecliptic. The effect of this change in the angle at which the axis is inclined to that plane is of course to lessen the duration of the long day and night at both poles. And if it were possible for the obliquity to entirely disappear, and the plane of the ecliptic to coincide exactly with that of the equator, there would then be no difference in the day and night in any part of the world during the year. But the limits within which the position of the axis, with reference to a fixed point in the heavens, can thus change are very moderate. After a vast succession of ages, the obliquity which just now is diminishing will again increase until it reaches its maximum of 24° 50′ 34″. This is a very small change in the angle of inclination after all, but Sir John Herschel thought that, if millions in place of thousands of years were taken into account, the maximum change in the obliquity might increase to 3° or even 4° on each side of the mean. With such a change as this the polar day and night would of course lengthen out, for the farther the axis bends over towards the sun the larger must be the area round the poles illuminated in the long summer day, and conversely the broader must be the extent over which in winter-time the polar night will prevail.

CHAPTER X.

CAUSE OF COSMICAL CHANGES OF CLIMATE—*Continued.*

Effect of variation of the eccentricity of the earth's orbit.—Sir J. Herschel's opinion.—Arago's.—Purely astronomical causes insufficient to afford a solution of the problem.—Mr. Croll's theory.—Changes of climate result indirectly from astronomical causes.—Physical effects of a high eccentricity of the orbit.—Extremes of climate at opposite poles.—Modifications of the course followed by ocean-currents.—Perpetual summer and perennial winter.—Physical effects of obliquity of the ecliptic.—Succession of climatal changes during a period of high eccentricity.

THE possible bearing that astronomical phenomena might have upon the climate of our globe has not infrequently engaged the attention of astronomers. Sir John Herschel was inclined to admit that variations in the eccentricity "may be productive of considerable diversity of climate," and might so operate during great periods of time "as to produce alternately in the same latitude of either hemisphere a perpetual spring, or the extreme vicissitudes of a burning summer and a rigorous winter." And he was also of opinion that, owing to the precession of the equinoxes and the shifting of the earth's axis by another movement, these strongly contrasted conditions would gradually be transferred from one hemisphere to another. Hence he thought it not improbable that some of the indications noted by geologists, of widely different climates having prevailed at former epochs in the northern hemisphere, might in part at least be accounted for.

Other eminent astronomical writers, among whom was Arago, came to quite a different conclusion, and were of opinion that no appreciable change of climate could possibly result from any variation in the ellipticity of the earth's orbit. The earth, they argued, receives the same total amount of heat in the aphelion as in the perihelion section of its orbit. And this, as we know, is due to the fact that our globe moves with greater velocity in perihelion than it does in aphelion. Thus it happens that, although the southern hemisphere in perihelion is turned towards the sun, and must receive *per diem* a larger share of heat than the northern hemisphere derives in the same space of time in aphelion, yet the perihelion section of the orbit is quickly travelled; so that greater nearness to the sun only serves to make up for the short time the earth keeps in that position, just as in aphelion the greater distance of the earth from the sun is exactly counterbalanced by the longer time our globe remains exposed to the solar rays. Nor does it matter, said some of the astronomers, to what extent the ellipticity of the orbit may vary—for it may reach the very highest degree of eccentricity, and yet the equal distribution of the sun's heat in perihelion and aphelion must continue invariable. Any difference in the amount of heat that might follow upon an increase of eccentricity, would, it was thought, rather take effect in bringing about a generally warmer climate for the whole world, than in producing a glacial period in one hemisphere and perpetual spring or summer in the other. And this conclusion was based on the fact that the total heat derived by the earth from the sun is inversely proportional to the minor axis of the earth's orbit.

Such were some of the arguments brought forward to show that cosmical changes of climate could not be due to variations in the eccentricity of the earth's orbit. Each of these arguments is strictly consonant with well-ascertained fact, and cannot possibly be gainsaid; and therefore it may at once be admitted that purely astronomical causes alone will not account for that wonderful alternation of extreme climates to which the geological record bears witness. But in a remarkable paper published in 1864, Mr. Croll clearly showed that, "although a glacial climate could not result directly from an increase in the eccentricity of the earth's orbit, it might nevertheless do so indirectly." At the present time the ellipticity of the orbit is such that the earth, when nearest to the sun or in the perihelion part of its course, receives in a given time one-fifteenth more heat than it does in aphelion. Now, this being the case, it is quite evident that, did the earth travel round the sun at the same rate in all parts of its orbit, or, in other words, were the seasons of equal duration, the southern hemisphere, which has its summer in aphelion would not only, as it does now, receive more heat *per diem* than the northern, but its annual proportion would also be greatly in excess. But, as the reader has already been reminded, this is prevented by the unequal pace at which our globe hurries on its way—the result being that both sections of the world receive the same yearly amount of heat. Mr. Croll points out that the present difference between the two hemispheres, as regards the proportion of heat derived from the sun in a given time, would be vastly increased when the eccentricity reached its highest value. If at a period of maximum eccentricity the winter of our hemisphere should happen in aphelion, he thinks that we should then be receiving nearly one-fifth less heat during that season than we do now, and in summer-time, of course, nearly one-fifth more. But if, on the other hand, our winter should, as at present, fall in perihelion, the effect of a great ellipticity he believes would be such as to annihilate the difference between summer and winter in the latitude of this country. And he grounds these conclusions upon purely physical considerations.

During a period of great eccentricity of the earth's orbit, the earth in aphelion would be rather more than eight and a half millions of miles farther from the sun than it is now, and the present long frigid winter of the southern hemisphere would then become still longer, and the cold much more intense. If it happened to be the northern hemisphere whose winter occurred in aphelion, of course similar climatal results would ensue, and the mean temperature of our winter would fall below the freezing point. Consequently, all the moisture precipitated in our latitude during that season would fall in the form of snow, and the British seas would be frozen over. Nor would the greater proximity of our hemisphere to the sun in perihelion avail to free these islands from their frost. It is true that the direct heat received in perihelion during the summer of a period of great eccentricity would exceed that which we now derive during that season by nearly one-fifth; but this intense heat, paradoxical as it may seem, would not give rise to a warm summer. The summers of North Greenland, we know, are colder than our winters, notwithstanding that the rays of the sun in that region are so strong as to melt the pitch on the sides of ships. Every one, indeed, has heard of the heat of the arctic sun, which shines day and night during the whole summertide. But despite the sun's power, the mean temperature of summer in North Greenland does not exceed one or two degrees above freezing point, and this is entirely owing to the presence of snow and ice. Were it not for these, the sun would heat the ground, and the ground would impart its warmth to the atmosphere, and the summer temperature would then rise to something like our own. It is only in this way, or from passing over warm water, that the air can be heated, for the direct rays of the sun pass through it without sensibly affecting it. Now in the polar regions the sun's heat is used up chiefly in melting the snow and ice, and not in warming the ground; so that comparatively little of the summer heat finds its way into the atmosphere by radiation. Such being the case, it is not difficult to follow Mr. Croll when he argues that the sun, during a period of great eccentricity, would not be able to give a warm temperature to that hemisphere whose summer happened in perihelion. An increased amount of evaporation would certainly take place, but the moisture-laden air would be chilled by coming in contact with the vast sheets of snow that had gathered during the long intolerable winter, and hence the vapour would condense into thick fogs, and cloud the sky. In this way the sun's rays would to a large extent be cut off, and unable to reach the earth, and consequently the winter snows would not be all melted away. Nor, supposing there were drenching rains during the summer, would these suffice to dissolve more than one-eighth part of the snow and ice—for, as Mr. Croll remarks, "it takes nearly eight tons of water at 58° Fahr. to melt one ton of snow."

The accounts given by voyagers who have sailed in the south polar seas are often highly interesting, as showing the great difference in climate between lands situated at the same distance from the poles in the two hemispheres. They all agree as to the intenser cold of the antarctic as compared with the arctic summer, and to the greater frequency of cold raw fogs in the southern polar regions. Captain Forster, of the *Chanticleer*, who spent several months making observations at Deception Island, mentions specially the thick fogs and strong gales which he encountered. The fogs indeed were so frequent and thick, that for ten days neither sun nor stars were seen, and the air was so intensely raw and cold that Lieutenant Kendal did not remember to have suffered more at any time in the arctic regions. And yet this was in January, the very midsummer of the south, and in a latitude corresponding to that occupied by the Faroe Islands, where the climate is much the same as in our own Shetland. Now, when we remember that this cold raw summer of the south happens while the earth is actually nearer to the sun than it is at the time the milder northern polar regions have their summer, we cannot but admit that mere proximity to the sun will not necessarily produce a warm season.

At the time of greatest eccentricity, when the earth would be nearer to the sun in perihelion than it is now, there can be no doubt that the heat received would be correspondingly increased. But during the long winter of aphelion—longer by thirty-six days than the summer of perihelion—such an accumulation of snow and ice would have taken place, that even the diminished distance between the earth and sun in summer-time would be powerless to effect its removal; and so it would go on increasing year by year, until all northern countries (winter happening in aphelion) down to the latitudes of these islands were swathed in a dreary covering of snow and ice. There would then be a glacial period over our hemisphere, while at the antipodes a very different condition of things would obtain. Supposing, as before, that the precession of the equinoxes had caused the summer of the southern hemisphere to happen at the time the earth was in aphelion, we should then have a climate for our antipodes exactly the opposite of that which, as

Mr. Croll has shown, a maximum eccentricity would confer upon the northern hemisphere. The heat received would be less in a given time, but then summer would be thirty-six days longer, while winter would be much milder, and correspondingly short, owing to the sun being nearer than in summer by more than eight millions of miles. The result of all this would be to equalise the seasons. There would be a long cool summer and a short genial winter, during which probably little, if any, snow would gather; and thus there would be an approach to what Herschel has called "a perpetual spring."

There is another set of circumstances, however, which, according to Mr. Croll, would help powerfully to increase the difference between the two hemispheres. Along with many other eminent physicists and geographers, he maintains that the great constant currents of the ocean are due to the action of the trade-winds pressing upon the surface of the sea. Now the trade-winds exist, as every one knows, by reason of the unequal temperature of the atmosphere at the equator and the poles. The air is heated under the equator, and rises to flow towards the poles, while cold currents set from the poles towards the equator to restore the equilibrium. At present all the constant oceanic currents appear to flow out of the Antarctic Ocean. In this way the wide Equatorial Current is only a continuation of the great Antarctic Drift-current, which, flowing north-east, enters the Indian Ocean, sending one branch by the west coast of Australia northwards through the

Indian Archipelago, and another stream westwards, so as to strike the east coast of Africa. Leaving the Mozambique Channel, this great current now doubles the Cape, and then continues on its course north-westwards along the African coast, until eventually, sweeping across the whole breadth of the Pacific, it divides, one stream flowing south along the coast of Brazil, and the other striking for the Gulf of Mexico, from which, when it issues, it takes the name of the Gulf-stream.

During such a glacial condition of things as would follow upon a great increase in the eccentricity of the earth's orbit, the air in the northern hemisphere (supposing the winter of that hemisphere to occur in aphelion) would be chilled down to a very much lower temperature than in the corresponding latitudes of the opposite hemisphere. And as such would necessarily be the case, it follows that the aerial currents flowing from the poles to restore the equilibrium which the upward set of the heated air under the equator had disturbed would be of unequal strength. The winds from the severe wintry north would sweep with much more vigour towards the equator than the opposite winds from the south pole. And hence Mr. Croll contends that with weaker winds blowing from the south the great Antarctic Drift-currents would be reduced in volume, while the subsidiary currents to which they give rise, namely the broad Equatorial and the Gulf-stream, would likewise lose in volume and force. And to such an extent would this be the case, that, supposing the outline of the continents to remain unchanged, not only would the Brazilian branch of the Equatorial Current grow at the expense of the Gulf-stream, but the Gulf-stream, he thinks, would eventually be stopped, and the whole vast body of warm water that now flows north be entirely deflected into the southern oceans. For the same reason also the currents of the Pacific, which carry so much warmth from the tropics to the north* would also be turned back. If such were the case, we can easily conceive that the reduction of temperature caused by the withdrawal from the north of all these great ocean-rivers of heated water would be something enormous. But so much loss to the northern hemisphere would be just so much gain to the southern, which would have its temperature raised to such a degree that, in place of a "perpetual spring" there might well be "perpetual summer" within the Antarctic Circle.

Besides an increased degree of ellipticity of the earth's orbit, there is also another astronomical cause which may have no mean influence upon cosmical climate. This is a change in the obliquity of the ecliptic. We have seen that the axis of rotation, owing to the action of the planets, is compelled

* The southern hemisphere has a cooler mean annual temperature than the northern, notwithstanding that both hemispheres receive from the sun exactly the same quantity of heat. This anomaly is usually explained by assuming that the southern hemisphere loses by radiation more heat than the northern, the winter of the former being colder and longer than that of the latter; but Mr. Croll points out, with much apparent reason, that the cooler temperature of our antipodes cannot be so accounted for, but is due rather to the constant transference of heat to the north by means of ocean currents, nearly all the great currents originating south of the equator.—*Philosophical Magazine*, Sept. 1869.

slowly to change the degree of its inclination to the plane of the ecliptic. At present this inclination is gradually growing less, and the effect of this is to shorten the long day and equally long night at both poles. Nor is it difficult to understand how, if this inclination disappeared altogether, and the axis of rotation then became perpendicular to the ecliptic, we should have a perpetual equinox; for the plane of our earth's equator would then exactly coincide with that of the ecliptic, and the sun's light being equally diffused over one entire half of the globe, at the same instant of time, day and night would necessarily be of equal duration in all parts of the world all the year round. But, as we know, the inclination of the earth's axis can vary only within comparatively narrow limits. Narrow though these limits be, however, it is, to say the least, highly probable that a change from a minimum to a maximum degree of obliquity could not take place without considerably influencing the climate of the poles. Mr. Croll has calculated that at the period of maximum obliquity, that is, when the poles in summer-time bend over towards the sun, so as to bring a wider area of the polar regions within the solar influence, there would be "one-eighteenth more heat falling at the poles than at present"—an amount of heat which, were there neither snow nor ice at the poles, would raise the temperature within the Arctic and Antarctic Circles by something like fourteen or fifteen degrees.

The precise time occupied in passing from a minimum to a maximum obliquity has not as yet been determined exactly; but during that prolonged period of great eccentricity which caused our glacial epoch, a maximum obliquity would no doubt be attained more than once. At this period there was, as we have seen, a vast accumulation of snow and ice in our hemisphere, consequent upon our winter occurring when the earth was farthest removed from the sun. Under such conditions an increase of obliquity could not, so long as the snow and ice remained, raise the temperature; for the extra heat derived from the sun in the lengthened summer day would in a large measure be consumed in melting the snow and ice. But as there would be one-eighteenth more snow and ice melted than at present, the polar ice-cap would be reduced, and consequently the rigours of glaciation in our hemisphere would be diminished. To some extent, therefore, a change of obliquity would tend to neutralise the effects of a high degree of eccentricity.

At the opposite pole, upon which a large eccentricity had conferred a perennially warm and equable climate, the effects of increased obliquity would be to remove any ice that still remained, and in this manner to increase the general warmth of the atmosphere. When, upon the other hand, the obliquity had reached its minimum, the immediate result of that would be to increase the snow and ice in the northern hemisphere, and to give a somewhat cooler temperature to the southern.

Thus, while we are considering the effect upon cosmical climate of a great increase of ellipticity of the earth's orbit, we must be careful to remember that an increase of obliquity of the ecliptic will ever have a tendency to modify this. Nor, indeed, is it at all improbable that, when the eccentricity is very high without being actually at its maximum, the increase of heat due to an increased obliquity may be quite sufficient, so long as it lasts, to prevent any excessive degree of glaciation in that hemisphere whose winter happens in aphelion.

We have already seen that a high degree of eccentricity would give rise to a whole series of physical changes—every one of which would tend to widen the difference of temperature between the opposite hemispheres. Now, if the obliquity of the ecliptic reached a minimum during our glacial epoch, as indeed it must have done more than once, the effect of great eccentricity and diminished obliquity combined, would be to intensify the glaciation of our hemisphere. The result of this would be to aid still more in the transference of all warm ocean currents from northern into southern seas. I have already referred to the enormous influence exerted upon climate by the presence of these great bodies of warm water: so enormous indeed is this influence that it appears in the highest degree probable that the mere removal, by whatever cause, of such currents from our hemisphere would be sufficient to induce the growth of glaciers in this country, while in the Antarctic Circle, to which the warm currents had been transferred, ice and snow would in large measure disappear.

I would remind the reader, however, that while many eminent physicists maintain that the constant currents of ocean are caused by the continuous impact of the trade-winds, which, by pushing forward in one direction the superficial strata of water, force these to drag forward in turn the strata that immediately underlie them, several writers have advocated other theories of their origin. It would lead me into too long a digression, however, were I to attempt the discussion of these various theories, even if this were the place for doing so. The evidence, and as it would seem, the weight of opinion also, appear to favour that theory which would assign a chief part in the origin of the great constant ocean-currents to the action of the winds. It does, to say the least of it, seem highly suggestive that the course of the main currents from the south should be just the same as that followed by the trade-winds; and that the south-east trades, being so much more powerful than the north-east trades, the currents from the south should likewise be stronger than those from the north.

Having shown the probable effect upon the climate of the globe that would ensue from a great increase in the present ellipticity of the earth's orbit, and how this effect would be modified by changes in the obliquity of the ecliptic, it is hardly necessary to point out at length how this astronomical *vera causa* appears to harmonize with the remarkable facts brought to light by geologists. It has already been stated that astronomers have ascertained the time required for the earth's orbit to pass from a minimum to a maximum eccentricity, and this, it appears, is very irregular.

It is only, however, when the eccentricity arrives at something considerable that the climatal effects we have been considering will become apparent. During the millions of years that have elapsed since the oldest rocks that we know of were deposited, the earth's orbit has frequenty attained a high degree of ellipticity. But by laborious calculations it has been found that the duration of such an eccentricity as would suffice to produce extreme conditions at the poles is very unequal. In the past three millions of years alone, Mr. Croll shows that there have been three such periods, separated by exceedingly irregular intervals, the periods lasting respectively for 170,000, 260,000, and 160,000 years. So that even in the shortest of these periods there would be time for the precession of the equinoxes to complete several revolutions. That is to say that our hemisphere, during this long cycle of great eccentricity, might experience several glacial periods and several periods of such genial climates as we have referred to above, and each of these periods would last for thousands of years. For, as we have seen, the equinoctial point takes something like 21,000 years to effect a complete revolution upon the ecliptic, in half of which time the seasons in the two hemispheres would of course be reversed—and the pole which had enjoyed continuous summer would then be doomed to undergo perpetual winter—these conditions being modified from time to time by changes in the obliquity of the ecliptic.

That our hemisphere has frequently undergone such extraordinary vicissitudes of climate the records of geology sufficiently attest, and some of the proofs referred to are roughly stated in the opening pages of this chapter. It is not necessary, however, to enter into details with respect to all the great changes of climate, whether glacial or the reverse, of which the solid rocks have preserved some relics.* We are at present concerned only with those excessive glacial conditions that were the result, as we have seen reason to believe, of the last great increase in the ellipticity of our earth's orbit, which began some 240,000 years ago and terminated about 80,000 years ago—embracing a period of 160,000 years. The cold was most intense about 200,000 or 210,000 years ago, that is about 30,000 or 40,000 years after the glacial period had commenced. Now, during the continuation of this vast age of high eccentricity, our hemisphere must have experienced several great vicissitudes of climate. Glacial periods, lasting for thousands of years must have alternated with equally prolonged periods of genial conditions; for the latter, no less than the former, are a necessary consequence of extreme ellipticity, combined with the precession of the equinoxes. And during all these changes the general outline of the continents has remained much as it was before the advent of the glacial epoch. Whatever influence upon climate, therefore, the relative distribution of land and sea may be allowed to have, it is quite certain no one can show that our glacial climate was induced by any peculiar arrangement of land and sea. It has been considered that

* See Appendix, Note A.

during the age of ice the land in the high latitudes of the northern hemisphere stood at a relatively higher level than at present. But there is no proof that the land attained either a much greater altitude or covered a much wider area than it does now; on the contrary, all the evidence goes to show that large tracts of the northern hemisphere which are at present in the condition of dry land were, at various times during our glacial epoch, submerged.

If, then, it be astronomically and physically true that extreme eccentricity of the earth's orbit, combined with the precession of the equinoxes, would confer upon our hemisphere long periods of continuous summer, separated by equally long periods of continuous arctic winter, we may next inquire whether there are any geological facts connected with the till itself, which, apart from any other considerations, would lead us to infer that our glacial epoch was in reality not one continuous age of ice, but a period interrupted by long ages of milder conditions, during which the ice disappeared from the low-lying parts of the country, and may even have vanished for a time altogether.

CHAPTER XI.

BEDS SUBJACENT TO AND INTERCALATED WITH SCOTTISH TILL.

Beds in and below till.—Seldom seen except in deep sections.—Examples of superficial deposits passed through in borings, &c.—Sections exposed in natural and artificial cuttings.—Examples.—Beds contorted and denuded in and below till.—Examples.—Fossiliferous beds in the till.—Examples.—Striated pavements of boulders in till.

THE reader who has accompanied me so far will remember that, while describing the till, I mentioned, in passing, that it sometimes contained beds of silt, clay, sand, and gravel.* The beds referred to seem to occur somewhat partially, the till in many districts not showing any such intercalations, but this seemingly partial distribution is more apparent than real. In the upper reaches of the valleys, where the top-covering of till is not often thick, the streams are able to cut down through this to the solid rocks, and thus expose any intercalated beds of gravel and sand or clay which the till may chance to contain. But where the valleys widen out into the broad undulating tracts of lower ground, there are few natural

* These, I need hardly say, are to be distinguished from those thin lenticular patches, layers, and irregular beds of sand, earthy clay, and gravel which are here and there enclosed in the till, and form part and parcel of that deposit as described at page 24.

sections to disclose completely the character of the drift. Many of the streams do not go down through the whole thickness of the till to the underlying rocks, so that we cannot always be sure from the sections seen that the till may not contain or overlie beds of sand, silt, and gravel. In such low-lying districts, too, railway-cuttings do not often go deep into the subsoils, so that we get comparatively little aid from them either. So long as we have only a partial exposure of the drifts, and not a complete section from the surface down through all the subsoils to the pavement of rock on which these rest, we are not entitled to assume that the whole drift-covering consists of till, merely because that deposit may chance to be the only kind of drift visible at the surface. The results obtained from a number of careful borings should render us cautious in this matter; for in several districts where the superficial covering might have been considered to consist wholly of till, that deposit appearing everywhere over the entire surface of the ground, the boring-rods have, nevertheless, after piercing the till, gone down through considerable depths of sand, gravel, and other materials. At the risk of tediousness I shall here jot down a few records of such borings, that the reader may compare them with the sketch-sections.

The two examples that follow show the occurrence of beds of sand and gravel underlying one single mass of till. The first is from the valley of the Lugar, near Old Cumnock, Ayrshire:—

	Ft.	In.
Strong blue till with stones	76	0
Brown sand, very fine	3	6
Gravel and stones with large "whin" boulders	7	0
Rock.		

The next is the record of a boring made at Woodhall, near Ormiston, Midlothian:—

	Ft.	In.
Surface soil	2	0
Clay and stones	4	0
Sand and channel (gravel)	6	0
Sandy shales, &c.		

The succeeding show a greater variety in the subsoils. The localities from which they are taken are given within brackets:—

[ELPHINSTONE, MIDLOTHIAN.]

	Ft.	In.
Surface soil	1	0
Clay	5	0
Clay and stones	4	0
Channel (gravel)	4	6
Sand and channel	5	6
White sandstone.		

[WOODHALL, MIDLOTHIAN.]

	Ft.	In.
Surface soil	1	0
Clay and stones	17	0
Sand and channel (gravel)	13	10
Sandstone	2	11
Channel (gravel)	2	7
Clay and stones	1	4
Sandy shales, &c.		

[FROM PIT AT ORBISTON, LANARKSHIRE, MOSSEND IRON COMPANY.]

	Ft.	In.
Surface soil	1	0
Red till and stones	36	6
Sand and gravel	4	0
Dark muddy sand	21	8
Brown sandstone in beds	28	0
Whinstone block	0	10
Sand and gravel	4	9½
Whinstone block	0	3
Sand and gravel	1	5
Light sandstone.		

[Boring, Dykehead, Larkhall, Lanarkshire.]

	Ft.	In.
Sandy clay and stones	24	0
Sand and gravel	5	6
Sandy mud	38	6
Tough clay and stones	33	0
Mud	7	0
Sand	11	0
Sandstone block	4	0
Sand and gravel	12	0
Sandy clay	3	0
Soft mud	30	0
Soft mud and beds of sand	5	0

[Boring, in same District as last.]

Surface and soft sandy clay	5	0
Soft clay	13	0
Sand	3	0
Mud	24	6
Gravel and sand	25	6
Stiff clay and stones	14	6
Sand	4	6
Sand and gravel	16	6
Sandy clay and stones	2	0
Mud	8	0
Muddy sand	17	0
Hard gravel	3	6
Stiff sandy clay and stones	10	0
Mud	22	6
Mud with broken "metals"	7	6
Carboniferous strata.		

I shall have occasion presently to refer to these "borings," and to adduce others; meanwhile enough have been given to show that shallow stream-sections, and other natural and artificial exposures, do not always tell us the whole truth. Any one going over the ground from which some of these borings are taken could not possibly have guessed that underneath the till, or other deposits he saw at the surface, lay deep beds of sand, gravel, mud, and intercalated masses of till.

But although stream-sections never yield such deep

exposures of drift as some of the above, yet the cuttings laid open by the rivers are often highly instructive. And so often do the river-cuts disclose the presence of sand, mud, and gravel intercalated amongst or underlying till, that we must look upon the occurrence of these beds rather as the rule than the exception. It must not, however, be imagined that the beds referred to always, or even often, attain the great thickness indicated in some of the borings given above. They are generally much thinner, and frequently absent altogether, when nothing save sheer till covers the underlying rocks. This is most commonly the case in the hilly districts, the subjacent and intercalated beds becoming more frequent, extending more continuously, and acquiring a greater thickness as they approach the lower levels of the country. Yet, even in these last-named districts, they seldom continue far without interruption, but ever and anon disappear, leaving the stony clay to form the whole of the subsoil down to the rock-head.

I shall now bring forward a few sections to illustrate the general aspect of the till and its associated

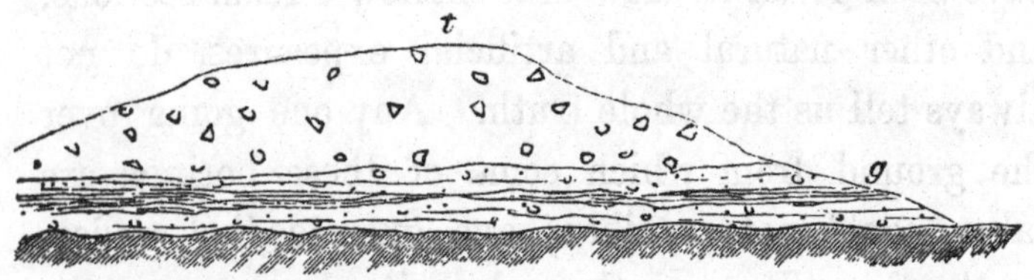

Fig. 17.—Till resting on stratified deposits, Douglas Burn, Yarrow.

deposits as presented to us in natural and artificial cuttings. The first I select, not only as an example of the occurrence of beds underneath the till, but also because it serves to show the general position occupied by the till in the valleys of our Southern Uplands.

Fig. 17 represents the section as seen in the left bank fronting the stream; Fig. 18 represents the same deposit as it would appear in a transverse section; thickness of deposits 10 to 15 ft.

Fig. 18.—Section across Douglas Burn, Peeblesshire.

A fine example of similar phenomena was pointed out to me by my colleague, Mr. B. N. Peach, in the north bank of the river Tweed, near Melrose. The annexed Fig. 19 will convey a general idea of the appearance of the beds. The rocks *r* were smoothed

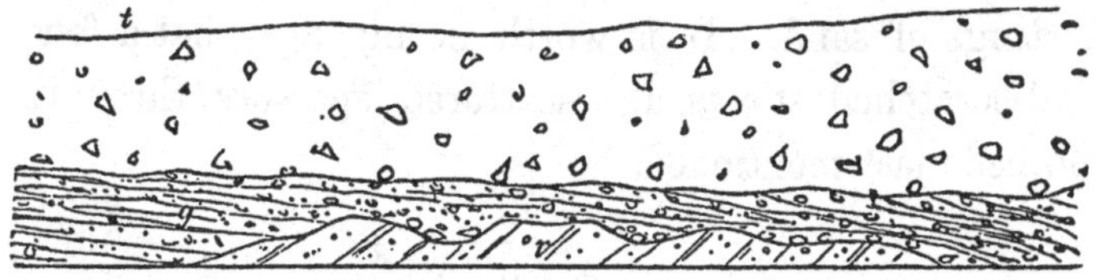

Fig. 19.—Till on river-gravel and sand, Tweed, Melrose (thickness, 30 to 40 ft.)

as if water-worn below the coarse shingle and gravel *g*, and from the arrangement of the gravel there could be no doubt of its fluviatile origin.

It frequently happens, however, as I have remarked above, that the stream-sections do not go down quite to the rock. In such cases, although we may sometimes surmise what the underlying drifts are, yet we never can be at all certain about the matter—so inconstant are they, so liable to change. In the sections that follow it will be observed that the subjacent rock is not seen, and therefore we cannot say whether the aqueous deposits that underlie the till form the bottom beds of the drift or not.

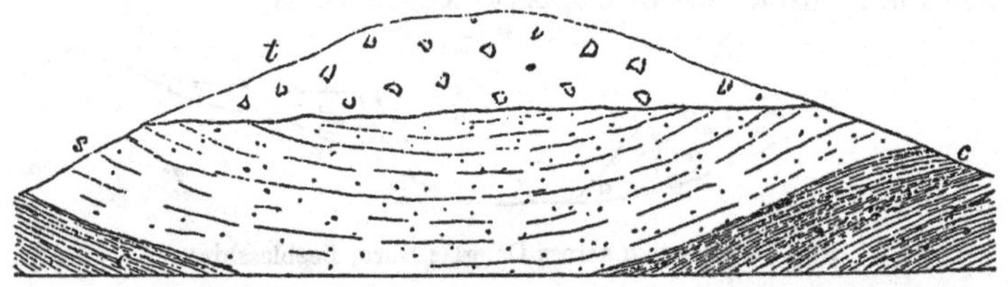

Fig. 20.—Till on stratified deposits, Glen Water, Ayrshire (thickness, about 30 ft.).

In this section (Fig. 20), a bed of strong tough till rests upon a fine yellowish white sand *s*, containing thin lines or laminæ of brownish clay. Underneath the sand comes fine clay, *c*, arranged in leaves, with partings of sand. It is worth noting also that a few well-scratched stones are scattered sparsely through the beds last mentioned.

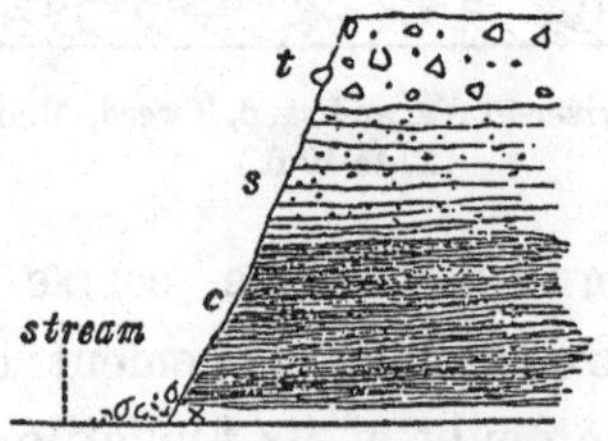

Fig. 21.—Till resting on sand, *s*, clay *c*, Garpal Water, Ayrshire.

In the sections now given it will be observed that the till rests upon a plain or level surface of sand, gravel, or clay, as the case may be. The junction-line, however, is not always or even often so regular. In Fig. 22 the till is represented as cutting down into beds of sand and gravel in a most irregular way—the lines of bedding in these deposits ending abruptly

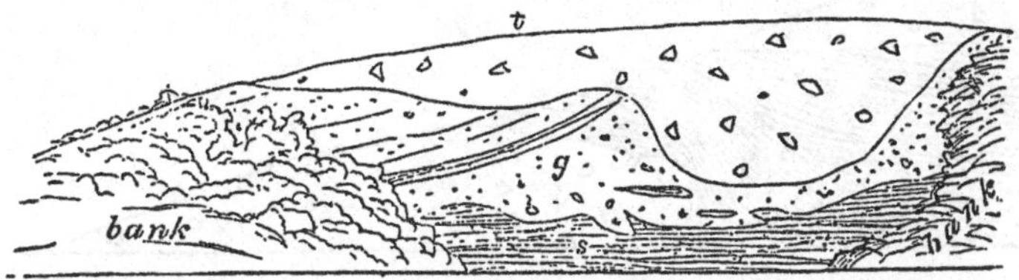

Fig. 22.—Till cutting into stratified deposits, river Clyde, near Covington.

against the till. A still better example of the same appearance was exposed during the progress of the excavations for the Peebles Railway at Neidpath Tunnel. Here a mass of tough till, with the usual scratched stones *t*, overlies a series of horizontal beds of clay, sand, and gravel, which terminate quite suddenly against the till. The clays were of that kind

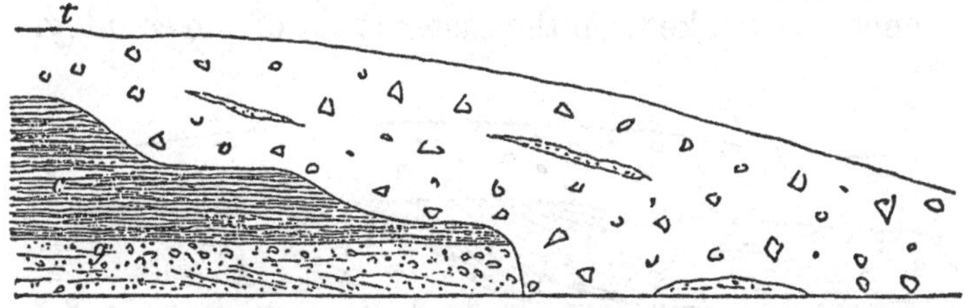

Fig. 23.—Till cutting down into stratified deposits, Neidpath, Peebles: thickness of drift, 40 to 50 ft.

which is termed "gutta-percha," exceedingly fine, and arranged in extremely regular layers or laminæ, underneath which were earthy gravel and sand.

Only one bed of till is shown in the above sketch-sections, but underneath the aqueous beds represented in Fig. 23 I have reason to believe that another deposit of till occurs. In the next illustration (Fig. 24), two beds of till are apparent. The intercalated beds

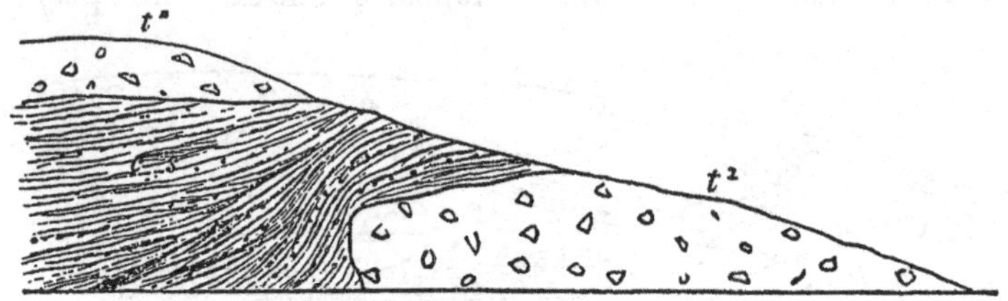

Fig. 24.—Stratified beds intercalated with till, Glen Water, Ayrshire (25 to 30 ft.)

here consist of sand and clay. They are capped by till, t^2, somewhat sandy, but quite unstratified and full of striated stones; a more tenacious mass of till, t^1, underlies the intercalated beds, which are standing nearly on end, and form a most irregular junction with the till upon and against which they rest.

Another example (Fig. 25) of somewhat similar phenomena I take from the eastern side of the country.

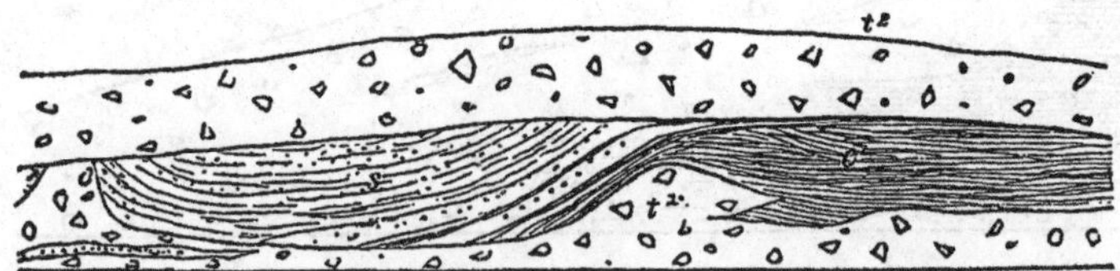

Fig. 25.—Stratified beds in till, Leithen Water, Peeblesshire.

In both these sections the crumpling up of the beds below the upper deposit of till is very marked. Other examples might easily be given, but from those already produced the reader will have a clear enough notion of what is meant by the contortion and displacement of the beds in the till.*

The sections to which I shall now refer are most interesting, inasmuch as they have yielded organic remains. Some years ago my brother described a section of till seen in the Slitrig Water, near Hawick. Dr. (now Professor) Young and myself saw the section at the same time, which is given by my brother as follows:—

	Vegetable soil.
	Boulder clay (*i.e.* till), 30 to 40 ft.
Stratified beds.	Yellowish gravelly sand.
	Peaty silt and clay.
	Fine ferruginous sand.
	Coarse shingle, 2 to 3 ft.
	Coarse stiff boulder clay (*i.e.* till), 15 to 20 ft.

It may aid the reader's conception of this succession if I give here a diagrammatic section across the

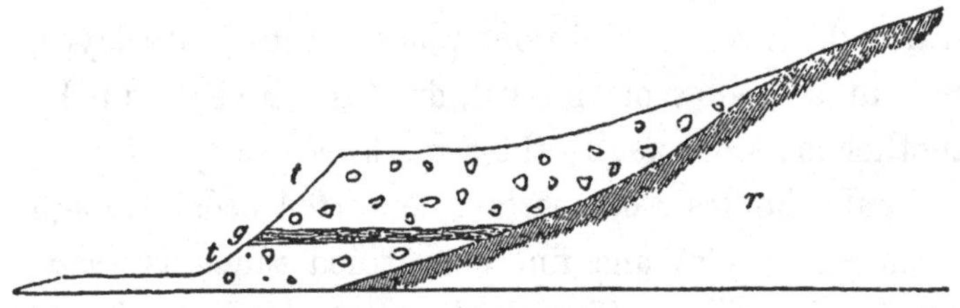

Fig. 26.—Fossiliferous beds in till, Slitrig Water, near Hawick.

deposits, which will show at the same time the position of the intercalated beds, and the mode in which the till occurs in the valley. "The cliff at

* For a further account of beds subjacent to or intercalated among till, the reader who is interested in the matter may refer to my brother's paper on the "Glacial Drift of Scotland" (*Glasg. Geol. Soc. Trans.*, vol. i. part ii.), in which he will find references to other papers descriptive of the same phenomena.

this locality" (I quote from my brother's paper) "is at least forty or fifty feet high, and consists of a stiff bluish clay stuck full of boulders. The bed of stones or shingle is well seen, even at a little distance, running as a horizontal band along the face of the cliff at a height of some fifteen or twenty feet above the level of the stream. On closer examination this zone proved to consist not merely of water-rolled shingle: over the lower stratum of rounded stones lay a few inches of well-stratified sand, silt, and clay, some of the layers being black and peaty, with enclosed vegetable fibres in a crumbling state." "So far as it was possible to ascertain the nature of the vegetable remains in the peaty layer, they appeared to be the rootlets of a kind of heath."

On the banks of the Carmichael Water, Lanarkshire, according to the same observer,* beds of sand, silt, and gravel, with, in one place, a thin peaty layer, rest in a hollow of the till, and are covered up by another mass of exactly the same kind of deposit.

Again, he describes certain contorted beds of tough gutta-percha clay and finely stratified sands as occuring in the till at Chapelhall, near Airdrie. These deposits varied in thickness up to twenty or thirty feet, and in them layers of peat and decaying twigs and branches have been detected. They were clearly overlaid and underlaid by tough stony till.

The aqueous beds intercalated with the till not infrequently appear to lie in basins or saucer-shaped

* *Trans. Geol. Soc. Glasg.*, vol. i. part ii. I have seen the sections described.

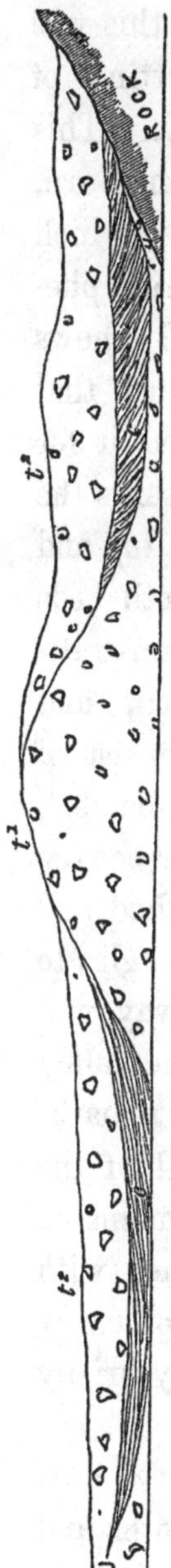

Fig. 27.—Fossiliferous deposits in till, Cowdon Burn Railway Cutting, Neilston, Renfrewshire.

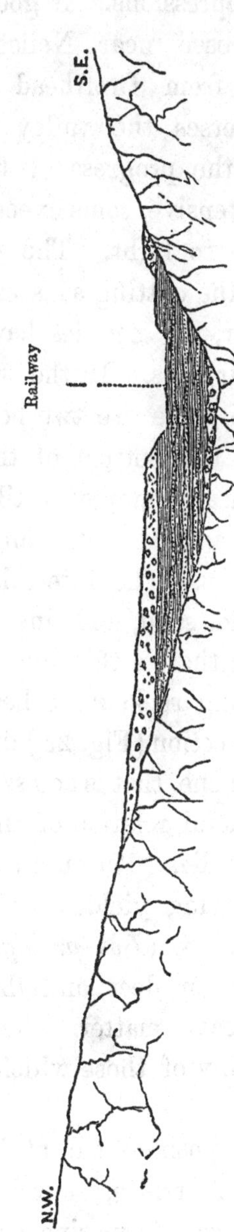

Fig. 28.—Section across the valley of Cowdon Burn, showing fossiliferous deposits resting upon till and covered by another overlying mass of till.

hollows or depressions. A good example of this was recently exposed near Neilston, in the cutting of the railway from Crofthead to Kilmarnock. This railway traverses the valley of the Cowdon Burn, and during the progress of the excavations, which were very extensive, some exceedingly interesting phenomena came to light. The section (Fig. 27) shows the face of the cutting as seen in 1868; since that date, however, the navvies have not improved it for geological purposes. In the woodcut *t* represents the till, of which there are two beds, one at the top and the other at the bottom of the section. Both beds are good typical examples of till, being quite unstratified, and crammed with angular, scratched, and polished stones. The intercalated beds *c* consist of silt, clay, mud, sand, and fine gravel, all well-bedded, and here and there a thin line of peaty matter occurs. The underlying rocks *r* are beautifully smoothed and striated. A section (Fig. 28) drawn at right angles to the preceding one, that is across the line of railway, will show the general relation of these drifts to the valley in which they lie. The intercalated beds are remarkable from having yielded an imperfect skull of the great extinct ox (*Bos primigenius*), and remains of the Irish elk or deer, and the horse, together with layers of peaty matter. The beds are precisely similar to many of those which I have already briefly referred to.

Below a deposit of till, at Woodhill Quarry, near Kilmaurs, in Ayrshire, the remains of mammoths and reindeer and certain marine shells have several times been detected during the quarrying operations. The first notice of these discoveries dates so far back as 1817, when Mr. Bald described* two elephant's tusks, as having been got at a depth of seventeen and a half feet from the surface. More recent investigations have shown that the mammalian remains obtained from this quarry occurred in a peaty layer between two thin beds of sand and gravel which lay beneath a mass of till, and rested directly on sandstone rock.† From this infraposition of the organic remains to the till some observers have inferred that these belong to pre-glacial times. But the results obtained during the examination of the ground by the Geological Survey show that the strata containing mammaliferous remains really occupy an intercalated or *inter-glacial* position, and that the cause of their sometimes being found to rest on the solid rock and not upon a lower mass of till is due to the irregularity of the surface on which the whole of the drift beds have been deposited; for where the level of the rock slopes down far enough it passes below the horizon of the stratified beds, and a lower and underlying till then makes its appearance.‡ Arctic marine shells occur in a layer overlying the mammaliferous deposit.

Remains of mammoth have been met with under somewhat similar circumstances at Chapelhall,§ near Airdrie, where they occurred in a bed of laminated sand underlying till. Reindeer antlers have also

* *Memoirs of Wernerian Society*, vol. iv. p. 64.

† *Quarterly Journal Geol. Soc.* vol. xxi. p. 213.

‡ *Memoirs of Geol. Survey, Scotland;* Explanation of Sheet 22, p. 29.

§ *Proceedings of Geol. Society*, vol. iii. p. 415.

been discovered in other localities, as in the valley of the Endrick,* about four miles from Loch Lomond, where an antler was found associated with marine shells near the bottom of a bed of blue clay, and close to the underlying rock—the blue clay being covered with twelve feet of a tough stony clay.

In the mass of the till itself fossils sometimes, but very rarely, occur. Tusks of the mammoth, reindeer antlers, and fragments of wood have from time to time been discovered in this position. They almost invariably afford marks of having been subjected to the same action as the stones and boulders by which they are surrounded.

Before leaving the intercalated beds of the true till it is necessary to call attention to another remarkable appearance sometimes presented by the till. The appearance to which I refer is what has been termed "a striated pavement," that is, a surface or level of till "where all the prominent boulders and stones have not only their original and independent striæ, but where they have subsequently suffered a new striation which is parallel and persistent across them all." Such pavements, as one would naturally expect, are more frequently exposed along a sea-coast or horizontal section than they are in the interior of the country. Nevertheless, some well-authenticated instances of the latter have been observed, where the pavements were only exposed upon the removal of overlying till.

* *Edin. New Phil. Jour.* (New Series), vol. vi. p. 105.

CHAPTER XII.

BEDS SUBJACENT TO AND INTERCALATED WITH THE SCOTTISH TILL—*Continued.*

Beds below and in the till indicate pauses in the formation of that deposit.—How the aqueous beds have been preserved.—Their crumpled and denuded appearance.—Their distribution.—Character of the valleys in which they occur.—The present stream-courses, partly of pre-glacial, inter-glacial, and post-glacial age.—Old course of the River Avon, Lanarkshire.—Pre-glacial courses of the Calder Water and Tillon Burn.—Buried river channel between Kilsyth and Grangemouth.

WE may now proceed to the explanation of the facts adduced in the last chapter. The reader has already seen that the till itself is a truly glacial deposit, due to the grinding action over the surface of the country of immense masses of glacier ice. But no one will doubt that its intercalated and subjacent beds of silt, sand and gravel have had a very different origin. They occur in such layers as could only have been spread out by the action of running water. Evidently, then, these strata are a very different kind of deposit from the till that encloses them, and it is equally self-evident that at the period of their formation the production of till must for a time have ceased, at least in those particular places where the stratified beds occur. And seeing that these intercalated beds are not confined to any one district, but are found in every part of the country where they have been searched for, it is reasonable to conclude that there were times when the great ice-fields that covered the country receded so far at least as to uncover the lowland tracts and valleys, and permit the accumulation in those regions of clay, sand, and gravel. Nor does it seem less reasonable also to conclude that after such a recession the ice again advanced and covered up the aqueous strata with thick deposits of stony clay. But here a difficulty will occur to the reader which it may be well to notice. How, it may be asked, could soft beds of sand, silt, and gravel escape being ploughed out by the ice-streams which are said to have deposited the overlying stony clay?

It has already been pointed out that the existence of the till itself is a difficulty of the same kind; and I have endeavoured to show that this deposit bore the same relation to the ice-sheet that river-detritus does to a river. There can be no doubt that in many places over which the ice-sheet passed till could not possibly accumulate, just as in the bed of a stream there are bare rocky slopes exposed to the full sweep of the water where detritus is not permitted to gather. It is no less certain that after till had been piled up in some places it was again and again ploughed out, and redistributed below the ice-sheet. Now we find that the intercalated beds of sand and gravel give unequivocal proof of having been subjected to great pressure. They are twisted, bent, crumpled, and confused, often in the wildest manner. Layers of clay, sand, and gravel, which were probably deposited in a nearly horizontal plane, are puckered into folds and sharply curved into vertical positions. I have seen whole beds of sand and clay which had all the appearance of having been pushed forward bodily for some distance, the bedding assuming the most fantastic appearance (see Figs. 20, 22, 23, 24, 25).

But the intercalated beds have not been crumpled only; they are everywhere cut through by the overlying till, and large portions have been carried away. Indeed, when we compare the bulk of these beds to that of the till, we must at once allow that they form but a small fraction of the drift deposits. Owing to the erosive power of the old glaciers, comparatively little of the intercalated sand, &c., has been spared; but enough is left to assure us of the former importance of the intercalated beds. The geological value of a deposit has not usually been measured by its bulk.

In exposed positions, such as hill-tops and hill-slopes, the till never contains intercalated beds; nor do these occur save as interrupted and fragmentary patches, in places that appear to have been open to the full sweep of the ice-currents. It is usually in positions sheltered in great measure from the pressure and grinding of the glaciers that the stratified beds of the till have been best preserved. But what, it may be asked, is meant by a position sheltered from the grind of the ice? Do not these sand and gravel beds occur exclusively in the valleys, and is it not just in such positions where the grinding action of the old glaciers was most powerful? If the ice-sheet covered the whole country, in what possible position could the sand and gravel beds be comparatively secure? Let me try to make this plain. In Scotland, as in other countries, the large rivers flow in broad open valleys, and are fed by lateral tributaries which issue from narrower and more confined valleys and ravines. The river Clyde, for example, which flows towards the north-west in a valley that gradually expands to a broad open strath, as it approaches its estuary, is joined from north and south by numerous streams, many of which run in deep narrow ravines, until they are just on the point of mingling their waters with the river. This appearance is very well seen in the neighbourhood of Hamilton. The Avon there winds through a deep cool ravine for several miles before it enters the Clyde, and the same is the case with the little tributaries of the Avon itself. The Calder from the north-east also makes its way towards the Clyde in a romantic glen, whose precipitous walls, like those of the Avon, are hung with greenery. Now during the glacial period, the ice-sheet, which followed the line of the principal valleys, must frequently have crossed the lateral and subsidiary valleys nearly at right angles. In the main valleys the glacier would exert its full influence, but it would not be able to do so in the narrow lateral valleys and ravines; the ice and till would merely topple into the glens referred to, and gradually choke them up, and the main mass of the glacier would then pass on over the whole. Here the analogy of running water again will help us. In a stream-course we see how the detritus accumulates in deep holes and pools, at the bottom of which the water is often well-nigh still, while a current is sweeping across at the surface.

In such narrow glens, then, any silt, sand, or gravel that had gathered during the temporary absence of the ice-sheet would not be so likely to be ploughed out when the ice returned, as the similar materials which had accumulated in those broader and more open valleys where the ice would have full freedom to move and exert its erosive power.

But it may be objected that it is precisely in the narrower ravines where at the present day we meet with no drift whatever, where in fact the streams flow between bare walls of rock; while, on the contrary, the broader and more open valleys show considerable depths of sand, silt, gravel, and stony clay. To this objection it may be answered that the narrow ravines in which so many of our streams flow have been formed almost without exception since the close of the glacial epoch. The ravines are in most cases new cuts excavated by the streams after the confluent glaciers had finally vanished.

How this has happened will readily appear when we remember that the work performed by the old glaciers was twofold. In many cases the massive ice-streams deepened the valleys that already existed—in certain regions digging out great hollows, to the nature and origin of which reference will be made in a subsequent chapter. But after having deepened valleys and widened glens, they very frequently buried these again more or less completely under vast piles of clay, sand, and boulders. Some valleys indeed they completely obliterated, so that when the ice finally melted away, and left the land once more exposed to the light of day, the streams and rivers could no longer flow in their old courses, but were compelled to form for themselves new channels.

It is quite true that, speaking generally, the present drainage-system is very much the same as that which obtained before the advent of the glacial epoch; nevertheless the course followed by each river and stream seldom agrees precisely with that along which the waters made their way in pre-glacial times. Sometimes the streams flow throughout nearly their entire length in new channels which have been cut in rock since glacial times, the older courses being still choked up and concealed under the clay and stones that were shot into them by the old glaciers. More frequently, however, the present river-courses are partly new, partly old. When, after the ice had disappeared, the water again began to make its way down the slopes of the land to the sea, it is self-evident that the direction of the streams would be determined by the configuration of the ground. But this, as we have seen, would not exactly agree with that of pre-glacial times. Many of the valleys had been levelled up, and in not a few cases long hills and mounds of drift appeared where deep dells had formerly existed. The chief features of the country, however, remained; the broad valleys and straths, although clothed with drift deposits, yet again received the tribute of the lateral streams, and formed the highways of the principal rivers. But the lateral feeders of these rivers, following the new slopes of the ground, have not always been directed into their old channels. Sometimes, indeed, they wander for miles away from these, and join the main stream either far above or far below where they formerly debouched. Even when, after cutting down through the drift deposits, they have happened to regain their old channels, they usually leave these again and again as they journey on, plunging ever and anon into deep rocky gorges, which, as I have said, have been chiselled out by them since the close of the glacial epoch.

To the wanderer along the course of some of the lowland streams nothing can be more striking than the sudden and complete change of scenery that ensues upon the passage of a stream from its new into its old channel. In the former the water frets and fumes between lofty walls of rock, which, seen from below, appear to rise almost vertically from the river's bed. In such a deep narrow gorge the stream may continue to flow for miles, when of a sudden the precipitous cliffs abruptly terminate, and the water then escapes into a broad vale with long sloping banks of stony clay, sand, and gravel. After winding about in this open glade for, it may be, several miles, the stream not infrequently leaves again as suddenly as it entered, and dashes once more into another dell, whose steep walls of rock shoot up and overhang the water as before. The broader and more open portions of the valley, where the sloping banks are formed of drift débris, is of course that part of the old pre-glacial channel which the present stream has re-excavated; while, as already indicated, the narrow and rocky gorges are entirely new cuts made by the stream during the ages that have elapsed since the glacial epoch.

In many cases the present streams seldom continue in their old channels for any distance. We often find them cutting across these nearly at right angles, and in this way fine sections of the old buried river-channels with their contents frequently appear in the rocky glens of the Lowlands, to some of which I shall refer presently. A glance at the woodcut (Fig. 29), which

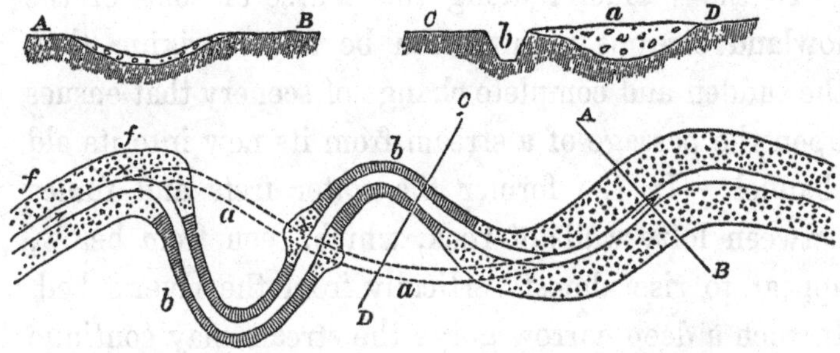

Fig. 29.—Diagram to show pre-glacial and post-glacial river-course. *a a* buried course; *b b* post-glacial channel; *f f* and dotted parts show re-excavated channel. Sections above indicate character of valley along the lines *A B* and *C D*.

is a diagrammatic ground-plan intended to illustrate the phenomena just described, will show how it comes to pass that a post-glacial channel may often cut a pre-glacial course, and yet the present stream coincide in direction with that of pre-glacial times. The buried course is represented at *a a*, and the new channel at *b b;* the thin line on either side indicating the tops of the cliffs, *b b*, and sloping banks, *f f*. From *f* to *f* it will be observed that the present channel coincides exactly with the old course, while at × × the latter is cut across nearly at right angles by the former.

In most cases the pre-glacial channels prove to be wider and sometimes deeper than the new cuts; consequently when a stream, after flowing for some distance in a post-glacial course, suddenly enters a pre-glacial channel, the softer character of the materials it has to excavate enables it to clear out a wider hollow than it has carved in the solid rock of its post-glacial course. A section across the valley at *A B*, compared with one drawn from *C* to *D*, shows the relative appearance of the old and new cuts. It will be noticed that not only are the old channels wider and sometimes deeper than the new—facts which indicate, of course, a greater age—but also that their sides are less precipitous. This latter appearance points to the long-continued action of frost, which, by splitting up and detaching the rocks upon a cliff, has a tendency to reduce all such steep river-walls to sloping banks; but the gently inclined slopes of many pre-glacial river-courses doubtless owe much of their character to the grinding action of the glaciers during the ice age. Not a few old river-courses, however, can be shown to be as steep and precipitous as any which have been formed since glacial times.

To the appearance presented by the deposits that fill up the buried river-channels, I must now direct attention. At p. 153 I have given the details of a boring made near Larkhall, for the purpose of testing the position of the underlying coal-seams. This boring and another near the same place pierce a great thickness of detrital materials which I have ascertained occupy a former course of the Avon—the present river flowing for a considerable distance in an entirely new channel excavated since the glacial epoch. The following figure gives a diagrammatic view of the choked-up channel as it has been ascertained by borings. A section of the old channel, where it is cut by the Avon, is seen on the banks of the river near Fairholm House, Larkhall; but owing to the incoherent nature of the silt and sand, these beds have slid forwards with the overlying clays and thrown everything into confusion.

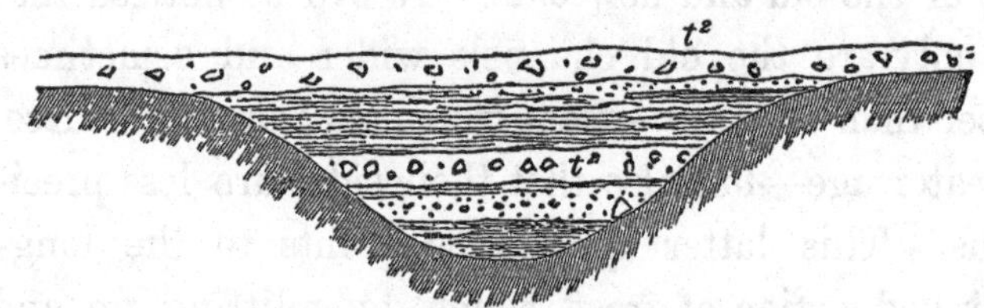

Fig. 30.—Diagrammatic section across pre-glacial course of the River Avon, Larkhall, Lanarkshire.

It will be remembered that the borings prove the existence of two masses of till, with intervening and underlying beds of silt, mud, sand, and gravel. The lower of these two deposits of till indicates the prolonged action of glacier-ice; yet the beds below retain a considerable thickness, notwithstanding their incoherent character. At this particular place, however, the ice-sheet crossed the gorge in which they lie, at an angle, and hence they must needs have escaped the powerful erosion to which they would have been subjected had the path of the ice coincided with the trend of the ravine. The presence of 40 ft. of silt, sand, and gravel above the till indicates a period of lessened cold, when the ice-sheet disappeared from this region and permitted the formation of such deposits. But after a time it would appear that the ice-sheet again overspread the country, doubtless sweeping out the silt, sand, and gravel from exposed positions, but sparing them in the narrow glens and gullies that intersected its path.

In the Scottish coal-fields such old stream-courses as I now describe are of common occurrence, and are locally known to the miners as "clay dykes" and "sand dykes," according to the prevailing character of the material that fills them. The coal is often worked quite close to the side of the buried ravine; after which, if the nature of the dyke will allow it, a mine is driven through the clay and sand, until the opposite face of the old glen is reached and the coal-seam found again. Sometimes, however, the dyke is charged with soft mud and running sand, and then it becomes impossible to mine, and a new pit must be sunk on the further side of the channel to get out the coal. Many accidents have happened from the breaking in of sand and mud into the pit-workings, when coal has been taken out too close to the "dyke."

A very good example of such a dyke, or buried stream-course, was exposed in the cutting made for the railway between Edinburgh and Holytown, quite close to the little village of Cleland, Lanarkshire. The cutting intersected the channel at right angles to its course, and thus a beautiful section of the old ravine, and the materials that choke it up, was obtained.* But the face of the cutting is now so "dressed," that the precise succession of the drift-beds cannot be readily deciphered. There cannot be a doubt, however, that a thick mass of till occupies the highest position, with beds of gravel and sand coming in below. The same old ravine is intersected by the Calder Water a little to the west of Wishaw House, where, however, the trees and brushwood somewhat obscure it.

It is remarkable that in the same neighbourhood there is another buried stream-course, which runs for a short distance parallel to the last-mentioned one, and is in like manner cut across by the present course of the Calder. It is well exposed on the banks of this stream a little below Coltness bridge, upon the side of the road leading from Wishaw. Both buried channels are filled with similar materials, chiefly sand with a little gravel—the whole being overlain with till. The coal-workings, which are very extensive in the district, enabled me, while carrying on the Geological Survey, to trace out all the windings of these remarkable sand-troughs; and by connecting the information thus obtained with what I was able to gather from natural sections, it became evident that considerable changes had taken place in the drainage-system of the neighbourhood since glacial times. The Tillon Burn, which is now a tributary of the Calder, was formerly an independent stream; while the Calder Water has forsaken its old channel, and at present flows for some

* See *Edin. Geol. Soc. Trans.*, vol. i.

distance in an entirely new course, after which it breaks into and continues along the pre-glacial or inter-glacial course of the Tillon Burn.*

I shall only refer to another example of pre-glacial and inter-glacial water-course.† This buried course has been traced from Kilsyth to Grangemouth, on the Forth, where it enters that estuary at the great depth of 260 ft. below the present sea-level. No trace or indication of this buried river-channel shows at the surface of the ground, and its existence would probably never have been discovered had it not been for the numerous borings and pits which have pierced it, for it cuts right through the coal-strata of that district. The nature of the deposits that fill up this old river-ravine is shown in the following section ‡:

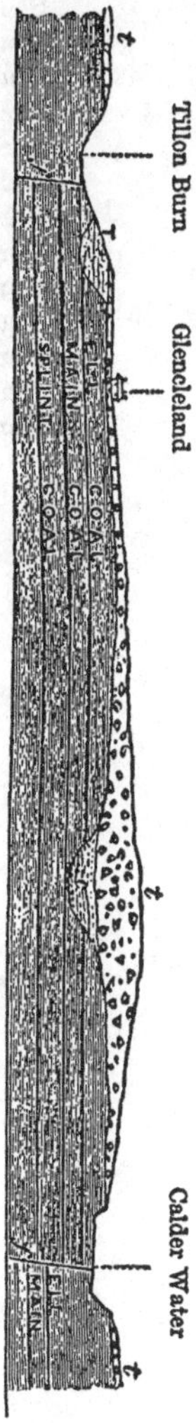

Fig. 31.—Section across carboniferous strata, showing pre-glacial stream courses (scale 6 inches to 1 mile; horizontal and vertical scale the same); *t* till; *c* pre-glacial course of Calder Water; T pre-glacial course of Tillon Burn; *f* dislocations, or faults, in coal-bearing strata.

BORING NEAR TOWNCROFT FARM, GRANGEMOUTH.

	Ft.
Surface sand	6
Blue mud	3
Forward	9

* See Geol. Survey Map of Scotland, Sheet 23.

† This "channel" is described by Mr. Croll in *Trans. Edin. Geol. Soc.*, vol. i. p. 330.

‡ See J. Bennie's "Surface Geology of the District Around Glasgow," *Glasg. Geol. Soc. Trans.*, vol. iii. part i.

	Ft.
Brought forward	9
Shell bed	1
Gravel	2
Blue mud	8
Gravel	3
Blue muddy sand	15
Red clay	49
Blue till and stones	20
Sand	20
Hard blue till and stones	24
Sand	2
Hard blue till and stones	40
Sand	7
Hard blue till	24
	234

CHAPTER XIII.

BEDS SUBJACENT TO AND INTERCALATED WITH THE SCOTTISH TILL—*Continued.*

Stratified deposits passed through in borings.—Probably in most cases of fresh-water origin.—Old lake at Neidpath, Peebles.—Lakes of inter-glacial periods.—Borings near New Kilpatrick.—Pre-glacial valley of the river Kelvin.—Origin of the deposits occupying that buried river-course.—Valley between Johnstone and Dalry.—Flat lands between Kilpatrick Hills and Paisley.—Condition of the country in pre-glacial times.—Loch Lomond glacier.—Ancient glacial lake.—Succession of events during recurrent cold and warm periods.

IT will be observed that while the hollows* in which these deposits occur have been described as old river-courses, yet nothing has been said as to the origin of the deposits themselves. Of the nature of the overlying till which is seen at the surface in some of the examples cited, there can be no doubt; but it may be objected that since we do not always see the intercalated and subjacent stony clays, we cannot be

* Similar pre-glacial river-courses are known to occur in other countries. An excellent example has been traced out in the underground workings of the Durham coal-fields, by Messrs. Nicholas Wood and E. F. Boyd (*British Association Reports* for 1863, p. 89; *Geologist*, 1863, p. 384). The buried river-courses of North America are also familiar instances. Professor Hitchcock has described these as "antediluvian river-beds;" that is to say, beds of rivers that existed before the last great submergence of North America. Logs and fragments of wood are often got at great depths in the buried gorges ("Illustrations of Surface Geology," &c., *Smithsonian Contributions*). Professor Newberry and others have more recently given interesting accounts of the like phenomena.

sure as to their character. Might not they be something else than glacial deposits? To this it may be answered that we know of no stony clay in Scotland which is not of glacial origin.* If the succession of strata disclosed by artificial borings were altogether peculiar and abnormal, we might have good reason for disregarding them; at all events, we could hardly be justified in drawing conclusions from them. But such successions are neither peculiar nor abnormal; on the contrary, similar sections occur at the surface which answer precisely in character to the accumulations passed through by the boring-rods. In the open-air sections the intercalated stony clays are true till, and the stony clays in the deep bore-holes must be till, or at all events, deposits having a glacial origin.

With regard to the aqueous strata which are associated with the stony clays, the almost total absence of fossils makes it in many cases difficult to decide whether they are of fresh-water or marine origin. In some cases, at low levels, it is not unlikely that they are partly one, partly the other. It must be admitted, however, that the absence of organic remains tells more against a marine than a fresh-water origin for the mud, sand, and silt that fill up the buried channels and hollows. It is unfortunate that a greater number of these should not have been exposed to the light of day; nevertheless, enough perhaps are actually exhibited to enable us to form a correct idea of those whose existence is only revealed to us by borings and

* In a subsequent chapter some account is given of the modified form of till which I term boulder-clay, and reference is likewise made to the occurrence of stones in certain shelly clays of marine origin.

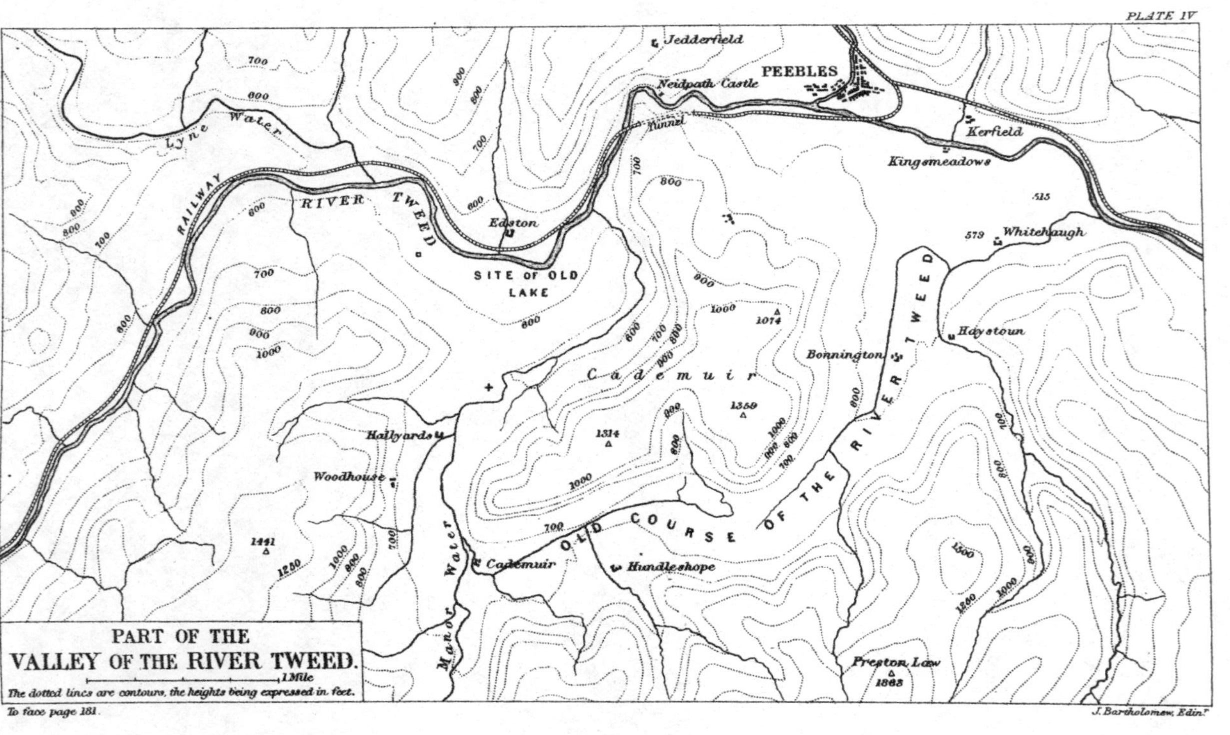

PART OF THE
VALLEY OF THE RIVER TWEED.

1 Mile

The dotted lines are contours, the heights being expressed in feet.

To face page 181.

J. Bartholomew, Edin.r

pit-workings. The opinion to which I incline is that the aqueous beds now filling up the old hollows and depressions of the land are in large measure of fresh-water origin (not by any means exclusively so); but if this be so, it seems certain that they cannot all have been laid down in the old ravines and valleys under conditions like those now obtaining in similar water-courses and depressions. No geologist would be inclined to admit that the great depths of fine sand, silt, and mud, which occupy the buried hollows, could possibly have been deposited by such streams as now flow in like places.

In many cases the deep aqueous drifts, as exposed in open section, have more of a lacustrine character than anything else. This was well seen in the section at Neidpath Tunnel, briefly described on page 157. The stratified beds in that locality appear to have been once very extensive—traces of the same deposits having been found in and below the till for some distance up the valley of the Tweed. At Neidpath there was a depth of from 40 to 50 ft. exposed, and the bottom was not seen. The beds partook of fluviatile and lacustrine characters (chiefly the latter), and appear to have been deposited in a lake-like expansion of the Tweed, at a time when that river flowed, not by its present course, but round by the back of Cademuir Hill. The accompanying sketch-map will serve to render this intelligible.

It will be seen that from Cademuir Farm to Bonnington there extends a broad flat hollow, the bottom of which at its highest level is only some 100 ft or so above the bed of the Tweed at Neidpath. Were the narrow glen at this latter place to be filled up, the Tweed would be dammed back, a lake would be formed, and thereafter both the Tweed and its affluent, the Manor Water, would flow round by the Cademuir hollow. That such actually was the course of one or both these streams at a comparatively recent geological date is proved by the fact that the Cademuir hollow is paved with river gravel, which could have come from no other source.

An examination of the present and the former course of the Tweed and the surrounding drift phenomena led me to conclude that the "gutta-percha clays" of the Neidpath section were deposited at a time anterior to the cutting-out of the Neidpath glen by the river; that, in short, the Tweed during pre-glacial and inter-glacial periods flowed by the Cademuir hollow; and that Neidpath glen has been to some extent at least hollowed out since the final disappearance of the ice-sheets. The alterations of surface, brought about by the massive glacier that cut through the gutta-percha clays and deposited the tumultuous mass of till above them, and the modifications of level induced by denudation in later glacial times, eventually compelled the Tweed to leave its old course and take the more direct route by Neidpath.

The occurrence of pre-glacial and inter-glacial lacustrine beds in the river-valleys is only what one might have reasonably anticipated. Whenever the ice-sheets retired an irregular surface of glacial drift would be exposed, in the hollows and depressions of which, lakes and pools would gather. Nay, in some cases the mouths of small lateral valleys would be closed up with drift, and thus, streams being dammed back, sheets of water would be formed in which fine sediments would accumulate. In like manner the retiring glaciers themselves would not infrequently become barriers to the drainage of small lateral valleys, and in this way produce true glacial lakes. Hence we need not be surprised at the frequent appearance of old lacustrine beds in the valleys. It would be much more surprising if they did not occur.

It is well known that in the valley of the Kelvin, near New Kilpatrick, a number of borings prove the existence in that district of a very great depth of superficial deposits. Two examples* of these borings may be given; they are as follows:—

	Ft.	Ins.
Sandy clay	5	0
Brown clay and stones	17	0
Mud	15	0
Sandy mud	31	0
Sand and gravel	28	0
Sandy clay and gravel	17	0
Sand	5	0
Mud	6	0
Sand	14	0
Gravel	30	0
Brown sandy clay and stones	30	0
Hard red gravel	4	6
Light mud and sand	1	8
Light clay and stones	6	6
Light clay and whin block	26	0
Fine sandy mud	36	0
Brown clay, gravel and stones	14	4
Dark clay and stones	68	0
	355	0

* For these and other borings, see Mr. Bennie's paper, *Glasg. Geol. Soc. Trans.*, vol. iii. part i.

ANOTHER BORING IN THE SAME NEIGHBOURHOOD GAVE—

	Ft.	Ins.
Soil	1	6
Muddy sand and stones	4	0
Soft mud	4	6
Sand and gravel	45	0
Sandy mud and stones	20	6
Sandy gravel and mud	52	4
Brown clay and stones	25	0
Sand and gravel	6	0
Brown sandy clay and stones . .	12	0
Sand	2	0
Brown sandy clay and stones . . .	4	0
Mud and sand	15	9
Sand and blue clay and stones . . .	7	9
	200	4

Mr. Croll has suggested that these deep drifts may occupy a pre-glacial bed of the Kelvin. If so, then this ancient buried channel must enter the Clyde at a depth of more than 200 ft. below the present sea-level. There is nothing abnormal in this. It has already been mentioned that an old buried river-channel enters the Forth near Grangemouth at a depth of at least 260 ft. below the sea; from which we must infer that at some period anterior to the filling-up of that channel the land stood at least 260 ft. higher than the present datum-line. In a subsequent page I shall have occasion to point out that the great sea-lochs of the western coasts are merely submerged land-valleys. Indeed, if we could but remove the superficial deposits from the surface of the Lowlands there can be no doubt that the sea would also reach a great way into the heart of these districts, penetrating sometimes for many miles by such valleys as that of the Clyde, the Ayr, the Stinchar, the Tweed, and other rivers.

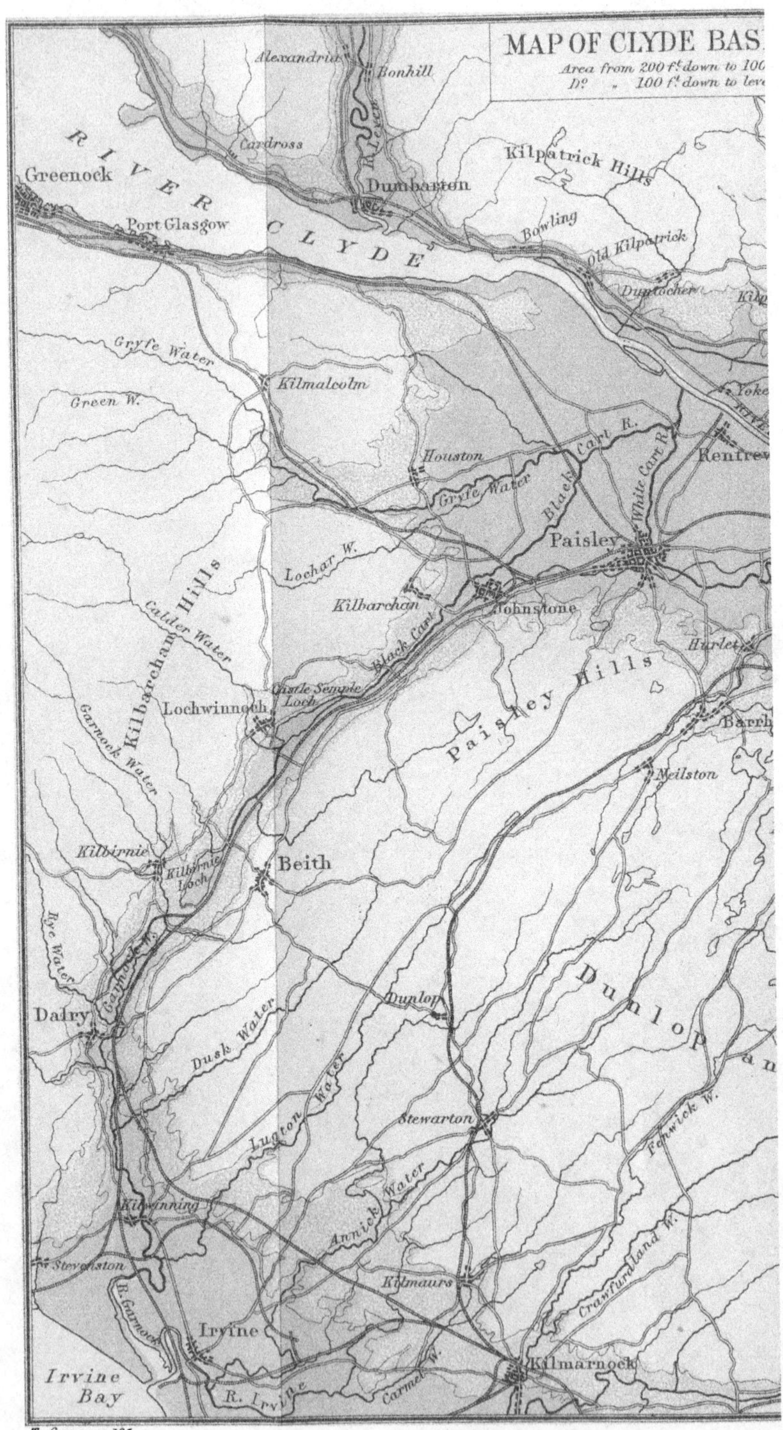

To face page 185.

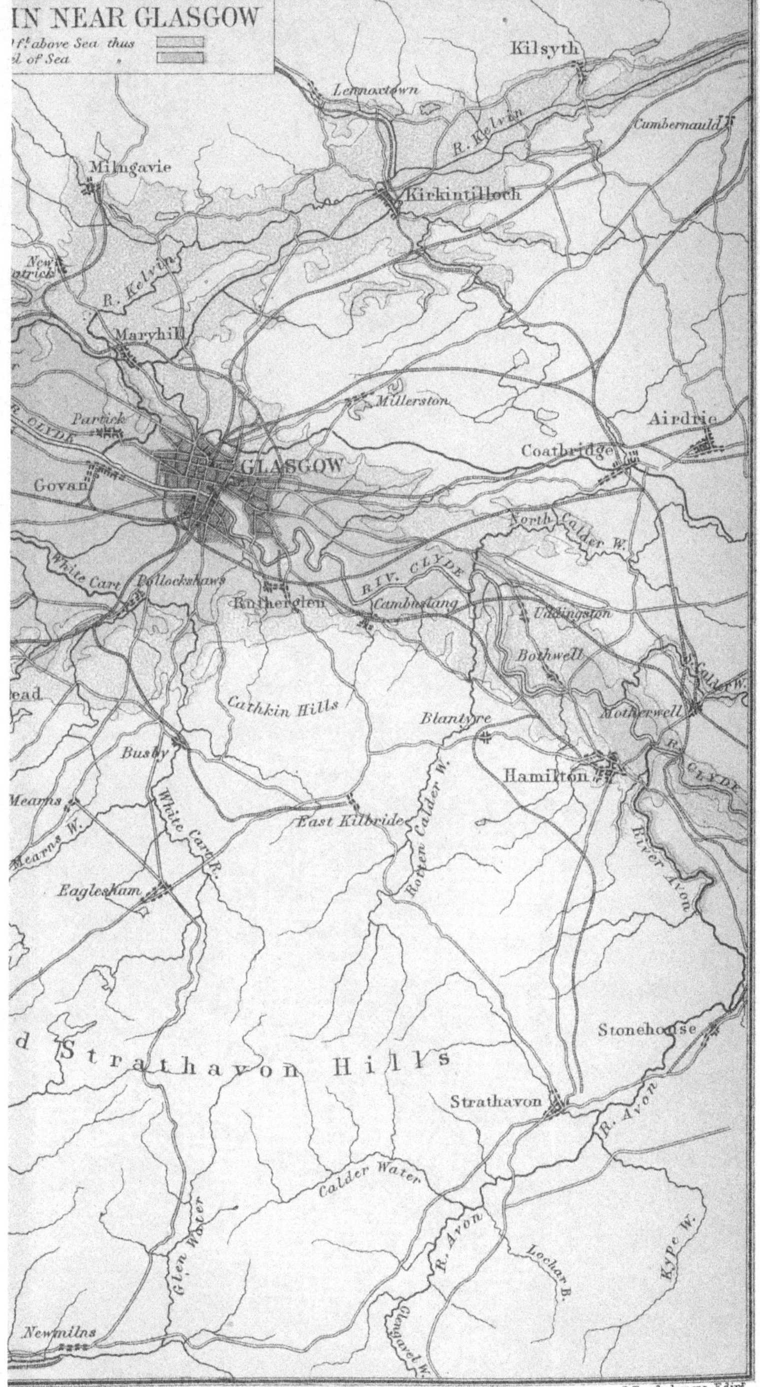
IN NEAR GLASGOW
ft. above Sea thus
el of Sea
Kilsyth
Lennoxtown
R. Kelvin
Cumbernauld
Milngavie
Kirkintilloch
R. Kelvin
Maryhill
Millerston
Partick
Airdrie
GLASGOW
Coatbridge
Govan
North Calder W.
Riv. Clyde
White Cart
Pollockshaws
Rutherglen
Cambuslang
Uddingston
Bothwell
Calder W.
Cathkin Hills
Motherwell
Blantyre
Busby
Hamilton
R. Clyde
Mearns
White Cart R.
East Kilbride
Rotten Calder W.
Mearns W.
River Avon
Eaglesham
Stonehouse
Strathavon Hills
Strathavon
R. Avon
Calder Water
R. Avon
Glen Water
Lochar B.
Kype W.
Glengavel W.
Newmilns
J. Bartholomew, Edin.

Of all these valleys that of the Clyde has yielded the greatest depth of superficial accumulations—these deposits in at least one place reaching the excessive thickness of 355 ft. No one can glance over the borer's sections given above without feeling assured that, whether or not the gravels, sands, and muds occupy a buried river-valley, they at least could not have accumulated underneath a river—they are either estuarine, lacustrine, or marine, or finally they may partake of a mixed character, and be partly of fresh-water and partly of marine origin. Their general resemblance to similar deposits exposed in other districts would incline me to consider them as for the most part of lacustrine origin, and an examination of the physical features of the district certainly tends to support this suggestion. As the question is an interesting one, it may be well to consider it in detail.*

Those who have travelled from Glasgow by Paisley and Johnstone into Ayrshire will remember that the railway in its course towards the latter place skirts the base of some rising ground which towards the south slopes up to form what we may call for want of a general name, the Paisley Hills. From the base of these hills a wide stretch of flat or gently undulating country extends northwards to the foot of

* Mr. D. Bell has advanced an explanation of these intercalated beds somewhat similar to that given in the text. He speculates on the damming-up of the Clyde at Bowling, the consequent formation of a lake, and the egress of the waters by Lochwinnoch and Dalry (*Trans. Glasg. Geol. Soc.*, vol. iv. p. 66). Mr. Bell's views and mine were arrived at independently; but since his have the priority of publication, he is justly entitled to claim the "copyright" of the glacial lake described in the text.

the Kilpatrick Hills, and westwards until it abuts upon the lower slopes of the Kilbarchan Hills. On north, south, and west, then, this plain is encircled by a screen of hilly ground. This screen, however, is breached in two places: at Bowling by the Clyde, and at Johnstone by the Black Cart Water. From Johnstone the railway runs up the valley of this water, passing Castle Semple Loch, Barr Loch (now drained) and Kilbirnie Loch, beyond which it follows the same hollow through the hills to the low-grounds south of Dalry. The surface of Kilbirnie Loch is very nearly on the same level as that of Castle Semple Loch, but there is a sluggish flow of water from the former to the latter. The head of Kilbirnie Loch, then, we may take as the watershed between the Black Cart Water and the streams that drain along the same hollow towards the south-west. But any one who examines the ground will have no difficulty in concluding that this long hollow has not been excavated by the streams that now flow in different directions along its bottom. They are quite inadequate. The whole appearance of the valley suggests strongly the idea of an old water-course that once drained along its whole extent from the north-east. That is to say that at some former period a river flowed from the valley of the Clyde across what is now the watershed of the Black Cart Water, and so on south-west into Ayrshire. This no doubt does at first sight seem an impossibility, but that it actually must have happened I shall now try to show.

Let me ask my reader to carry his mind back to

the pre-glacial period—to that far distant past before ice and snow filled the valleys, and ere yet any accumulation of glacial deposits covered the surface of the country. In those days the Clyde and its affluents certainly flowed at a considerably lower level than they do now. Could we clear away all the superficial deposits that rest in the basin of the Clyde between Glasgow and the sea, we should find, not as at present the broad undulating plain that stretches from the base of the Kilpatrick Hills to the heights behind Paisley, but a deep valley dotted with rocky knolls and ridges. We should find also that at least one deep lateral valley, carrying the drainage of the Campsie and Kilpatrick Hills, entered that of the Clyde from the north-east. The bed of the pre-glacial Clyde at or near Bowling must lie buried at a depth of more than 200 ft. below the level of the sea. Even at Glasgow the old channel of the river is not less than 80 ft. under the same datum-line.* Now when the river was flowing at these levels it need hardly be said that the land must have stood relatively higher than it does now; at all events it is certain that our shores extended much farther out to sea.

* In a series of borings made for the Clyde Trustees at Mavisbank Quay the rock was attained at 70·48 ft., 77·60 ft., and 80·23 ft. below high water mark in three bores respectively. At Stobcross, on the opposite bank of the river, the borings, at a distance of two or three hundred feet from the water, reached the rock at a nearly similar depth from the same datum-line. But when the borings were continued at the distance of a few hundred feet further from the river, the rock-head appeared at a less depth from the surface. The superficial deposits were thus shown to thin off towards the north, but they did so in a very irregular manner. For these details I am indebted to the kindness of Mr. Deas, Mem. Inst. C.E., Engineer to the Clyde Trust.

Let us, then, conceive that while the Clyde valley is in the state I have described, glacial conditions of climate supervene, that snow and ice begin to thicken in the mountain valleys, and great glaciers to creep out and deploy upon the low-grounds. One immense stream of ice fills up Loch Lomond (or the place where Loch Lomond was afterwards to be), and flowing onwards through the vale of the Leven, advances across the bed of the Clyde until it abuts upon the opposite slopes of the Kilbarchan Hills. This invasion results of course in damming back the Clyde, and a lake accumulates over the low-grounds which are encircled by the Kilpatrick, Kilbarchan, and Paisley Hills. But as the Clyde continues to flow and the surface of the lake to rise, the water must eventually find a channel of escape. Now supposing the valley of the Black Cart Water to have existed at this time, it is evident that as the lake rose it would penetrate this valley until it reached the watershed, over which a river would pour south-westward into Ayrshire. It is highly probable, however, that the great hollow now occupied by Kilbirnie and Castle Semple Lochs and the Black Cart was not so strongly pronounced in preglacial times. Be this, however, as it may, it is evident that the old Clyde, swollen in summer-time by melting ice and snow, must have swept through the notch or breach* in the hills with great force, and this condition

* The reader will, perhaps, understand my meaning better, if I merely state here that the work I conceive to have been done by the Clyde was simply the denudation, or wearing away, of the *col* between two valleys, and the subsequent deepening and widening of those valleys.

of things continuing for a long period, as it must have done, the hollow along which the water flowed would be widened and deepened. But before the river had commenced to widen and deepen this secondary course, the lake from which it flowed would cover a large tract of country and stretch along the base of the Kilpatrick Hills as far at least as Kilsyth. In this manner the old lateral valley mentioned above as being probably the pre-glacial channel of the Kelvin would be completely submerged, and so also would be the bed of the Clyde up to and beyond Glasgow.

The area covered by the lake would then necessarily become an area of deposition. Gravel, sand, mud, and silt would accumulate upon the bottom, the finer sediments settling down in the deeper parts, and thus all the drowned river ravines would have a tendency to silt up. The process would be a gradual one, and sometimes it might even be interrupted by some local recession of the Leven glacier, which would lower the surface of the great glacial lake and allow the streams to re-excavate a portion of their beds. But these and other obvious considerations I need not stop to point out.

Let us now further conceive that the arctic cold continues to increase, that the snow and ice grow in depth and breadth until even the Kilpatrick and Campsie Hills are overflowed by massive confluent glaciers descending from the Highlands, and the whole of the great central valley of Scotland—the broad lowland country—is brimful of ice, forming one wide and far-stretching mer de glace. The effect produced in the old lake-bed of the Clyde basin by such an advance of the glaciers would no doubt be most destructive. Over broad areas the soft and incoherent masses of sand, mud, gravel, &c., which had gathered upon the bottom of the old lake would be ploughed up and intermingled with the other débris continually gathering and being pushed onward underneath the ice-sheet. But it might well be that in deeper hollows and in such ravines as intersected the path of the ice, some portions of the lake deposits might escape and receive a covering or cap of ground-moraine or till.

But a glance at the "borings" given above (pp. 183-4) will show that the buried hollows and ravines may contain more than one stony clay separated by considerable depths of aqueous deposits. These stony clays probably indicate just so many incursions of the ice-sheet; the intermediate beds of silt, sand, and gravel, may point on the other hand to periods when the ice vanished from the low-grounds and crept back to the mountain valleys. Every time the Leven glacier advanced and choked up the Clyde valley, a lake would form over the region under review, and fresh-water beds would be deposited; every time the ice-sheet covered the whole country fresh masses of stony clay would accumulate in protected hollows and ravines; every time the ice-sheet retired the lake would reappear, until the Leven glacier finally drew back from the valley of the Clyde. It is therefore unnecessary to call in the aid of the sea to explain for us the occurrence of those beds of gravel, sand, and mud which are intercalated with the stony clays in the buried valleys and depressions of the Clyde basin.

I have spoken of the outlet of the ancient glacial lake of the Clyde having been by the valleys of the Black Cart and the Rye. It is quite possible, however, and even highly probable, that the discharge at some periods may have been to the north-east by Kilsyth into the basin of the Forth. In attempting to restore the physiography of the land during successive stages of the glacial epoch, we have to bear in mind that after every descent of the glaciers very considerable changes would be effected here and there upon the configuration of the ground. In one place the level would be lowered by the abstraction of rock —in another it would be raised by the accumulation of superficial deposits. And it might quite well be that, owing to some such changes, the Clyde lake might during certain inter-glacial periods drain into the Forth—for even now the difference of height between the watershed at Kilsyth, and that of the Black Cart is only some 50 ft. or so.

The actual certainty that such great disturbances of the drainage-system must have taken place during the glacial epoch has not received so much attention from glacialists as it deserves. Yet no one who gives the subject any consideration can fail to see how disturbances of the kind must have occurred in many other valleys besides that which I have selected for illustration. Long before the land-ice increased to such an extent as to overflow the Kilpatrick Hills, the Campsies, and the Ochils, many of the streams that drained into the Forth must have been dammed back by the great glacier which occupied the principal valley, and which in all likelihood extended many miles below Stirling, before the hills referred to were overtopped and buried. In attempting to read the records of the glacial epoch such considerations as these ought not to escape us; had they always been fully realised perhaps we should have been less liable to set down every thick bed of inter-glacial silt, sand, or gravel, to the action of the sea.*

It is quite certain, however, that marine deposits do sometimes occur intercalated among true morainic accumulations, and to these we shall refer presently. Meanwhile I must direct attention to some examples of inter-glacial fresh-water deposits which have already been briefly described.

* I am reminded here of those deposits of sand and gravel covered with morainic débris, that occur abundantly in the low-grounds of Switzerland, and which geologists have shown to be of fresh-water origin, but which, had they occurred in similar positions in our country, some of us might have had small hesitation in assigning to the action of the sea, so difficult is it for a speculative islander to escape the influence of his geographical position.

CHAPTER XIV.

BEDS SUBJACENT TO AND INTERCALATED WITH THE SCOTTISH TILL—*Continued.*

Inter-glacial deposits of the Leithen Valley, Peeblesshire.—Other examples of similar deposits.—Crofthead inter-glacial beds.—Climatal conditions of Scotland during inter-glacial ages.—Fragmentary nature of the evidence not conclusive as to the climate never having been positively mild.—Duration of glacial and inter-glacial periods.—Inter-glacial shelly clay at Airdrie.—Arctic shells at Woodhill Quarry, Kilmaurs.—Recapitulation of general results.

AT p. 158 Fig. 25 shows certain stratified deposits of gravel, sand, and clay, which betray all the usual characteristics of river detritus. They rest, as will be observed, upon a very uneven surface of till, and show a considerable degree of confusion. The Leithen (Peeblesshire), on the banks of which they occur, flows in one of those deep, narrow, but softly outlined valleys that form so familiar a feature in the Southern Uplands. The stream has its source at a height of 1,750 ft. above the sea, and after a course of nine miles joins the Tweed, at Innerleithen, some 500 ft. above the same level. Its drainage-area does not exceed 22 square miles; its watershed, however attains a height of upwards of 2,000 ft., and the highest point in this hilly tract of south-eastern

Scotland is only some 750 ft. higher. The significance of these details will appear presently.

Both lower and upper masses of till shown in the section, are crammed with well striated stones and boulders, and are in all respects typical deposits. From what has been said in regard to the origin of till, it is quite clear that at the time the lower mass was being accumulated, the whole drainage-area of the Leithen must have been filled with ice to overflowing. And not only so, but the same must have been the case with all the valleys in the Southern Uplands. This follows, as a matter of course, as any one may convince himself, by studying the Ordnance Survey maps of the region under review. It is quite inconceivable that the Leithen valley, which is by no means one of the highest in that district, should alone have been filled with glacier ice, but the presence of a deep ice-mass here necessarily implies the like presence of large glaciers in all the main dales and hopes of these Uplands.

The presence of the intercalated river gravel and sand indicates plainly that an interruption to this arctic condition of things took place. Before these river deposits could be laid down, the ice must have vanished from the Leithen valley, and if such was the case with this mountain-valley, we are driven to conclude that the great ice-sheet could not then have covered any portion of the Scottish Lowlands. Glaciers may have lingered still in the higher valleys of the country, but it is obviously impossible that a great ice-sheet could exist while upland streams like the Leithen had freedom to flow.

The aspect of the upper deposit of till in the section testifies to the disappearance of the mild conditions under which the river accumulations were formed, and to the return of an intensely arctic climate. The stones and boulders in this till are all well blunted, scratched, and smoothed, thus indicating the former presence of a depth of ice sufficient to overflow the heights overlooking the valley, and so to prevent the introduction of sharply angular and unpolished blocks amongst the glacial débris gathering underneath.* The contortion and confusion of the river deposits was no doubt caused by the pressure of this second ice-flow.

A similar train of reasoning applies to the sections shown in Figs. 24 and 26, and, therefore, they need not to be more particularly referred to. Referring back to Fig. 27, page 161, another interesting section will be seen. Here beds of clay, sand, silt, &c., are represented as completely enclosed in till. These strata showed lines and layers of peaty matter, and yielded, during the railway operations, an imperfect skull of the great urus, and some remains of the Irish deer and the horse. The beds are clearly of lacustrine formation, and their position proves that after a mass of till had been deposited by some great ice-sheet, a milder climate ensued, when streams once more flowed down the valleys, and lakes occupied the hollows and depressions of the land, which was tenanted by oxen, deer, and horses. Finally, the

* See *ante*, p. 86.

presence of the overlying mass of till indicates that this mild period passed away, and was succeeded by severe arctic conditions, when thick ice again streamed over central Scotland.

Precisely the same inference as to change of climate is to be drawn from the other instances of inter-glacial beds with organic remains, mentioned at pp. 159, 160; the stratified deposits in all these instances point to oscillations of temperature—to periods when the great ice-sheet disappeared from the low-grounds, and shrunk into a series of local glaciers among the mountain regions.

But as I have endeavoured to show, the same conclusion must be arrived at, even supposing the inter-glacial deposits had never yielded any organic remains whatever. These, however, are extremely valuable, inasmuch as they enable us more fully to realise the nature of those physical conditions that characterized the inter-glacial periods. Combining the evidence we learn that not only did the great ice-sheet sometimes retire from the low-grounds, and give place to lakes and streams and rivers, but also that, during such periods of milder conditions, a vegetation like that of cold temperate regions clothed the valleys with grasses and heaths, and the hill-sides with birch and pine. Reindeer wandered across the country, while herds of the great white ox, the horse, the Irish deer, and the woolly-coated mammoth frequented the grassy vales. If one might draw conclusions from the aspect of the few fossil remains disinterred from the deposits in the till, he might compare Scotland during the inter-glacial periods to that tract of country which extends along the extreme southern limits of the Barren Grounds of North America—a region where a few firs and other hardy trees cover the drier slopes, and where carices and grasses grow luxuriantly enough in the sheltered valleys—those favourite breeding-places of the rein-deer, which roam over the dreary deserts to the north. Whether during any of the inter-glacial periods the climate was ever mild enough to melt away all the ice and snow from the highland valleys, the record does not say. What positive evidence we have points rather to the existence of local glaciers in the higher valleys—to moderate summers and severe winters—during such inter-glacial periods as we have any certain records of.

And yet we might be committing a grave error were we to assume that Scotland, during inter-glacial times, never enjoyed milder conditions than now obtain in the forest regions and barrens of North America. We must ever bear in mind that the inter-glacial deposits are the veriest fragments. They have been preserved, for the most part, only in sheltered hollows from the ravages of the great ice-plough; and the interrupted and patchy portions that remain are mere wrecks of what must once have been, in the broader valleys, widespread and continuous deposits. Every renewed descent of the glaciers upon the low-grounds would tend to effect the removal of these accumulations; and it may well be that of several inter-glacial periods, not a single representative deposit now remains. Even during the inter-glacial periods themselves, the streams and rivers would help to clear away and redistribute those beds of sand, gravel, and silt which the glaciers had spared—just as in our own day the streams are gradually excavating and washing away the materials that fill up old pre-glacial and inter-glacial ravines and water-courses. Moreover, we must not forget that if really warm climates ever did supervene during inter-glacial times, every such warm period must have been followed successively by temperate, cold-temperate, and arctic conditions; and these last would consequently be the most fully represented of the series.

So far then as the Scottish glacial drifts are concerned, there is no evidence whatever to show that the inter-glacial periods might not have been warm enough at times to cause all the ice and snow to disappear from the country. Whether we shall ever obtain any decisive evidence on this head, will probably depend upon the assiduity with which the inter-glacial deposits are examined. The till was for many years looked upon as a deposit destitute of all traces of life, and only a few hammerers continued, Micawber-like, to hope for something turning up. I believe that mammalian remains have been oftener obtained from the beds in the till during shaft-sinking and other mining operations than geologists are aware of. While carrying on the Geological Survey of the Scottish coal-fields, I have frequently heard of "bones" and "horns" having been met with by the workmen in sinking through the deep drifts. These relics, unfortunately, have almost invariably been lost or mislaid; but there can be little doubt, from the descriptions that were given to me by intelligent overseers, that the relics were true fossils, and still less doubt that these fossils were obtained, not in recent alluvial but in glacial deposits.

We may, perhaps, never learn how many great changes of climate took place during the accumulation of the till and its associated deposits. This arises from the fact, already adverted to, that during every period of intensest cold, when the country was covered with a more or less thick sheet of snow and ice, the loose materials, which in the preceding age had gathered in river-valleys and in lakes, would almost inevitably be subjected to excessive denudation. That the records of mild inter-glacial periods should be at the best but fragmentary, is no more than one might have expected. The wonder is not that they should be so interrupted, but that any portion whatever has been spared. Owing to this interrupted mode of occurrence, it is obviously impossible to correlate them, bed with bed. The inter-glacial deposits in one place may or may not be contemporaneous with the similar accumulations in some other locality. There were more changes than one during the formation of the lower drifts; and hence, in the case of isolated inter-glacial beds, we never can tell whether they ought to be referred to an earlier or a later stage of that period.

The disappearance of a mer de glace, which in the Lowlands of Scotland attained a thickness of nearer

3,000 ft. than 2,000 ft., could only be effected by a very considerable change of climate. Nor, when one fully considers all sides of the question, does it appear unreasonable to infer that the comparatively mild and genial periods, of which the inter-glacial beds are memorials, may have endured as long as those arctic or glacial conditions which preceded and followed them. We have a difficulty in conceiving of the length of time implied in the gradual increase of that cold which, as the years went by, eventually buried the whole country underneath one vast mer de glace. Nor can we form any proper conception of how long a time was needed to bring about that other change of climate, under the influence of which, slowly and imperceptibly, this immense sheet of frost melted away from the Lowlands and retired to the mountain recesses. We must allow that long ages elapsed before the warmth became such as to induce plants and animals to clothe and people the land. How vast a time, also, must have passed away ere the warmth reached its climax, and the temperature again began to cool down! How slowly, step by step, the ice must have crept out from the mountain fastnesses, chilling the air, and forcing fauna and flora to retire before it; and what a long succession of years must have come and gone before the ice-sheet once more wrapped up the hills, obliterated the valleys, and, streaming out from the shore, usurped the bed of the shallow seas that flowed around our island! Finally, when we consider that such a succession of changes happened not once only, but again and again, we cannot fail to have some faint appreciation of the lapse of time required for the accumulation of the till and the inter-glacial deposits.

It will be observed that hitherto we have done no more than make a brief reference to the occurrence of undoubted marine beds in the till. To avoid confusion, I have delayed considering these intercalated marine deposits until now.

The only marine beds of the kind which can be certainly recognised as such, occur, with one exception, below the level of 100 ft., or thereabout, above the sea. The exception referred to is that of a clay with arctic shells at Airdrie, which has been ascertained to occupy an interbedded position in the till, at a height of 512 ft. This clay occurred as a mere local patch of no extent, dividing the upper and lower mass of till for only a few yards. Small and insignificant although it be, it nevertheless enables us confidently to assert that at the time of its deposition, the land stood more than 500 ft. below its present level.

In our attempts to realise the physical conditions that obtained at various periods during the glacial epoch, it is necessary, as I have tried by some examples to show, that we should keep in view not only the height above the sea, but also the geographical position of the deposits. Airdrie, where the shelly clay was got, lies nearly in the centre of that broad tract of lowland country, which is drained by the river Clyde and its tributaries. At the time when a free ocean flowed over the site of Airdrie, it is evident that a very large part of the Scottish Lowlands would be submerged. The sea would wash the base of the Kilsyth Hills in the north, and towards the south would extend up the Clyde valley as far, at least, as Hazelbank, a few miles below Lanark.

Now such being the case, a very little consideration will suffice to convince us that however cold the climate may then have been, Scotland could not by any possibility have presented such an intensely arctic aspect as when the ice-sheet was at its thickest. The till that underlies the shelly clay speaks to us of a time when the great central valley of Scotland brimmed with glacier-ice. Under no other condition could a *moraine profonde* gather in the position of that till at Airdrie. But the presence of the clay with shells shows that a great physical change supervened, and it seems reasonable to conclude that at the time this shelly clay was deposited the greatest intensity of glacial cold had passed away. But if the accumulation of the lower mass of till at Airdrie implies the former existence of one great confluent ice-sheet in Scotland; then, in like manner does the overlying mass of till compel us also to conclude that after a comparatively mild period had endured for some time another mighty ice-sheet again overflowed the land.

Above the height of 500 ft. I am not aware that any marine deposits occur intercalated with the till. The interstratified beds, wherever I have seen them at this and greater heights, are of fresh-water origin; and similar fresh-water beds occur at all levels down to the neighbourhood of the sea. Indeed, if the inter-glacial beds that fill the buried valleys and depressions in the lower reaches of the Clyde basin be, as I have tried to show they very likely are, of fresh-water formation, then fluviatile and lacustrine beds of inter-glacial age occur even below the present sea-level.

The evidence, however, entitles us to assume that, during the accumulation of the till and its associated stratified beds, there was at least one period of considerable submergence, and, for all that we know, there may have been several. At all events, there can be no doubt that oscillations of level did take place. It will be remembered that overlying the mammalian remains at Woodhill Quarry, near Kilmaurs, marine arctic shells occur, clearly indicating that after the deposition of the fresh-water beds a subsequent submergence of more than 100 ft., or so, took place. When we get down to levels of 100 ft., or thereabout, above the sea, marine beds are frequently found associated with stony clays, either in a subjacent or intercalated position. In most cases, however, it will be found that these stony clays differ in some respects from the typical till; but before saying anything further about them, it may be well to summarise, in a few words, the general results at which we have now arrived.

In concluding what we had to say about the changes of climate that were likely to result from astronomical causes, we found that a full consideration of the question led us to infer that glacial periods, lasting for thousands of years, must alternate with equally prolonged periods of genial conditions every time that the orbit of the earth reaches a high degree of eccentricity. Now if I have succeeded in making the geological evidence at all clear, it must be apparent that the tale told by the glacial deposits strikingly confirms the truth of the astronomical and physical theory which we considered in Chapter X.

We have found that there is abundant proof to show that the accumulation of a *moraine profonde* by one great ice-sheet was interrupted several times; that the ice-sheet vanished from the low-grounds, and even from many of the upland valleys, and that rivers and lakes then appeared where before all had been ice and snow. We have also learned that during such mild inter-glacial periods, oxen, deer, horses, mammoths, reindeer, and no doubt other animals besides these, occupied the land. Moreover, we have ascertained that the land itself experienced at least one considerable submergence, and that, during the period of submergence referred to, the sea that covered the drowned districts of the country was tenanted by molluscs of northern habitats. Finally, we are assured that no definite conclusion as to the climate of inter-glacial times can be drawn from these organic remains. The deposits which contain them are much too fragmentary to enable us to say, with anything like certainty, that inter-glacial Scotland never enjoyed a milder climate than is now experienced in cold-temperate regions. But, as we shall see in the sequel, a study of the glacial deposits in other countries throws no small light upon this difficult question.

CHAPTER XV.

BOULDER-CLAY BEDS OF SCOTLAND.

Stony clays of maritime districts.—Character of boulder-clay as distinguished from till.—Traces of bedding, &c.—Water-worn shells.—Crushed, broken, and striated shells.—Shell-beds.—Boulder-clay passes into till —Boulder-clay of Stinchar valley; of Berwick cliffs.—Boulder-clay and associated deposits of Lewis; of Caithness and Aberdeenshire.—Boulder-clay of Dornoch Frith; of Wigtonshire and Ayrshire.—Origin of boulder-clay.—Scotland resembled Greenland during the formation of this deposit.

WE have now traced the history of the glacial epoch down to a critical period. We have seen how when the cold was most intense a thick crust of snow and ice smothered all Scotland, filled up our seas, and overflowed even the Outer Hebrides: and we have also seen how these arctic conditions were ever and anon interrupted by the advent of mild inter-glacial climates. We come next to examine those deposits which mark the final retreat of the great ice-sheets underneath which the till accumulated.

Towards the close of last chapter mention was made of the frequent appearance in maritime districts of certain stony clays. It is very often difficult, nay in isolated sections it is sometimes quite impossible, to distinguish these stony clays from true or typical till. Nevertheless, there are certain marks by which a careful observer will learn to know them. They consist usually of a coarse aggregate of sub-angular and blunted stones and boulders scattered confusedly through a somewhat sandy or earthy clay. Many of the included stones are well glaciated, and do not differ in any degree from those that occur in the till. Nevertheless, a large number are often quite angular and show rough unpolished faces, and big blocks and boulders are usually more common than in the till. Occasionally, too, these big stones have travelled far, and can be shown to have come from regions whence it is hard to see how they could have been borne by glaciers. As a rule the deposit is quite unstratified, but traces of bedding are not uncommon; the stones being not infrequently arranged in lines, while bands of sand, clay, and gravel come in here and there. In some of these bands comminuted water-worn fragments of shells are occasionally met with, and well-preserved shells of arctic species have also been detected; and in places similar remains, often broken and crushed and striated, are scattered somewhat plentifully through the clay itself. Not infrequently, however, all these characteristic marks disappear, and the deposit is not to be distinguished from true till—the one stony clay passes insensibly into the other.

The *boulder-clay*, as I shall term this deposit to distinguish it from till, appears to be confined to maritime districts. But since it frequently yields the same outline or configuration as the till it is often difficult to tell, in the absence of sections, whether the drift-covering belongs to the latter or the former deposit. It is quite possible therefore that here and there boulder-clay may extend for some distance up the valleys from the sea into the heart of the country. It cannot do so, however, in any force, else it could hardly have escaped detection. It attains its greatest thickness in the lower reaches of the broad open valleys where these approach the coast-line, and appears gradually to pass into till or to disappear altogether as we follow it inland. In some maritime districts it does not apparently occur; nor would it seem to have been met with at a greater height than 260 ft. above the sea.

In the annexed cut both the till and the boulder-clay are present. The till *t* is a tough, tenacious,

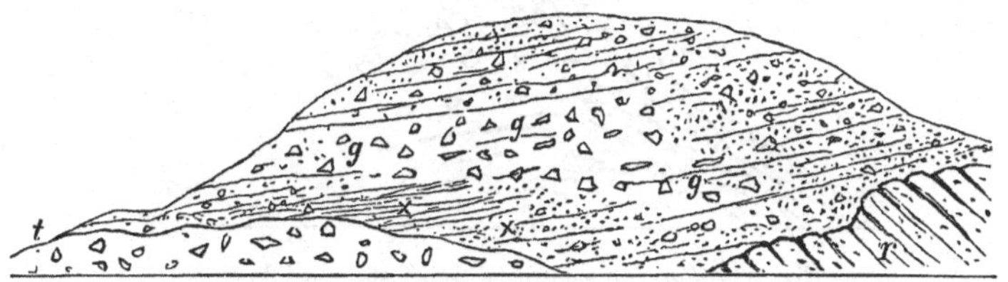

Fig. 32.—Till overlaid with boulder-clay; River Stinchar.
r, Rock. *t*, Till. *g*, Boulder-clay. ×, Fine gravel, &c.

brown clay, quite unstratified, and stuck full of finely-scratched and smoothed stones. The boulder-clay, on the other hand, is more or less distinctly bedded in parts, the bedded portions being indicated by the faint lines. At × it consists of thin layers of fine gravel, earthy clay, and sand, with stones scattered throughout, some of which are striated. Towards the top the gravel is much coarser, not so well bedded, and contains many angular blocks and stones, some of them smoothed and scratched. At *g* we see a stony clay, looser and earthier than the subjacent till, and only faintly stratified. The stones, while usually smoothed and scratched, are, upon the whole, not so distinctly marked in this way as the stones in the till.

This section may be compared with the following, taken from the opposite side of the island. The till is shown at *t;* it is a reddish clay, stuck full of

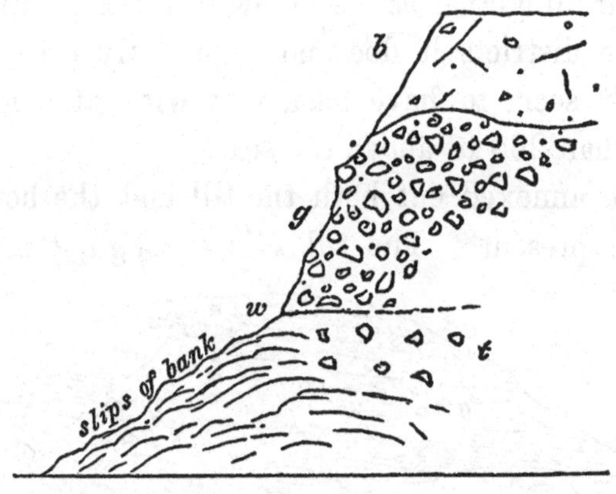

Fig. 33.—Section seen in sea-cliff at Berwick.

scratched stones, but owing to the copious percolation of spring-water (at *w*) between it and the overlying looser deposits it is rendered soft and incoherent. The superjacent beds consist of coarse bouldery shingle *g*, in a sandy clay matrix. The stones look waterworn, but are angular and blunted, and some are glaciated. The base of this shingle-bed is not well seen, and consequently the junction line between it and the till is not distinctly visible. This arises from the action of the springs, which are continually producing little slips of the bank—the succession, however, is sufficiently distinct. Overlying the coarse shingle occurs a dark reddish brown clay, with a few scattered stones, which are in other places more closely aggregated. They are generally ice-marked, but some show no traces of glaciation. Broken shells occur here and there, but appear to be rare. In other parts of the cliff beds of sand are intercalated with *g* and *b*, and sometimes the whole of the drift deposits overlying the solid rocks consist of a red boulder-clay, often very coarse, overlaid with beds of sand and clay. The till is only seen in one or two places, and is apparently very thin.

The next sections I take are from a much less known part of the country. Fig. 34 represents two

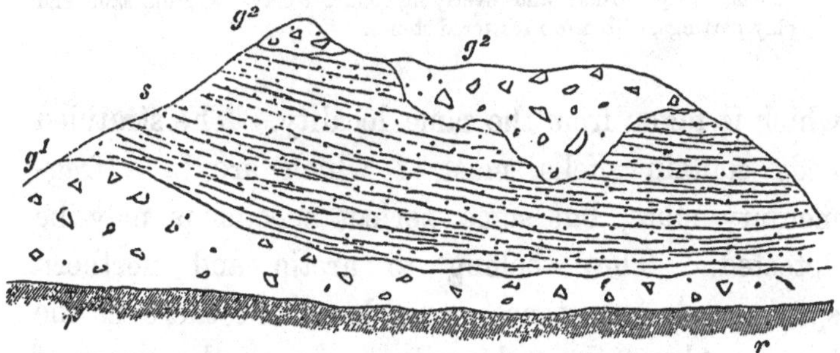

Fig. 34.—Boulder-clays and associated deposits, Traigh Chrois, Island of Lewis. *r*, Solid rock. g^1 g^2, Lower and upper boulder-clay. *s*, Sand, gravel, and clay.

beds of boulder-clay, with intermediate deposits of sand, gravel, and clay. The lower boulder-clay bed rests upon gneissic rocks. It is a dark greyish brown sandy or earthy clay, quite unstratified, with numerous blunted stones and boulders. These consist almost exclusively of gneiss, of a kind which does not preserve striæ, consequently only a few of the finer grained, harder, and more compact stones show any striations. But the most remarkable feature of this stony clay is the presence of broken arctic shells, which occur in an irregular manner through the mass. The upper surface of the boulder-clay is denuded—a character better shown in Fig. 35,

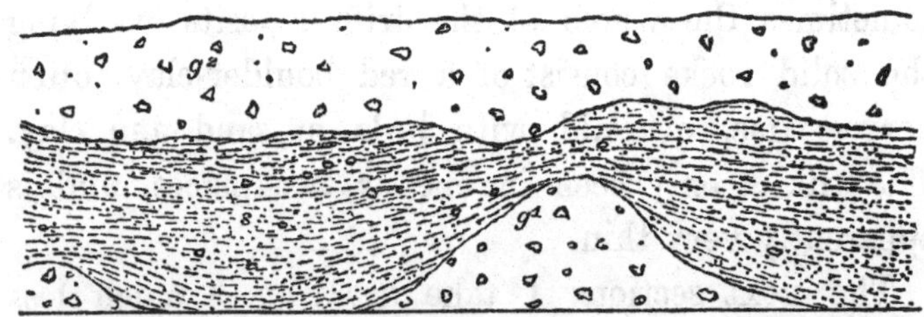

Fig. 35.—Bounder-clay and associated deposits, Traigh Chrois, Island of Lewis. g^1 g^2, Under and overlying boulder-clay. *s*, Fine sand and clay partings, with a few scattered stones.

which is taken from the same locality. The stratified beds contain shells, most of which are in a fragmentary state, but some perfect specimens may be detected. They belong to arctic and northern species. A few stones occur here and there in the beds. Above these deposits comes another mass of unstratified boulder-clay of a reddish brown colour, and somewhat more sandy in texture than the underlying mass. The stones it contains are quite similar to those in the lower bed. No shells appear to occur. It will be observed that this upper clay cuts down into the beds upon which it rests.

Fig. 36 shows the same deposits, with the exception

of the lower clay, which, however, would no doubt appear if the section had cut deeper. It will be

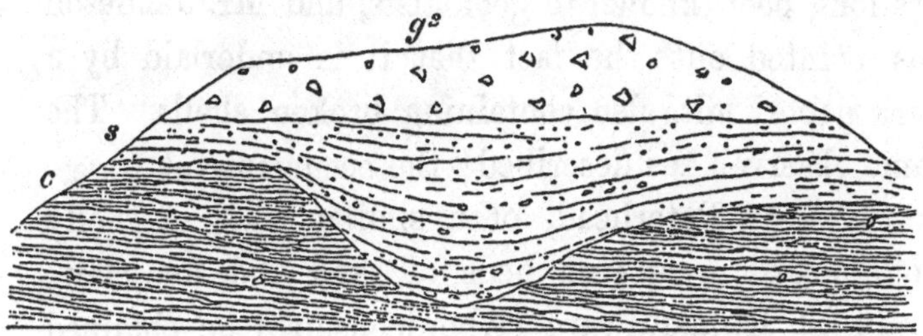

Fig. 36.—Boulder-clay overlying stratified deposits, Traigh Chrois, Island of Lewis. g^2, Boulder-clay. *s*, Sand and gravel. *c*, Clay and silt.

noticed that an unconformity occurs in the stratified beds,—the clay and silt (*c*) having been denuded before the sand (*s*) was deposited above them.

From an examination of all the sections in north of Lewis I found we had the following succession of deposits, beginning with the upper bed :—

1. Boulder-clay.
2. Sand, gravel, and coarse shingle, with rolled and angular fragments of shells.
3. Dark bluish-brown clay and silt, with shells, many of them mere fragments.
4. Boulder-clay with broken shells.
5. Till.

The last-mentioned deposit nowhere appears in any of the coast sections underneath the lower boulder-clay, but it is of common occurrence throughout Lewis, and I satisfied myself, on evidence which need not be detailed here, that it is the oldest member of the glacial series in that island.

A nearly similar succession of deposits occurs in Caithness and the maritime districts of Aberdeenshire.

In the former county an unstratified stony clay containing broken shells of arctic and northern forms has long been known to geologists, and Mr. Jamieson has pointed out* the fact that it is underlaid by a dark pebbly silt, also containing broken shells. The same observer has described† the occurrence at Invernettie, near Peterhead, of a similar mass of shelly boulder-clay overlying beds of sand and fine clay; again, at King-Edward (Aberdeenshire) he obtained the following interesting section:—‡

	Ft.
1. Stratified sand and gravel (no fossils) . . .	10 to 25
2. Unstratified boulder-clay, with shell fragments in lower part	20 to 30
3. Fine sand, in some places rich in shells . . .	1 to 2
4. Fine silt, with whole shells, excavated to a depth of 10 ft.; bottom not reached.	

My colleague, Mr. R. L. Jack, tells me that he has got from a gravel at Gartness railway station (Stirlingshire) well-rolled fragments and a few perfect specimens of boreal and arctic shells. The gravel occurs at a height of 140 ft. above the sea-level, and is part and parcel of a wide-spread deposit of fine sand, gravel, laminated clay, and boulder-clay. The beds rest on till, and reach 260 ft. as an upper limit.

Further illustrations of the mode of occurrence of boulder-clay might be adduced, but for the most part they would only be repetitions of those already given. It seems to occur pretty generally in most maritime regions. It may be seen on the shores of Dornoch Frith, at Tain, where it shows lines of

* *Quart. Journ. Geol. Soc.* (1866), p. 265.

† *Op. cit.* (1858), p. 518.

‡ *Op. cit.* (1866), p. 275. The section is given in descending order.

stones and thin bands of fine clay. Again, in the extreme south it has been observed at various places along the coast of Wigtonshire, where it contains thin layers of clay, &c., with arctic marine shells. Broken and comminuted shells also appear in boulder-clay near Ballantrae, &c., in Ayrshire.

During the formation of the till the whole of Scotland was, as we have seen, buried underneath snow and ice, and such being the case, the surface of the great mer de glace could not be covered with stones and débris, as are the puny glaciers of the Alps. The moraines gathering below and in front of the great ice-sheet would therefore contain few or no rough unpolished stones. But when, owing to the melting of that ice-sheet, mountain-tops and ridges began to stand boldly up above its general level, and so by slow degrees to separate it into a series of local glaciers, it is evident that long trains of blocks and rubbish would begin to sprinkle the surface of the glaciers. This débris slowly carried downwards, would eventually topple over the terminal front of the ice, and mingle with the glacial mud and stones which were being extruded upon the sea-bottom. Thus there would be mixed together in one and the same deposit heaps of scratched stones, with quantities of rough blocks and débris; and it is plain, moreover, that the action of the sea upon the moraine rubbish-heaps in front of the glacier would tend occasionally to give a stratified arrangement to the detritus, sometimes even sifting the materials and forming beds of clay, sand, and gravel. In such beds it would only be natural that shells of arctic species should sometimes become entombed.

During such a condition of things the rocky parts of the coast between separate glacier valleys would no doubt be fringed with a belt of ice, just as at present is the case in Greenland, and this ice-foot occasionally breaking off would float away with the rocky rubbish and débris which alternate freezings and thawings had detached from the cliffs and showered upon it. In this manner rock-fragments from a distant part of the country might be dropped upon the bed of the sea, and so get intermingled with moraine matter brought down by glaciers from the interior.

It is well known that in alpine countries the glaciers sometimes advance beyond their usual limits, and the same in all likelihood is the case with the great glaciers of Greenland. It is, therefore, not unreasonable to infer that during our glacial epoch the ice-rivers may have been subject to similar fluctuations. It is easy, therefore, to see how, during such a temporary advance, the submarine heap of clay, stones, sand, and gravel lying in front of a glacier would be pushed forward and thrown into confusion. The shells would often be crushed and broken, and the deposits themselves would become intermingled with the *moraine profonde*.

The section from the south of Ayrshire (Fig. 32) exhibits not a few of the very features which a little consideration would naturally lead one to expect. We there see scratched and glaciated stones confusedly intermingled, and sand and gravel passing, as it were, into a sandy clay with scratched stones, that closely resembles true till.

The sections from Lewis are also highly instructive. They show us how the sea-bottom, tenanted by various species of mollusca, was liable to be invaded by the advancing and retreating ice—which ploughed up the deposits of sand and clay and incorporated these with its *moraine profonde*.

Such, then, I believe to have been the origin of boulder-clay: it appears to have been deposited at or near where the great glaciers terminated in the sea. Hence it is sometimes partially stratified, at other times contains beds, bands, and layers of sand, gravel and clay, in which shells or fragments of shells are not infrequently detected. Again it passes into a regular stony clay which it is difficult, or even impossible, to distinguish from the till of the interior. In short, it is the conjoint production of the ice-sheet and the sea—consisting partly of true *moraine profonde* and partly of the nature of a submarine terminal moraine. And I have sometimes thought that the frequent absence of even fragments of shells from the aqueous beds of the boulder-clay may not improbably be accounted for in this way. We know that even in arctic regions rivers and torrents escape freely all the year round from the glaciers that terminate on shore, and it can hardly be otherwise with those which shed their icebergs in deep water. Now the presence of any large body of fresh water constantly flowing out from beneath the ice could hardly fail to keep away marine organisms from that portion of the sea-bed immediately in front of the glacier. And it seems, therefore, far from improbable that this may be one reason why the stratified portions of the boulder-clay do not more frequently contain shells and other exuviæ.

During the accumulation of the boulder-clay the land would appear to have stood lower* relative to the sea than it does now; but what was the extent of this submergence we cannot tell, the upper limits of the boulder-clay not having been precisely determined. No boulder-clay, however, has yet been noticed higher than 200 or 260 ft. above the sea, and generally it occurs at much lower levels.

The conditions that brought about the accumulation of the boulder-clay are thus seen to have differed to some extent from those that induced the deposition of the till. During the formation of the latter the ice was much thicker, and extended farther into the sea. It may even have been that the land itself stood at a greater elevation, as it certainly appears to have done in pre-glacial times.† But in the later glacial periods the ice-sheets were most probably of less extent than in the earlier stages of the great cycle. When the

* At p. 109 those great movements of elevation and depression to which the earth's crust is liable have been referred to subterranean action. The sea, however, may rise upon the land without any movement of the land itself. M. Adhémar and Mr. Croll have pointed out that a vast ice-cap (such as that which covered so many northern regions during the cold periods of the glacial epoch) would of itself cause a rise of sea in the ice-covered area, by displacing the earth's centre of gravity. But to what extent this rise would take place is uncertain. It is highly probable that some of the minor submergences of which we have evidence may be due to this cause; but the great depressions to which allusion will be made in Chaps. XVI. and XVII. are more likely to have been caused by movements of the earth's crust.

† See *ante*, p. 184.

boulder-clay which is found superimposed on the till of our maritime regions began to be thrown down, all the valleys were filled with ice, but not to overflowing. The watersheds and elevated ridges now broke up the mer de glace into a series of separate glaciers, and it is to this period that some of the crossing sets of rock-striæ described at p. 91 must be referred. When the mer de glace became less continuous the glaciers would sometimes change their direction, being less impeded in their course by the pressure of neighbouring ice-masses. During the former periods of great confluent ice-sheets the condition of Scotland closely resembled that of the antarctic polar continent, but at the time the boulder-clay was being accumulated it more nearly approached to the present aspect of Greenland. Great glaciers reached the sea and presented steep faces of ice to the swell of the Atlantic, but between the glacier-valleys were long stretches of rocky coast-line fringed with a thick shelf of ice like that which flanks the shores of many regions that border on the dreary Arctic Ocean. Coast-ice and bergs floated about, and the bottom of the sea was tenanted by arctic molluscs.

CHAPTER XVI.

UPPER DRIFT DEPOSITS OF SCOTLAND.

Morainic débris and perched blocks of Loch Doon district; of Stinchar valley; of northern slopes of Solway basin; of Rinns of Galloway; of the Clyde and Tweed valleys; of the Northern Highlands.—Erratics; carried chiefly by glaciers; have travelled in directions corresponding with the trend of the rock-striæ.—Condition of Scotland during the carriage of the erratics that occur at high levels.—Sand and gravel series.—Kames.—Terraces.

WE come now to consider the nature and origin of the upper drifts—those deposits, namely, which belong to a later date than the till and boulder-clay. They consist, as will be seen presently, of very diverse materials, and the mode of their formation has long been the subject of contention. It is a matter of no little moment, as the sequel will show, that we should ascertain what was the precise sequence of events that followed upon the deposition of the boulder-clay; and this we can only do by carefully considering the evidence in detail.

Turning attention first to the hilly districts—to the great uplands of the south—we find that the till, in a sorely denuded state, is frequently covered by a coarse earthy débris of angular fragments and large blocks and boulders. This débris ascends to great heights upon the sides of the mountains, and may be traced far down the valleys, even into the low-grounds beyond. Thus, for example, the mountains that hem in Loch Doon (Ayrshire) are sprinkled with loose angular and subangular stones, some of them striated, and with immense numbers of large boulders of grey granite which do not belong to the hills upon which they rest, but have travelled outwards from the central mountain region. The angular débris, as we trace it down the valley, appears to become thinner and thinner, until, when we reach the low-grounds about Dalmellington, it cannot be distinguished. But the grey granite boulders are more easily detected, and appear here and there on the hill-sides for several miles further down the valley. They are not, however, confined to the immediate slopes of the valleys of that district, but are scattered promiscuously over all the hill-tops up to a height of 1,700 ft. In the valley of the Stinchar, which drains the same great mountain tract, similar appearances may be noted; angular rubbish and large boulders are scattered over the hill-slopes down as far, at least, as the village of Colmonell, and they even appear near the very top of Beneraird (1,400 ft. above the sea). The same facts have been observed by my colleagues on the Geological Survey in the valleys that drain towards the Solway Frith. Everywhere great boulders are distributed over the valley-slopes and hill-tops and even over the low-grounds beyond. Thus the low-lying Rinns of Galloway, according to Mr. D. R. Irvine,* show numerous loose boulders of grey granite which have come from

* See *Mem. Geol. Surv. Scot.*, Exp. of Sheet 3, par. 39.

the hills of Cairnsmore to the north-east. In all these districts, in short, there is abundant evidence to show that both the angular débris and the boulders or *erratics*, as they are termed, have radiated outwards from the central knot of mountains down all the principal valleys to the low-grounds. We meet with the like phenomena in the valleys of the Clyde and the Tweed. Loose earthy and clayey rubbish, containing some scratched stones, and large erratics sprinkle the sides of the hills up to considerable heights, and this for many miles down the course of those valleys. In the Tweed valley, for example, such débris appears in decided masses as far down as Drummelzier. The deep glens of the Highlands present us with similar phenomena. The mountain-slopes are everywhere sprinkled with loose earthy rubbish, in which a few faintly glaciated stones sometimes occur, and large erratics occur up to all levels, even as high as 3,000 ft., according to Mr. Jamieson. But, for reasons which will be afterwards given, neither the evidence derived from the highland glens nor that obtained from the high valleys of the Southern Uplands is of any use to us in our present inquiry. It is the loose earthy angular gravel and rubbish and the large erratics that are met with at the lowest levels, and furthest removed from the centres of dispersion, with which we are at present chiefly concerned.

The angular rubbish is unquestionably of morainic origin—it answers in every respect to the rude débris which gathers on the surface of an alpine glacier and is shot over the end of the ice to form terminal

COOLIN MOUNTAINS, SKYE. Erratics resting on glaciated rocks in foreground. (By H. M. Skae.) *To face p.* 220.

moraines. It speaks to us, then, of a time when all the mountain valleys were yet filled with ice—with massive local glaciers, the direct descendants of the great ice-sheet that produced the till and boulder-clay. After the boulder-clay had been deposited the ice-sheet continued to retire until at last it no longer reached the sea, but deposited its moraines upon the land. It still covered a large part of the Lowlands, but such hills as the Lammermuirs, the Pentlands, and the Ochils, now rose above the level of the mer de glace, while in the Highlands and the Southern Uplands the ice was restricted, for the most part, to the valleys. It was only under conditions such as these that the morainic débris, sprinkled over the hill-tops and occurring far down the valleys, could have been deposited. At a long subsequent period local glaciers again occupied the higher valleys, but, as I shall afterwards point out, they were mere dwarfs when compared to the gigantic ice-streams that existed at the period at present under consideration.

Now let us glance for a little at the testimony of the erratic blocks, and see how this agrees with the evidence yielded by the ancient morainic débris.

Erratics are of all shapes and sizes—occasionally reaching colossal proportions, and containing many hundred cubic feet. Some are rounded, others only partially so, and very many are angular and subangular; not a few also show one or more scratched surfaces. In certain districts they are exceedingly abundant. I have already described how they occur in the valleys of the Southern Uplands and are plentifully scattered over the hill-tops up to considerable heights in that region, and how common they also are in the low-grounds that sweep out from the base of these hills. Reference has likewise been made to their frequent occurrence in similar positions in the Highlands.

In the intermediate Lowlands they are not wanting, but large numbers have disappeared in the progress of agriculture, so that they are not so plentiful as they used to be. But upon the slopes of those more or less isolated hilly tracts which rise up between the Highlands on the one hand and the Southern Uplands on the other, they are often abundantly met with. Along the northern flanks of the Ochils, for example, we find them thickly strewn. They consist in that district of such rocks as mica-schist, gneiss, granite, &c., all of which have evidently been derived from the highland mountains to the north and north-west. Boulders of similar rocks also make their appearance still further south. They are met with here and there in the low-lying parts of Fife, and Mr. Maclaren has described the occurrence of a large mass of mica-slate at a height of 1,020 ft. on the Pentland Hills—the nearest rock from which it could have come lying fifty miles to the north or eighty miles to the west. Boulders of highland rocks have also been noted on the northern slopes of the Lammermuir Hills. They likewise occur in considerable numbers on the crests of the trappean heights that rise between the valleys of the Clyde and the Irvine. In the south-west of Scotland, as already indicated, those undulating and hilly districts that roll out from the foot of the Galloway mountains, are studded with innumerable boulders which have radiated outwards from the central heights. Even the islands that lie off the coasts are dotted over with loose boulders or erratics, which can frequently be shown to have travelled great distances.

Erratics rest on bare rock, till, and angular débris alike, and they are also found on the slopes of certain hillocks of gravel and sand in the Lowlands. The position they occupy in the mountain valleys is often precarious—perched at a great height on some narrow ledge or jutting point of rock, where it would seem as if a slight push might send them bounding to the bottom.

As a general rule they prove to have been carried from higher to lower levels; the rocks of which they once formed a part stand at a greater elevation than those upon which they now repose. But to this rule there are exceptions; for loose boulders are occasionally found at a considerably greater height than the rock from which they have been broken.

What, then, do we learn from the erratics? How do we account for the scattering of these far-travelled blocks over, we may say, the whole face of the country. Some of them, it is evident, must have crossed wide valleys and considerable hills before they came to a final rest. The highland boulders on the Pentlands and the Lammermuirs, for example, after crossing Strathallan or Strathearn, traversed either the Campsie or the Ochil Hills, and passed athwart the broad vale of the Forth before they finished their journey. By what agent were they transported? The answer is—by a colossal glacier. So in like manner would I account for the presence of the numerous grey granite boulders that strew the slopes of the Galloway mountains, and are found distributed far and wide over the low-grounds at their base; for the boulders that cluster so numerously along the northern face of the Ochils; for the perched blocks that occur up to great heights in the glens and valleys of the Highlands; and for those that dot the surface of Orkney and Shetland and the islands of the Hebrides. But, as I shall point out in Chapters XVII. and XVIII., icebergs and ice-rafts have also had something to do with the distribution of the erratics.

It is a fact that most, if not all, the erratics have travelled in directions that coincide with the trend of the rock-striæ. Thus in Lewis we get boulders of Cambrian sandstone on the beach at Barvas (on the west side of the island), and at the Butt, which have evidently travelled either from Stornoway, or the mainland; now the low-grounds of Lewis are glaciated from south-east to north-west. In Aberdeenshire and Forfar all the erratics have streamed outwards from the mountains. It is the same in Perthshire, Argyle, Ross, and other highland districts. In the southern counties the same rule holds strictly true. The erratics lying loose upon the ground have moved in the identical direction followed by the till of the same regions—a direction which it need hardly be said coincides with that of the underlying rock-striations. Indeed, when the till is carefully searched it not infrequently yields fragments of the same rocks as those of which the erratics lying loose at the surface are composed. Thus, as mentioned in a previous page, fragments of highland rocks are got in the till of the Ochils, the Campsies, and the Paisley Hills. I have seen also bits of mica-schist in the till at Reston in Berwickshire.

Let us for a moment recall the appearance presented by Scotland during the accumulation of the till, and then consider what would be likely to result upon a gradual change of climate. When the cold was at its climax one great sheet of snow and ice enveloped the whole country, above which perhaps only the tips of some of the higher mountains appeared. As this ice-sheet melted away, and the great confluent glaciers withdrew from the bed of the shallow seas, of course the surface of the ice must have been lowered in proportion, for melting would take place atop just as at the extremities. As a consequence of this, more and more of the mountain districts would peer above the waste, and frost would then rupture the rocks—and rubbish, and débris, and great blocks falling upon the ice, would be carried outwards in the direction of its flow. In this way long trains of erratics would be travelling north, south, east, and west, from the Highlands, and in similar directions from the Southern Uplands. As the ice continued to melt, hills like the Pentlands would begin to rise above the level of the ice, and form islands in a great mer de glace; and just as the retreating tide will strew the beach with the waifs of ocean, so the ice-current, as it pressed upon and slowly crept away from these desolate islands, would leave upon their frozen shores the wreckage of the distant mountains. The surface of the ice still sinking, erratics would be left stranded on mountain-slopes and hill-sides at ever decreasing levels, until at last, the mer de glace having shrunk beyond the reach of the waves, erratics toppled over the terminal fronts of the glaciers upon the Lowlands themselves. And so the process would continue until the glaciers had shrunk back into their mountain valleys.

It would thus appear that the erratics belong to different stages of the glacial epoch. Those that lie upon the islands, and, at great heights on such hills as the Pentlands, are, in all probability, as old as the boulder-clay of the maritime districts, while those occurring at lower levels, and nearer to the mountains, must date to more recent times.

There is an apparent difficulty in accounting for the transportation of erratics from lower to higher levels. It is evident, however, that these could not have been carried on the surface but must have been pushed on below, or carried along engorged in the ice.* But if erratics and the morainic débris described above, were scattered over hill-tops and strewn along mountain-slopes by gigantic glaciers, it is evident that much moraine matter must also have been left upon the Lowlands, as the ice gradually drew back

* Erratics of the same character occur in Sweden. See *postea*, chapter XXVI.

to the mountain valleys. Have we any trace of such terminal moraines, and, if not, what has become of them? To answer these questions we must consider now the upper drift deposits of the lowland districts.

In the districts referred to, the till and boulder-

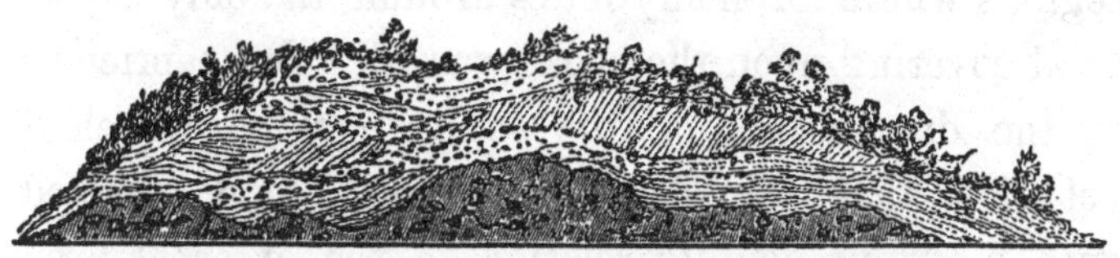

Fig. 37.—Sand and gravel resting on denuded surface of till, railway cutting, near Douglas Mouth Bridge, Lanarkshire.

clay are overlaid in many places by great masses of distinctly water-worn materials. These deposits occur at all levels, from the coast up to a height of more than 1,500 ft. above the sea. The most characteristic form assumed by them is that of rolling mounds, cones, and ridges, all of which consist, for the most part, of gravel and sand. To such an extent, indeed, is this the case that the whole group is often spoken of as the "sand and gravel series." It does not cover nearly so large a tract of ground as the till. In the higher mountain regions many miles of country may be traversed without discovering a single patch of sand or gravel. As a general rule the deposits belonging to this group are confined to lowland districts, where they appear at first sight to be distributed in a most arbitrary manner. Occasionally we may follow them for miles, when all at once they will die out, and then we may not meet with them again until we have passed into a quite different district. Again, they may be represented by only one or two mounds, often widely separated, with no trace of sand or gravel in the intervals. I have not infrequently come upon a solitary and isolated mound of sand and gravel, in regions where for many miles around the only superficial covering upon the rocks was till. But capricious as the distribution of the series may be, we shall yet find that this drift is arranged and grouped with a certain definite relation to the external form or contour of the country.

The sands and gravels have, as I have just said, a tendency to shape themselves into mounds and winding ridges, which give a hummocky and rapidly undulating outline to the ground. Indeed, so characteristic is this appearance, that by it alone we are often able to mark out the boundaries of the deposit with as much precision as we could were all the vegetation and soil stripped away and the various subsoils laid bare. Occasionally, ridges may be tracked continuously for several miles, running like great artificial ramparts across the country. These vary in breadth and height, some of the more conspicuous ones being upwards of four or five hundred feet broad at the base, and sloping upwards, at an angle of 25° or even 35°, to a height of 60 ft., and more above the general surface of the ground. It is most common, however, to find mounds and ridges confusedly intermingled, crossing and re-crossing each other at all angles, so as to enclose deep hollows and pits between. Seen from some dominant point, such an assemblage of *kames*, as they are called, looks like a tumbled sea—the ground now swelling into long undulations, now rising suddenly into beautiful peaks and cones, and anon curving up in sharp ridges that often wheel suddenly round so as to enclose a lakelet of bright clear water. Fine examples of sand and gravel hills are seen in Lanarkshire, at Carstairs and Carnwath. They are also well developed in Haddingtonshire, near Cockburnspath; in Berwickshire, at Dunse, and north of Greenlaw; in Roxburghshire, at Eckford; and another fine set is seen in the valley of the Tweed, at Wark and Cornhill, Northumberland. At Leslie and Markinch, in Fifeshire, a similar series occurs, and like accumulations appear more or less abundantly throughout the lowland districts.

Not infrequently the slopes of the kames are carpeted with fresh green turf, in strong contrast to the more sombre-hued vegetation of the surrounding clay-covered tracts. The local names in the country sufficiently attest this peculiarity. "Green Hills" are of very common occurrence, and I have usually found the name restricted either to kames or to certain little projecting bosses and cones of friable, decomposing, igneous rocks. When the kames are composed of large stones, the vegetation on their slopes becomes coarse and poor, and the fresh green grass then gives place to clumps of broom or stunted gorse. This is more usually the case with the sharper ridges and peaked cones, these being made up chiefly of coarse gravel and shingle. The gentler undulations consist for the most part of fine sand and gravel, and hence, in a rough way, the slope of a kame and the character of the vegetation that clothes it serve as a kind of index to the nature of the materials lying below.

Almost all the isolated solitary mounds that I know of are made up of fine sand, and some of the best examples of these occur in Fifeshire. A small one, quite close to Dunfermline, is locally famous under the name of Mont Dieu. According to old story this drift mound owes its origin to some unfortunate monks who, by way of penance, carried the sand in baskets from the sea-shore at Inverkeithing. A similar tradition accounts for a conical hill of fine sand at Linton, in the valley of the Kale Water, Roxburgshire; of this hill it is said that "two sister nuns were compelled to pass the whole sand through a riddle or sieve as a penance for their transgressions, or to obtain pardon for a crime of a brother." *

Another mound of the same material (Norrie's Law), a few miles north from Largo, in Fifeshire, is noted as the burial-place of some great worthy of past times. Who he was does not appear, but no doubt he must have been "a superior person," for he was buried in a suit of silver armour, most of which, unfortunately, found its way to the melting-pot soon after its discovery by a farmer. Other isolated cones, in various parts of the country, have often been described by

* *History and Antiquities of Roxburghshire.* By A. Jeffrey, vol. i. p. 41.

local antiquaries as tumuli, apparently for no other reason than that they resemble these in external appearance. It is not unlikely, however, that such cones may occasionally have been used as burial-places. It is certain, too, that some of the bolder ridges and mounds have been fortified and utilised for purposes of defence, the ditches scooped in their sides being still apparent. Protected by strong palisades of wood, and surrounded as most of them probably were by dense forest, one can easily see how an abrupt kame or steep cone might be made a very formidable fortress in the days of spears and arrows.

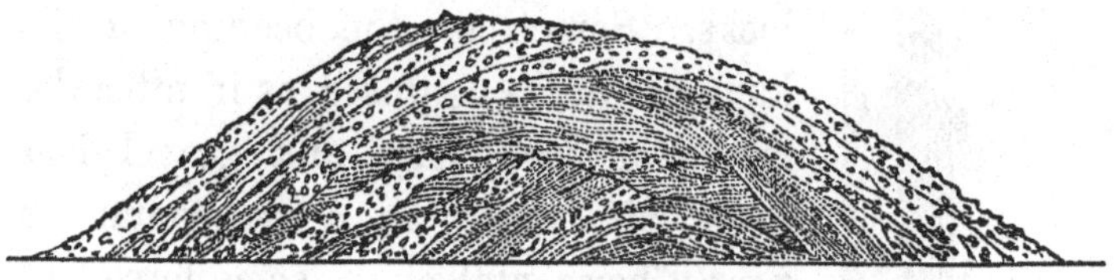

Fig. 38.—Section across kame, Douglas Railway, near Lanark.

The deposits of which the kames are composed are usually stratified, and, in some of the finer-grained accumulations, very beautiful examples of false or diagonal bedding frequently occur. But in many cases the coarser heaps of gravel and shingle do not exhibit any traces of stratification, the stones being piled up in dire confusion. It is remarkable, however, that the gravel-stones, whether small or large, are almost invariably well-rounded and water-worn. This is at moderate elevations; but at the highest elevations reached by the kames this water-

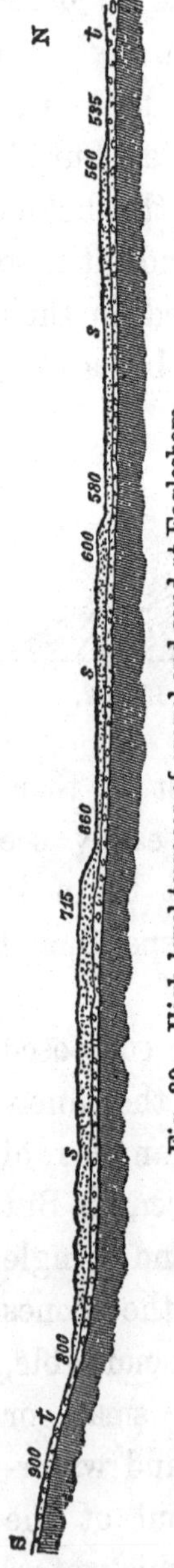

Fig. 39.—High-level terraces of gravel and sand at Eaglesham.

worn aspect becomes less conspicuous, or vanishes altogether. Occasionally, too, we come upon a large stone, or boulder, embedded in the sand and gravel, but this is by no means a common occurrence; on the contrary, when erratics are found associated with kames, they almost invariably repose upon the tops and slopes of these hillocks. Sometimes the bedding of the kames is much confused, as if, after the layers of sand, clay, and gravel had been laid down in a horizontal or nearly horizontal plane, some force had squeezed and pushed them out of place, twisting, folding, and crumpling them up; but this also, as far as my experience goes, is not a common occurrence.

Associated with the sand and gravel, we here and there come upon deposits of silt and clay which have occasionally been worked for brick-making. These beds are usually finely laminated. But neither in them, nor yet in the sand and gravel, have any organic remains been discovered.

In some hilly districts we find the slopes of the ground fringed, at uniform levels, with shelves and terraces of sand and gravel. These shelves and terraces generally dip outwards and downwards, with a gentle inclination from the high grounds against which they abut. Occasionally, several such platforms occur in succession, and, when seen in profile, resemble giant staircases. The deposits of which they are composed are usually more or less well bedded, and consist of gravel and sand, with which clay is sometimes intermingled. The deposits, however, are not always well waterworn; sometimes, indeed, they consist of only angular or subangular stones, and a kind of earth or earthy sand and clay. None of them have as yet yielded any organic remains.

A very fine example of the phenomena described occurs at Enoch near the village of Eaglesham, about twelve miles south-west from Glasgow. In the accompanying section, the general outline of this series of terraces is shown; *t* represents the till, *s* the sand and gravel, and the figures indicate the height in feet above the sea-level.

Having now given an outline-sketch of the sand and gravel series, we may proceed to inquire into the origin of these deposits. In doing so I shall supplement the facts already adduced by further evidence, which, for convenience' sake, has been reserved to this place.

CHAPTER XVII.

UPPER DRIFT DEPOSITS OF SCOTLAND—*Continued.*

Distribution of the kames.—Kames of Carstairs and Douglas Water; of the Kale and the Teviot; of the Whiteadder valley; of Kinross-shire; of Leslie and Markinch, in Fifeshire; of the Carron Water; of Perthshire and Forfarshire; of the valleys of the Don, Dee, Ythan, Deveron, Spey, and Findhorn; of the vale of the Ness; of the Beauly Frith.—Sand and gravel along flanks of hilly ground between rivers Clyde and Irvine.—Absence of kames, &c., in Lewis and Caithness.—Evidence of transport down the valleys.—Passage of well-rounded gravel into angular gravel and débris at high levels.—Sand and gravel overlying morainic débris.—Denudation of older drifts in sand and gravel districts. —Evidence of river action.—Fossils in ancient river-gravels, &c.—Isolated cones and ridges of gravel and sand, and high-level terraces.—Evidence of action of sea.—Sand and gravel drift of Peeblesshire.—Limits of submergence of Scotland.—Summary of conclusions.

IF there have been many opinions held concerning the mode of formation of till, there have been just as many theories offered in explanation of the origin of kames. Some writers have insisted that these find their nearest analogues in the terminal moraines of alpine glaciers, others have maintained their fluviatile origin, while a yet larger number of observers unite in ascribing them to the action of the sea. With each of these apparently conflicting opinions I partially agree. Neither, taken alone, contains the whole truth, but when the three are combined it seems to me that we have a simple and natural explanation of the phenomena; and I shall now attempt to point out the line of reasoning which has led me to this belief.*

It appears to hold generally true of all the larger areas of the kame deposits that these occur in valleys at or near where the rivers escape from the confined mountain glens or upland dales to enter upon the broad low-grounds. And not only so, but the extent of the gravel beds seems frequently, if not always, to be in direct proportion to that of the drainage-area in which these occur. When this last is very extensive the kames almost invariably attain a vast development. On the other hand, when the river system is comparatively insignificant, so likewise are the deposits of sand and gravel that cumber the ground where the main valley begins to open out upon the Lowlands. A few examples will illustrate my meaning.

If the Ordnance Survey map (sheet 23) be examined, it will be observed that the river Clyde, after leaving the hilly country through which it flows in a general northerly direction, suddenly turns to the west, near where it is crossed by the Caledonian Railway, so as to skirt for some miles the base of the Uplands from which it has just escaped. Immediately to the north of this westerly part of the river's course (in the neighbourhood of Carstairs and Carnwath) the ground rises in a very gentle incline, and undulates away to north and north-east for several miles, at a general level of little more than one hundred feet above the

* The results I have arrived at do not differ greatly from the views held by Mr. Jamieson.

river. It is precisely here where we encounter a widespread series of kames, cones, mounds, and banks of gravel and sand. Standing among these hillocks and turning towards the south we look right up the valley of the Clyde and into the great Uplands. But the gravel deposits in this case are not restricted to the low-grounds on the northern slopes of the river valley. Considerable heaps of gravel and sand may be traced up to and even beyond Lamington, at heights which could not possibly be reached by the present river, and these deposits are clearly a continuation of the similar accumulations near Carstairs.

Some four or five miles below Carstairs the Clyde receives on its left bank the tribute of the Douglas Water, a considerable stream, draining a large area. It takes its rise on the slopes of Cairn Table, at a height above the sea of 1,500 ft., and reaches the Clyde after a course of fifteen or sixteen miles. In the lower reaches of its valley we find numerous heaps and mounds of gravel and sand, and the same deposits are cut through by the Clyde opposite the mouth of the Douglas Water. These deposits, however, are by no means so extensive as those in the neighbourhood of Carstairs.

Another considerable assemblage of mounds, hillocks, banks, and undulating flats of sand and gravel occupies a similar position in the valley of the Kale Water, between the base of the Cheviot Hills and the River Teviot, near Eckford. This is a rather striking example of the phenomena under review. The Teviot here runs north-east, skirting the outlying spurs of the Cheviots, which are seen rising up boldly in the south. After receiving Jed and Oxnam Waters, the river suddenly wheels away from the hilly ground and makes directly north for the valley of the Tweed, being joined about a mile below Eckford by the Kale Water. From this point the whole valley of the Kale, up to where the stream escapes from the Cheviots, at Morebattle, is more or less covered with gravel and sand, which rises into banks and mounds, and extends in broad undulating flats. Similar deposits are seen opposite the junction of the Kale with the Teviot on the west bank of the latter river. None of these accumulations could possibly have been formed by the present streams; they are not only too extensive, but they occur also at too great an elevation.

We find similar appearances characteristic of the Lammermuir districts. The Whiteadder Water, for example, after leaving the Lammermuir Hills, enters upon a low-lying undulating country, which is thickly strewn with sand and gravel over an area many square miles in extent; and the great bulk of these deposits is strictly confined to the drainage-area of the water. Along the northern flanks of the same hills similar phenomena recur, the low-grounds being plentifully coated with gravel and sand opposite the mouths of the larger upland valleys.

On the opposite slopes of the Forth Basin excellent examples are not wanting. Considerable accumulations of gravel and sand extend along the low-grounds of Kinrosshire, opposite the valleys that open from the Ochils. Again, if we follow the course of the

Leven we shall find that, shortly after leaving the Loch, it flows through a great series of mounds, hillocks, and banks of gravel and sand, which are especially well seen at Leslie and Markinch.

Reference may also be made to the great sand and gravel heaps that occupy the low-grounds at the base of the Kilsyth Hills, within the drainage-area of the Carron Water.

Mounds, irregular banks, and extensive sheets of the like materials attain a vast development along that great trough that lies between the Perthshire Highlands and the Ochil Hills. It will be found throughout this wide tract that the sand and gravel bulk most largely opposite the mouths of the large mountain valleys.

In Forfarshire, and indeed along the whole north-east of Scotland, we invariably find that the greatest gatherings of gravel and sand are collected in similar positions, and that they frequently ascend the larger valleys for long distances. Vast deposits, for example, crowd the valleys of the Don, the Dee, and the Ythan, and the same appearances are repeated, as Mr. Jamieson has shown, in the rivers that drain north-east into the Moray Frith, as for example the Deveron, the Spey, and the Findhorn. In the neighbourhood of Inverness like masses of water-worn materials form conspicuous objects in the Vale of the Ness, and at the head of the Beauly Frith the lower reaches of the Farrar exhibit similar accumulations.

In all the cases now cited, and many more might be given, the extent of the gravel and sand deposits, which frequently assume the form of cones, peaks, and ridges, is invariably proportionate to the drainage-area of the valleys in which they occur. The same fact becomes more conspicuous when we limit our attention to any well-defined hilly region of the central Lowlands. Take, for example, that broad, undulating hilly district, which, beginning at the Frith of Clyde, extends south-east along the borders of Renfrewshire, Ayrshire, and Lanarkshire, until it gradually falls away into the valley of the Clyde, near Strathavon. This wide district forms the water-shed between the Irvine and numerous small tributaries of the Clyde.

Along both flanks of this wide, hilly, and moorland district it holds generally true that all the larger accumulations of gravel and sand are disposed at or near where the streams leave the hills and enter upon the low-grounds—the extent of these deposits being in proportion to that of the drainage-area and the height of the watershed. Valleys draining a limited district of low elevation have no marked accumulations of gravel and sand at their mouths; on the other hand, valleys draining from lofty sheds invariably contain in their lower reaches extensive deposits of gravel and sand, which are frequently heaped up into mounds, and the larger the valley the bulkier the accumulations of water-worn materials. These phenomena are well illustrated by the valleys of the Dusk, the Lugton, the Crawick, and the Crawfurdland Waters, and by the Avon and its tributaries—the Glengaber, the Calder, the Kype, and the Locher Waters.

We may now glance very briefly at some of the low-lying districts in the extreme north of Scotland. Some reference has already been made to the glacial phenomena of Lewis. The northern portion of that district may be described as a wide undulating moorland, no portion of which rises higher than a few hundred feet above the sea. It is drained by several inconsiderable streams flowing from the central axis of the island to north-west and south-east. In the extreme south there rises a bold line of mountains, against which the moorlands somewhat suddenly abut. But from this mountain district only one large valley opens upon the low-grounds to the north, and this hollow is entirely occupied by Loch Langabhat.

Now, nowhere in Lewis, neither in the undulating moorlands to the north, nor along the slopes of the mountains in the south, do any heaps of gravel and sand occur. Not a single trace of kames is to be met with in any part of that region. A similar absence of gravel mounds and heaps, as Mr. Jamieson has remarked, characterizes the great flats of Caithness.

Enough, perhaps, has now been said to give the reader some notion of the mode in which the larger accumulations of gravel and sand are distributed throughout the country. A word or two may now be added in regard to certain appearances presented by the deposits themselves.

I have said that in the case of the larger valleys and glens the gravel beds frequently ascend these hollows for some distance. When this happens these deposits almost invariably give evidence to show that they have been carried along by a force acting in a direction down the valley. This is most conspicuous in the long flat-topped banks and irregular terraces with an undulating surface, but it may also be sometimes noticed in the lower or undermost portions of well-marked ridges, or typical kames. Abundant evidence on this head will be found in the valleys of the Tweed and some of its tributaries; in the upland districts traversed by the Clyde; in Annandale, and many other drainage-areas in the south of Scotland. Mr. Jamieson, several years ago, recorded the fact in reference to the great valleys of the Spey, the Findhorn, and other rivers in the north.

Another peculiar appearance has next to be noticed. When the gravel beds are traced far up the valleys, they are frequently seen to pass into a kind of earthy angular débris, and the same kind of angular earthy stony rubbish is often found to form the upper limits of the sand and gravel series when these deposits are followed up the sides of the valleys. This angular débris is not to be confounded with that coarse rubbish which the frosts are yearly sprinkling over steep mountain-slopes. It forms in places distinct mounds, which are usually quite unstratified. Such angular gravels are well developed along the foot of the Moorfoot Hills. They bear a close resemblance to the moraine rubbish which accumulates in front of an alpine glacier, and a close search among them will sometimes detect a glaciated stone or two. But for many reasons, which will presently appear, it is often extremely difficult to say what relation the kames and mounds of the low-grounds bear to the angular rubbish and perched blocks of the mountain regions. These kames, as we have seen, form part and parcel of vast gravel and sand deposits which frequently ascend the valleys for great distances, until, as just stated, they pass into a kind of loose morainic débris. Now the sides of such valleys are usually sprinkled with moraine matter, and dotted over with erratics, from the head of the valley down to the low-grounds. When this is the case it is clear that the gravels and sand occupying the bottom of the valley must be of more recent date than the coarse débris that hangs upon the mountain-slopes on either side. The gravel beds of the Tweed at and above Drummelzier, for example, are clearly of more recent formation than the morainic débris over which and against which they lie. But the more typical assemblages of kames usually occur upon the low-grounds just beyond the mouths of the valleys in places where there is as a rule no morainic débris to be seen. Cases, however, do occur where well-marked kames run along hill-sides that are sprinkled with moraine matter and dotted with erratics, in such a way as to show that these last were deposited before the kames.

One other fact remains to be mentioned. In all the valleys that contain gravel and sand in any quantity both the till and the morainic débris have suffered extensive denudation—often all that is left being a few large boulders scattered here and there along the sides of the valleys. Nor can we be in doubt as to the direction from which the denuding agent came; for while the till and morainic débris have been stripped from the faces of such knolls and projecting rocks as look up the valley, the same deposits are found sheltering in the rear.

Now putting these various considerations together the conclusion seems forced upon us that all those accumulations of water-worn materials whose peculiar distribution has now engaged our attention, owe their origin to currents that once flowed down the valleys. And not only so, but we must also admit that those currents were proportionate in size to the extent of each particular valley-system in which such accumulations are found. In short, we can only, as I think, account for the appearances described by attributing the deposition of the greater areas of gravel and sand to river-action. But if so, then the rivers must have greatly surpassed in volume and breadth their present puny representatives. It is impossible to conceive that the masses of gravel and sand occupying the lower reaches of the upland valleys, and some of the highland glens, could be laid down by rivers like the present, even although these were to continue in constant flood. Some great change has taken place since the old gravel beds were deposited —the amount of water circulating in the valleys has in some some way vastly diminished—some of the rivers have even ceased to flow in their old courses and are now working out for themselves new channels —driven from their former beds by the huge heaps of detritus which they themselves, at some early period, carried down from the mountain regions.

The explanation appears to be simply this. The great ice-sheet underneath which the till accumulated had, after depositing the boulder-clay, continued to retire until, as already described, it was reduced to a system of gigantic local glaciers. In summer-time such streams and rivers as flowed in glacier-valleys would be vastly swollen by the water derived from melting snow and ice. Great currents would sweep down the valleys, carrying with them the angular débris derived from terminal moraines and from freshets rushing down the slopes of the hills. As this débris was hurried along, it would gradually be rounded by attrition, and eventually pass into good gravel. At the same time the till and ancient morainic débris over which the river rushed would be denuded and washed away from exposed positions. As the valley widened the river would also expand and begin to deposit material; if, however, the valley continued comparatively narrow until where it suddenly opened into the low-grounds, then the river would suffer but little gravel to gather in its course, but would sweep everything onward until it escaped from the hilly regions, when it would at once expand and throw down the major portion of its burden. One may still see upon a small scale how this process was carried on, by examining the behaviour of such little mountain brooks as are liable, upon every thunder shower, to be converted into roaring torrents. When these torrents are swollen with rain, they rush impetuously downward, often completely filling the deep gullies in which they flow. Arrived at the base of the mountains, they immediately spread out and deposit heaps of stones, débris, and coarse sand. In course of time long sloping banks are thus formed, which expand in fan-shape from the foot of the gullies. When but little water is flowing the brooks employ themselves in cutting courses through that thick débris which only sudden floods could have enabled them to carry.

The larger areas of gravel and sand are, therefore, strictly analogous in origin to the heavy masses of gravel and coarse sand that strew the beds of alpine valleys. Those who are familiar with the appearances presented by such areas as that of the Aar, the Rhone, and other well-known Swiss districts, or by the glacier valleys of Norway, as that of the Justedal, or the smaller but even more interesting ones of Fondalen, will be the first to recognise the close similarity of the Scottish gravel beds to those characteristic of glacier regions. Like these latter the Scottish deposits are quite unfossiliferous. Some time ago, however, I obtained from a thin bed of sand in the series numerous small bones, which Professor Huxley determines to be those of frogs and water-rats. Thus at the time the gravel beds were forming the country could not have been a totally uninhabited desolation. But the further consideration of the physical conditions under which these deposits were accumulated will perhaps be better understood after we have concluded our examination of certain phenomena connected with the gravel beds, to which we have as yet only alluded.

In our general description of the sand and gravel series mention was made of certain isolated cones and solitary ridges, and of shelves or terraces that here and there fringe the slopes of the hills. The reader will also remember that much of the sand and gravel opposite the mouths of the upland valleys is arranged in the shape of peaked cones and ridges, and that the material of which these curious hillocks is made up not infrequently shows beautiful diagonal bedding.

No one can examine a typical kame, such as that near Greenlaw, in Berwickshire, without feeling assured that it owes its shape to the mode in which its materials were heaped up. In not a few cases, however, the rapid undulations so characteristic of sand and gravel areas are clearly due to denudation. The deposits have not originally been heaped up as mounds, but are only the remnants of larger deposits—of broad terraces or flats—through which more or less deep hollows have been worn by the action of water. The abrupt manner in which the layers of sand and gravel are cut off by the slopes of the so-called kames, shows that many of them have been so formed.

But while thus admitting that many mound-shaped hillocks of gravel and sand are only the denuded remains of what were once continuous flats of fluviatile origin, still there are appearances connected with the more typical assemblages of kames, cones, and mounds which can hardly be explained by what we know of rain- and river-action. To account for some of the phenomena we are apparently compelled to call in the agency of the sea. The deep circular depressions, surrounded on all sides by smoothly-rounded cones and banks, and often occupied by lakelets or peat mosses, cannot possibly be due to the action of rivers; the well water-worn character of the gravel precludes all chance of the deposits being morainic, and the generally undisturbed appearance of the bedding shows that the mounds have not been caused by glaciers advancing upon and pushing up before them pre-existing beds of gravel and sand; although, as Mr. Jamieson has suggested, some of the kames may have been formed in this way. Even with this admission, however, we can hardly escape crediting the sea with the formation, or, at all events, with the shaping-out of many and even most of the characteristic kames.

At lower levels than 900 ft. they frequently exhibit diagonal or false bedding. This appearance points to the action of shifting currents in a somewhat shallow sea. It is well known, of course, that diagonal bedding also occurs in undoubted freshwater deposits, but only in a very partial manner. In some gravel and sand areas, however, it is so widespread and common that we are justified in considering it rather as an indication of marine than of fluviatile action.

Again, when we take into consideration the fact that isolated cones and ridges sometimes occur in zones that run along an exposed hill-face, far removed from any valley, or are dotted here and there over extensive moorlands away beyond the reach of any possible river; and when, moreover, we note that strings of gravel ridges and mounds may sometimes be followed up one valley across the dividing col into a totally different drainage-system, we cannot but conclude that ordinary river-action is out of the question as an explanation of the phenomena. In the present state of our knowledge, we appear to have no alternative but in such cases to admit the marine origin of such kames.

The same assumption is necessary to explain the occurrence of those elevated shelves or terraces which here and there fringe the slopes of the hills. The shelves of gravel at Eaglesham, for example, appear to be ancient sea-beaches. It is true that they contain no trace of shells, but still, when we consider their position upon an exposed hill-face where no temporary lake could possibly have been formed by any glacier, the probability of their fresh-water origin seems to be out of the question. The water that spread them along the slopes of the Eaglesham hills must have stretched far away for some sixteen miles at least, until it abutted upon the flanks of the Kilpatrick and Campsie hills. The highest of the terraces does not reach beyond 800 ft. above the level of the sea. Similar terraces, however, have been met with at greater elevations. I have traced them on the Moorfoots up to a height of 1,050 or 1,100 ft., and these, like the Eaglesham beds, seem equally to require the agency of the sea. Still further south high-level shelves of gravel and sand have been detected by my colleague, Mr. H. M. Skae, in Nithsdale, at a height of 1,250 ft. above the sea. But he agrees with me that these may possibly have been deposited in a temporary lake during the retreat of the great ice-sheet, and while a massive stream of ice yet occupied the broad vale of the Nith. Several instances of the occurrence of stratified accumulations of sand, gravel, clay, and silt, at considerable elevations in the Northern Highlands are given by Mr. Jamieson,* and at one time he was inclined to consider these as of marine origin. But he now thinks it doubtful if they are, seeing that "similar stratified beds are frequently found in Alpine districts which have been occupied by glaciers," and which there is no reason to believe have been submerged since those glaciers melted back.

But there are yet other considerations which seem to render it extremely probable that many of our kames have been shaped out by the action of the sea. I shall try to make this plain by describing briefly one district where they are typically developed.

In Peeblesshire shelves and terraces of gravel, which can hardly be other than marine, occur at a height of 1,050 or 1,100 ft. Let us then for the moment assume that this level indicates the degree of submergence experienced in the south-east of Scotland, and then glance at the peculiar distribution of the kames in the valleys of the district referred to. These do not occur in all the valleys alike—some they quite choke up, from others they are entirely absent. Nor at first sight is there anything remarkable in the valleys themselves to account for this anomaly. They may be wide or narrow, deep or

* *Quarterly Journal of Geological Society*, 1865, p. 177.

shallow, winding or comparatively straight. But when the height of the various watersheds and dividing cols is compared with that of the high-level shelves, the kames are then found to be restricted to valleys whose cols are either at or below 1,100 ft.—that is, the level reached by the gravel terraces. If now we, in imagination, depress the Peeblesshire uplands until these terraces are washed by the waves, we shall find that all the valleys which are occupied by kames form connecting straits between opener spaces of sea. Those valleys, on the other hand, whose cols and watersheds are higher than 1,100 ft. would, on a submergence to this extent, become long fiords winding into the heart of the hills. Now, in such valleys, kames are either entirely absent or occur only at the lower ends, where the valleys expand into the more open districts beyond.

Putting these curious facts together we seem to gain additional assurance that some kames at least have been shaped out by the sea. If the land, for example, were to be covered by the ocean up to the level of 1,100 ft., certain valleys, as we have seen, would be converted into straits, and some of these straits could hardly escape being swept by strong currents. Those who have cruised much among the Western Islands will, no doubt, have a vivid recollection of the strong and rapid tides that sweep to and fro along the narrow straits. With such currents flowing through the submerged valleys of Peeblesshire, ancient river gravels, moraines, and till would be eroded, redistributed, and heaped up into banks and ridges. Nor is the aspect presented by the fiord valleys of Peeblesshire less suggestive. In such valleys no current-action could take place, hence the till remains very much as the old streams left it, and the ancient valley gravels are not worked into kames. One has only to examine the work done by the sea in the western sea-lochs or the long fiords of Norway to become convinced that Neptune, when imprisoned in such deep, narrow ocean valleys, is powerless. The denudation in a *cul de sac* of this kind is inadequate to produce stratified heaps of well-rounded pebbles, the débris upon the shores of a sea-loch not being more water-worn than the shingle that fringes the margin of a confined inland lake. Along the Norwegian fiords we see wooden erections—landing-stages, warehouses, baths, &c.—resting upon long piles that are driven into the bed of the sea. These stand well, for there is no wave-action to speak of, the water merely rising and falling with the tide. Upon such a coast, even when it shelves, well-rounded pebbles and fine sand are exceptional. The mountain-slopes either plunge at once into deep water, or if there be any beach at all, that is most usually strewn with great angular blocks and coarse débris, detached from the rocks at the coast-line, more by the action of frost than by the power of the waves.

Taking all these matters into consideration, it seems to me probable that this south-east part of Scotland was not submerged to a greater extent than 1,100 ft. or thereabout. If, however, we are to take kames as tests of submergence, then we have evidence to show that in the western districts a greater degree of depression was attained—my colleague, Mr. R. L. Jack, having detected kames in the Fintry Hills, at a height of 1,280 ft. above the sea.

Some account of the erratics has already been given: it only remains to add that wandered boulders are found now and again reposing upon the tops and slopes of the kames—an appearance which, as we shall afterwards observe, is by no means confined to Scotland. Occasionally, too, they may be detected embedded in the kames; but this is a somewhat rare occurrence—their most usual position is upon the outside.

I shall now briefly summarise the results arrived at from a study of the morainic rubbish, perched blocks, and gravel and sand.

When the great ice-sheet was beginning to deposit the boulder-clay which is now met with in the maritime districts, the higher hills of the central Lowlands stood above the level of the mer de glace like islands in a frozen ocean. At the same time the mountain ridges of the Highlands and the bold hills of the Southern Uplands rose up so as to separate the ice-sheet into a series of gigantic local glaciers, which, however, still coalesced to form one mighty stream upon the broad Lowlands. Frost shivered the rocks and loosened out great blocks, which eventually toppled down upon the ice below, and, along with heaps of angular rubbish, were slowly carried away. Sometimes the stones and boulders fell into crevasses or between the ice and the rock of the mountain-slopes, and so got ground and polished on one or more sides: but they always travelled farther and farther off from their parent mountains. The tops of the lowland hills, peering above the ice, caught some of the wanderers as they drifted past, but many were borne out to the terminal front of the ice and dropped into the sea, where they mingled with the scratched stones that were being pushed out from underneath the glaciers.

As the ice continued to melt, erratics and angular débris were stranded at ever-decreasing heights upon the mountain-slopes and hill-sides, and at last the ice drew back from the sea and the glaciers then dropped their rubbish upon the land. Great streams of water escaping from the melting ice swept the morainic matter down the valleys, and angular stones and rubbish, as they were pushed along, became rounded by attrition, and arranged by the rivers in great flats of gravel and sand. Thus, ever as the glaciers withdrew, the angular débris that gathered in front of them was ploughed down and distributed over the bottoms of the valleys by the swollen rivers, the perched blocks at great elevation on the sides of the valleys and upon the slopes of the lowland hills, still remaining to indicate the heights formerly reached by the glaciers.* There being no great river-valleys draining from mountain regions in the low-grounds of

* I would again remind the reader that some of the crossing of striæ upon the rocks was in all probability caused during this recession of the ice-sheet. As the glaciers ceased to be confluent, their courses would in many cases be modified. A good example is given in *Trans. of Geol. Soc. Glasg.*, vol. iv. p. 223.

Lewis and Caithness, the absence of gravel deposits from such districts is easily accounted for.

To what extent the ice was eventually reduced we have no means of ascertaining, neither do we know much of the climatal conditions which at this period obtained in Scotland. All that we can safely assert is that the ice disappeared entirely from all the Lowlands, and drew back into the deeper mountain valleys. Of the plants and animals which at this time may have clothed and peopled the land, we know next to nothing. I have here and there in the gravel and sand-beds detected some vegetable matter, but in too decomposed a state to enable me to say what it was. In one section, however, near Carham, on the Tweed, I obtained from a bed of sand in the series numerous remains of water-rats and frogs. It would be hard to believe that these were the sole denizens of the land; as yet, however, they are all that we have got to show.

After such conditions had lasted for a longer or shorter period, the land gradually sank into the sea. As it slowly went down the waves and currents ploughed up and redistributed much of the old glacial accumulations and river deposits. Broad terraces of gravel and sand were cut into, and their materials winnowed and re-arranged. Here and there also ridges of gravel and cones of sand were heaped up in places where no sand and gravel existed before—the sea using up for the purpose the till and morainic rubbish.

To what depth the submergence extended has not

yet been ascertained. The position of the kames on the Fintry Hills would seem to indicate a depression of 1,238 ft.; but in the south-east of the country the depression did not perhaps exceed 1,050 or 1,100 ft.

At some period during the submergence, probably as that approached its maximum, the climatal conditions would appear to have become severer. While the kames were being shaped out little or no ice could have been floating in the sea, otherwise erratics should have occurred plentifully inside the kames, but the presence of the wandered blocks, which are strewn over the tops and slopes of the gravel, seems to show that after the formation of these hillocks and ridges, rafts of ice carried seawards blocks and angular stones. It must, therefore, be often extremely difficult to distinguish such ice-floated erratics from those perched boulders which were left behind upon the land by the retreating glaciers. When the stones are proved to have travelled in directions which the ancient glaciers never followed, we may be sure that these at least have been dropped from floating ice; it is singular, however, that this can very seldom be shown to be the case, for nearly all the erratics have been carried outwards from the mountains and distributed along the paths of the ancient glaciers. For example, the boulders of grey granite which occur so abundantly in the deep valleys of the Galloway mountains and are scattered over the hills up to a height of 1,700 ft., are also widely sprinkled over the low-grounds at the foot of the mountains. They appear as far north as the valleys of the Girvan and the Doon, and, along with other erratics derived from the same knot of mountains, appear to be distributed over all the low-grounds bordering on the Solway Frith. Blocks of the granite of Criffel are also widely scattered along the maritime districts of the north-west of England. But, as we have seen, it was precisely in these directions that the gigantic glaciers moved. If all, or even if any proportion, of the erratics derived from the Southern Uplands had been transported by bergs and rafts during the period of submergence, surely we might have expected to meet with them in the Lowlands to the north. Why, for example, should they not occur in the north of Ayrshire, in Lanarkshire, and even in the Lothians? The fact is, explain it as we may, that the erratics from the Southern Uplands never appear beyond the districts which are known to have been occupied by the great glaciers of the south.

The same fact holds true in regard to the boulders which have travelled from the Highlands. These are scattered over all the regions which are proved to have been covered with ice that flowed from the Highlands, but they never appear within the districts which, during the periods that preceded submergence, were continuously occupied by the great glaciers of the Southern Uplands. Yet if rafts and bergs did carry them south, why is it that we do not meet with erratics from the Highlands in the extreme south of Ayrshire, and indeed along the whole front of the Southern Uplands. We get them on the Pentland

Hills, and we find them on the Lammermuirs, but both rock-striations and till assure us that the highland ice rubbed the northern slopes of these hills as it swept east and south-east into the bed of the German Ocean. But not a single trace of any highland erratics occurs within those districts of the south-west and south which are plentifully sprinkled with the grey granite boulders from the Galloway mountains.

An impartial consideration of these facts, and of the phenomena connected with the retreat of the gigantic glaciers as described above, has led me strongly to suspect that we have hitherto greatly exaggerated the carrying powers of floating ice during the period of submergence. At most the only erratics which we can be at all sure were carried in this way, are those that sprinkle the tops and slopes of the kames, and this occurs only at comparatively low levels—at 900 ft. or less.*

* See remarks on the erratics of the "Northern Drift," in chapter XXVI. Before leaving the subject of the Scottish erratics, I may refer to some observations by Principal Forbes "On the First Appearance of Stones (erratics) on the Surface of Glaciers" (*Occasional Papers*, &c., p. 202), as perhaps throwing some light on the occurrence of boulders at considerable heights above the rocks from which they have been derived. He accounts for the appearance of certain erratics at the surface on the terminal slope of the Glacier du Nant Blanc and the Rhone glacier by inferring that they have actually been "introduced into the ice by friction at the bottom of the glacier, and forced upwards by the action of the *frontal resistance*," &c.

CHAPTER XVIII.

UPPER DRIFT DEPOSITS OF SCOTLAND—*Continued.*

Shelly clays of maritime districts.—Position of these deposits with respect to older drift accumulations.—General character of brick-clay sections.—Organic remains.—Ice-floated stones and boulders.—Crumpled and contorted beds.—General inferences.

THE deposits which we are now about to consider are memorable in the annals of geological discovery. Mr. Smith, of Jordanhill, was the first to introduce them to notice, and the phenomena as described by him at once convinced the most sceptical that an arctic climate had really at one time characterized our country. The deposits referred to occur more or less abundantly at many points along the seaboard, especially where the shore shelves sufficiently to give rise to a flat beach. Thus the low flats that fringe the margins of the Forth and the Clyde consist in large measure of these deposits, with a more or less thick covering of re-arranged or recent accumulations lying upon them. The low ground upon which Glasgow is built, and which, as we trace it westward, widens out on either side of the river Clyde, especially south, by Paisley, Johnston, and Houston, so as to form a broad expanse many square miles in extent, is composed, for the most part, of fine sand, silt, and brick clay, the lower portions of which deposits all belong to the glacial series. Of like nature are the under portions of those wide terraces of sand, silt, and clay, through which the river Forth flows for several miles before joining its estuary. Along the borders of that estuary similar deposits continue, and are occasionally exposed when the upper or more recent accumulations are thin or wanting, as at Kirkcaldy, Elie, and Portobello. The same appearance recurs upon the coasts of the Frith of Clyde, and many of the sea-lochs in that region. Brick-clays occupying a like position are found in several localities south from the Clyde, as at Stevenston, Monkton, Girvan, Ballantrae, and Stranraer. North of the Clyde they have been detected here and there in some of the fiords, and they occur also in the Outer Hebrides. On the east and north-east coast they have not been so frequently observed, but deposits of this age appear at Montrose, and like accumulations are well developed in the low-grounds of Aberdeenshire that border on the sea. It is highly probable, indeed, that if the recent shingle, sand, and silt were removed from the flat beaches that skirt a large part of the coast-line, the deposits which we are now about to consider would be found more or less continuous along the shore. There can be little doubt at all events that in some places they cover the bed of the sea; for the fossil shells which they contain, and to which I shall presently refer, have not infrequently been brought up in dredges.

None of these deposits has ever been detected at high levels or in the interior of the country. It is true that beds of brick-clay, loam, and silt are of common occurrence there; but these beds, however much they may sometimes resemble those I now refer to, yet cannot be confounded with them. The superficial brick-clays in the interior of the country are, for the most part, unfossiliferous; but when they do contain fossils, these invariably prove to be the remains of terrestrial and fluviatile organisms. Now the brick-clays of maritime and low-lying districts more frequently contain fossils, and these are all without exception marine. The greatest elevation to which fossiliferous deposits, apparently belonging to this age, have been traced is in Aberdeenshire, where Mr. Jamieson met with them at a height of 300 to 360 ft. above the sea.

In the Clyde district, where these deposits were first studied, they have been identified by means of their fossils at a height of 125 ft. above the sea.* The flat ground that stretches up the valley of the Clyde from Glasgow to near Hamilton appears to be composed for the most part of these accumulations, covered up, however, over wide areas with considerable depths of river-sand and gravel. At Uddingston they yielded sea-shells. But fossils are certainly by no means of common occurrence in the brick-clays at this level. It is not until we descend to the level of 30 ft. above the sea that organic remains begin to be plenti-

* I may remind the reader that the shells got at Airdrie, at a height of 512 ft. above the sea, rest upon and are overlaid by till, and therefore do not belong to the period of submergence during which the later glacial-marine clays, and gravel and sand of the Clyde, the Forth, and Aberdeenshire were deposited.

fully present. It is from the brick-clays occupying this level that Scottish geologists have made their largest and most varied collections, and it was from an examination of these very clay-beds that Mr. Smith of Jordanhill was enabled, as I have said, to demonstrate that an arctic climate had at a comparatively recent date characterized the country. But before we proceed to inquire into the nature of the evidence he adduced, it is essential that we first ascertain what relation the clay-beds bear to the other glacial deposits which have already engaged our attention.

When the bottom of the clay-beds is reached, they are found resting sometimes upon solid rock, sometimes upon an irregular and hummocky surface of till. But in many places we do not see the basement beds of the shelly deposits. We know only that in deep borings and other mining operations the clay-beds frequently alternate with silt, loam, sand, and sometimes gravel; but whether the beds of gravel and sand, which are often passed through between the clay-beds and the till, belong to the kame series, or are part and parcel of the shelly series, we cannot tell. Yet there can be no doubt that the shelly clays are of more recent date than the kames and the re-arranged river-gravels described in the last chapter. We find, especially in the basin of the Forth, that while the kames come down in many places to the margin of the great plains or terraces of shelly clays and associated deposits, they never in any case overlie these; on the contrary, they appear to extend below, and are overlapped by the latter. A careful examination of the physical features of the Forth and Clyde basins will convince one that this is the true succession of the superficial deposits. Proceeding inland from the margin of the Frith of Forth, we first pace over extensive flats or gently undulating ground which numerous sections have shown to be composed chiefly of deposits belonging to the shelly-clay group. After leaving these plains, we find that the ground where it begins to ascend more rapidly, has for a subsoil either till or gravel and sand—the latter in many places assuming the form of cones and irregular tortuous banks. The same is the case with the Clyde. At the lower levels we have the usual wide stretches of fine sand, silt, and brick-clays, which as we trace them inland are seen to abut upon the till. No true kames, however, come down in the Clyde district to the level of the brick-clays. At Montrose and in some part of Aberdeenshire the brick-clay beds are overlaid with heavy gravels; but these gravels, as we shall afterwards see, belong to a later date than the sand and gravel of the kames series. The annexed diagram will serve to illustrate the succession as seen in the basin of the Forth.

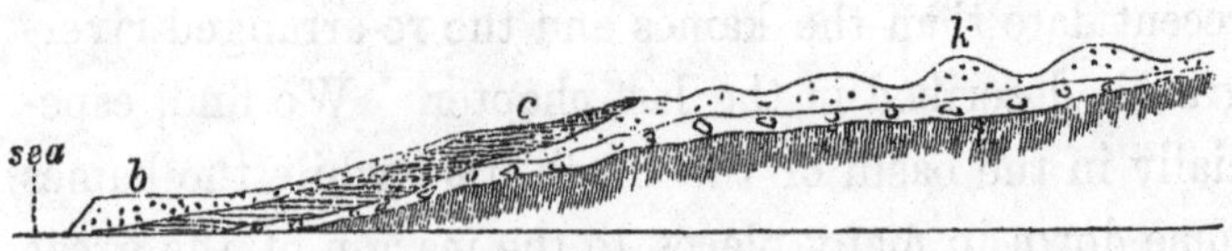

Fig. 40.—Diagrammatic view of drift deposits of the basin of the Forth. *b*, Recent beach deposits; *c*, brick-clay series; *k*, kames series; *t*, till and boulder-clay.

At the top of a good section of the brick-clay series we commonly get sand and fine gravel—sometimes loam and silt, below which come beds of clay of variable thickness. These clays are usually exceedingly fine grained, and are often arranged in thin leaves or laminæ. Fine exposures of such deposits were seen at the old College of Glasgow during the recent railway operations. Below the clays, and sometimes intermingled with them, occur occasional beds of mud, sand, silt, and, but only rarely, gravel. Irregularly scattered throughout all these deposits, a few angular, subangular, and smoothed stones and boulders are not infrequently met with—here and there crowding thickly together; and not a few of the stones and boulders referred to exhibit glacial scratches. But these glacial markings are seldom so well-marked as in the case of the stones and boulders of the till; many of the fragments indeed bear no trace whatever of glaciation.

The following section,* taken at the Kilchattan brick-works in Bute, shows the general character of the beds at and below the level of 30 ft. above the sea:—

1. Vegetable soil.
2. Sand and gravel, well stratified, false-bedded, passing down into a sandy clay with gravel, 10 or 12 ft.
3. Red clay without stones or shells, becoming dull olive-green in lower part, 1 to 2 ft.
4. Bed of fine dark clay, full of *Tellina proxima*, &c., many of the shells retaining both valves, 2 ft.
5. Finely laminated brown and reddish brick-clay without stones or shells, 15 to 18 ft.
6. Hard tough red boulder-clay with striated stones; its upper surface hummocky and irregular.

---

* See *The Glacial Drift of Scotland*, by A. Geikie, where this section and others are given; also papers by Messrs. Crosskey and Robertson (in *Trans. of Glasg. Geol. Soc.*), who are the best authorities on the subject of the Clyde brick-clay deposits.

The brick-clays of the Clyde and Forth basins at higher levels than 30 ft. are comparatively destitute of fossils. These, however, have occasionally been got at heights of 100 ft. and 125 ft. above the sea.

A large percentage of the fossils derived from the brick-clay beds are northern and arctic forms.* And since these clearly occupy their natural position —having lived and died and become entombed just where we now find them—there can be no doubt whatever that the sea in which the brick-clays and associated deposits were accumulated was considerably colder than the water that now laves the shores. In short, it seems clear that the climate during the life-time of these shells must have approximated in severity to that of Greenland. This is the certain conclusion to which we are brought by a study of the molluscs and other organisms yielded by the clay-beds; but it is a conclusion which, even in the absence of fossils, we could not fail to have arrived at from an examination of purely physical evidence alone. I have mentioned the fact that occasional stones and boulders are here and there sparsely scattered through the fine clay-beds. A little consideration will suffice to assure us that the gentle currents which disseminated the impalpable mud and silt over the bed of the sea could not possibly have carried along at the same time stones and boulders. These have clearly been dropped into their present position. We can often satisfy ourselves that this is so by closely examining the fine laminæ upon which the stones and boulders lie. Fig. 41

* See List of Fossils from Scottish Glacial Deposits in Appendix.

represents a boulder of sandstone about 2 ft. in diameter, which occurs all alone in a fine deposit of laminated clay. It will be observed that the laminæ below the stone are bent down as if by pressure from

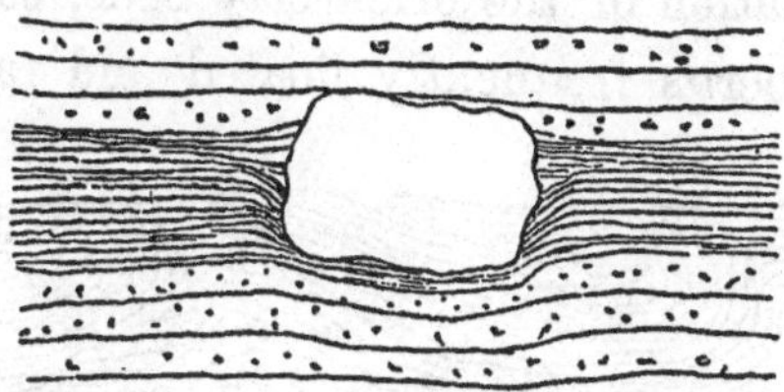

Fig. 41.—Boulder in stratified deposits, near Uddingston, Lanarkshire.

above—showing that the stone fell with some force upon what was then the bottom of the sea or estuary. The upper part of the boulder which projected above the level of the bottom was then gradually buried by the increasing sediment, as one may see by the mode in which the laminæ curve up, and at last sweep over the wanderer. We can hardly doubt that the carrying agent in this case was floating ice.

Some of these boulders have frequently been floated

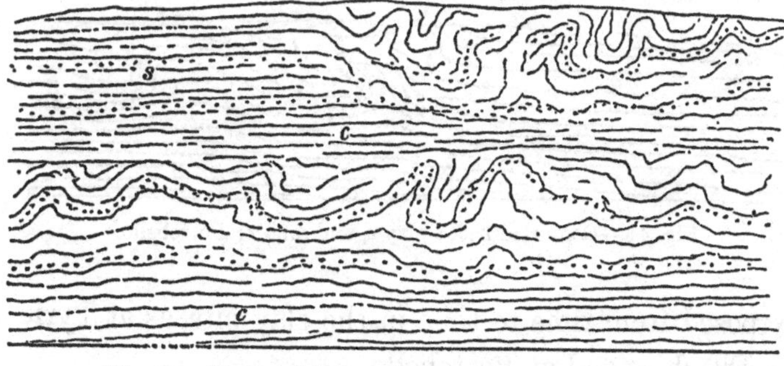

Fig. 42.—Contorted beds: clay *c*, and sand *s*; Portobello.

for considerable distances. Thus in the Portobello

brick-clays Mr. J. Bennie has obtained fragments and boulders of chalk which may have crossed the German Ocean from Denmark.

Again, we have strong evidence to show that during the accumulation of the brick-clay beds, coast-ice and perhaps icebergs frequently floated and ran aground

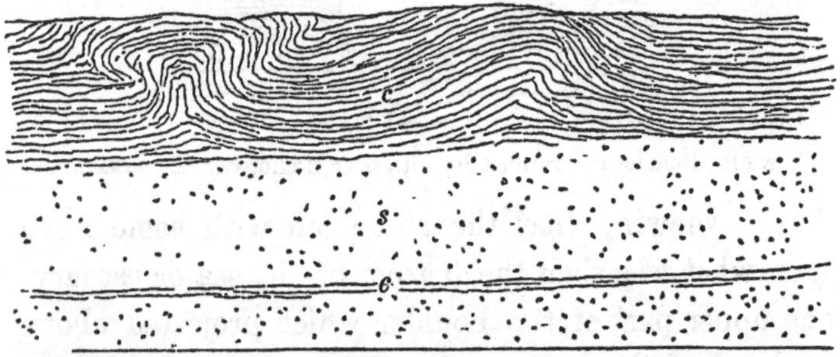

Fig. 43.—Contorted beds: clay *c*, and sand *s*; Leith. [Depth of cutting, 6 ft., J. Croll.]

in the seas. In many places the beds are confusedly twisted, crumpled, and contorted—the laminæ being bent violently over, now in one direction, now in another. Excellent examples of these appearances are

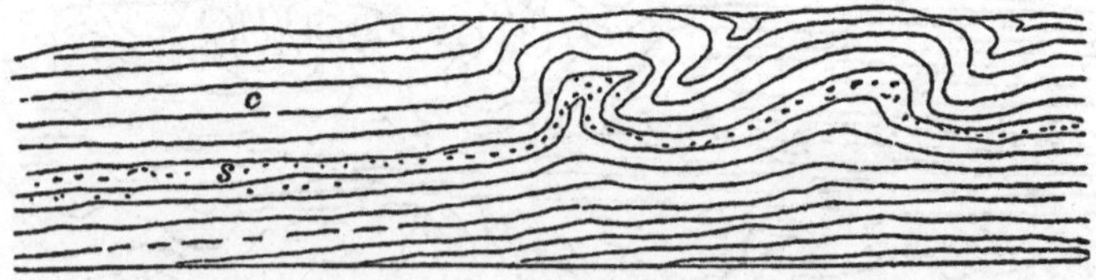

Fig. 44.—Contorted beds: clay *c*, and sand *s*; Portobello.

exposed from time to time in the clay-pits, as at Tyrie in Fifeshire and at Portobello.

I have in a previous chapter described the crumpling and contortion of the beds in the till. It will be remembered that the character of these contortions and crumplings plainly pointed to the exertion of force in one determinate direction. The beds, in short, were shown to be curved over in the direction followed by the till and the rock-striations, indicating the violent pressure of glacier-ice. But the case is widely different with the contortions visible in the maritime brick-clays. These are exceedingly irregular, and are just of such a character as we should expect would

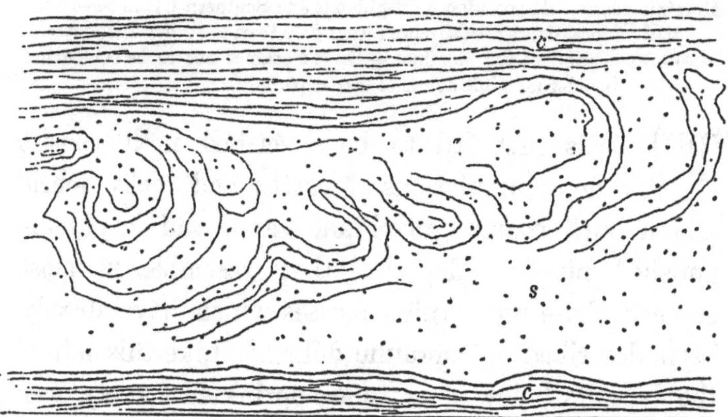

Fig. 45.—Contorted beds: clay *c*, and sand *s*; Portobello. [Depth of cutting, 6 ft., J. Croll.]

result from the grounding of ice-rafts and bergs. The rough sketches give a general idea of the appearances presented. One can hardly doubt that the submarine banks of sand and mud off the American coast must present very similar appearances after they have been bumped and crushed, and pushed forward by the bergs which are every now and then running aground, and stirring up the sediment in their frantic gyrations.

CHAPTER XIX.

UPPER DRIFT DEPOSITS OF SCOTLAND—*Continued.*

Morainic débris.—Its position in Highlands and Southern Uplands.—Essentially a local deposit.—Of more recent date than the other drift deposits.—Denudation of the moraines.—These moraines not to be confounded with the older morainic débris of earlier periods.

THE beds that fall to be described next consist of rude accumulations of earthy and rocky débris mixed with coarse gravel and sand, and large unpolished blocks — deposits that resemble in most respects those unstratified masses which have already been described as morainic débris. Like this latter, the deposits I now refer to are sprinkled loosely over the mountain-slopes, but in many places they assume a more or less distinct shape, so as often to form rather striking objects in a landscape, rising as they sometimes do in the throats of rugged mountain glens into abrupt concentric ridges and mounds—the convex faces of which invariably look down the valleys. These mounds are exclusively a mountain formation, never by any chance occurring in the Lowlands, but being strictly confined to deep highland glens and the high valleys of the Southern Uplands.

The origin of the débris is sufficiently obvious. The stones and blocks of which it chiefly consists have all without exception been derived from the hill-slopes overlooking the valleys. Each particular heap is but the accumulated waste of the glen in which it occurs. In this and other details the coarse débris is strictly analogous to the moraine matter of the Swiss valleys. The mounds and concentric ridges of the Highlands and Southern Uplands can only be terminal moraines, and point to a time when snow-fields covered the higher districts of the country and sent down streams of ice into the mountain glens.

That these glaciers really belong to the closing period of the Great Ice Age, is proved by the fact that in the Southern Uplands they have sometimes scooped out the moraine matter of earlier times from the bottoms of the valleys, but have left it untouched at heights on the hill-slopes which the later glaciers were unable to reach. Yet not a few of these latest glaciers were of considerable importance, as one may judge from the size and position of the moraines. Even the most extensive, however, were but pigmies when compared to those of the earlier cold periods. In the wild district of Galloway, which nourished at this time a number of fine glaciers, one can see that none of these ever got out from the mountain valleys to deploy upon the low-grounds. In the Peeblesshire uplands again, the only well-marked moraines are those of Loch Skene, and a few other valleys in its neighbourhood. All the evidence, indeed, goes to show that the last local glaciers in the south-east of Scotland were of little importance.

In the West Highlands where the sea washes the very base of the mountains, and where long fiords penetrate often for many miles into the heart of the hills, the moraines of these latest local glaciers come down, as we might naturally have expected, to very low levels, in some cases to only a few feet above the more recent raised beaches to be described in the sequel. In the central Highlands, however, they are confined to the heads of the great glens and their tributary valleys. Whenever the valleys begin to open out the moraines disappear. From which we may conclude that during the last local glacier period only a comparatively small portion of the country was covered with perennial snow and ice.

The moraines, as already mentioned, often form well-marked ridges. In many cases, however, these ridges are imperfect, broken down, and denuded. Frequently indeed the morainic heaps are well-nigh obliterated and their materials scattered about over the bottoms of the valleys. We find also that this débris has often been rearranged and partially stratified, and when followed down the valleys it gradually becomes indistinguishable from coarse river detritus. We have, in short, a counterpart of the appearances presented by the superficial deposits of alpine valleys. Larger streams than now flow down its glens must have existed in Scotland during this last glacier period. To these no doubt is due in chief measure the obliteration of the moraine mounds, and the accumulation of much of that coarse gravel which overhangs in broad terraces the present streams and rivers.

The reader will now perceive that the conditions which obtained in Scotland at the date referred to were closely analogous to those that characterized the final disappearance of the confluent glaciers in the times anterior to the great submergence; with this difference, however, that the ice-streams of the latest local glacier age were utterly insignificant in comparison with those of the earlier period. Yet the morainic débris of these two widely-separated periods has been frequently confounded, and consequently extravagant and erroneous ideas have been formed of the size of the latest Scottish glaciers. The morainic débris of the older period is scattered far and wide over the lower reaches of the mountain glens and upland valleys; it even spreads over considerable hills, as is well seen in the districts of Loch Doon and the Merrick Mountains; its perched blocks recline upon the slopes of isolated hill-ranges in the Lowlands; while the gravel and sand then carried down by the rivers, covers extensive tracts at the base of the high-grounds, sweeping out from the confined mountain-valleys, and accumulating in vast heaps over the low-lying tracts beyond. The moraines of the latest glaciers on the other hand are confined to the deeper glens and higher valleys, while the contemporaneous gravels, although frequently extensive, are yet comparatively insignificant.

They sweep out, however, from the mountain-valleys in the north and north-east so as to overlie the glacial-marine beds described in the last chapter, as for example in Aberdeenshire, and in Forfarshire at Montrose.*

* See papers by Mr. Jamieson in *Quart. Jour. of the Geol. Soc.*, 1865, p. 181; and by Dr. Howden in *Trans. of Edin. Geol. Soc.*, vol. i. p. 138.

CHAPTER XX.

UPPER DRIFT DEPOSITS OF SCOTLAND—*Continued.*

Parallel Roads of Glen Roy, old lake-terraces and not raised sea-beaches. —Valley dammed up by local glacier.—General summary of conclusions regarding succession and origin of the upper drift deposits.

IN a previous chapter I have pointed out that during the retreat of the great glaciers from the Lowlands, it frequently happened that glacial lakes formed in lateral valleys in the mountainous districts, and even at comparatively low levels. At that period the main valleys continued to be occupied with massive ice-streams, while the glaciers had either disappeared from or become greatly reduced in the smaller lateral and tributary valleys. In these last, therefore, the streams were dammed back by the ice of the main glaciers and thus gave rise to glacial lakes. Now it would appear that during the latest local glacier period precisely the same phenomena appeared. In the district of Lochaber it is well known that the slopes of certain valleys are fringed with successive horizontal shelves of coarse angular shingle and débris which are believed to indicate the margins of ancient glacial lakes. These are the famous "Parallel Roads."

In Glen Roy there are three distinct shelves, 856 ft., 1,065 ft., and 1,149 ft. respectively above the level of the sea. At one time these shelves were thought to be old sea-beaches, and this continued to be the general belief even after Agassiz had suggested their lacustrine origin. The later observations of Mr. Jamieson, however, would seem to have convinced most geologists at last that the glacial lake theory is the true explanation of the phenomena. A massive glacier descending from Glen Treig filled up Glen Spean, and thus formed a barrier to the escape of water from Glen Roy. Along the margins of the lake thus formed angular shingle and débris collected—derived in great measure, no doubt, from the degradation of the rocks under the influence of frost. As the icy barrier decreased, either by gradual melting or by sudden rupture, the lake was lowered, and thus another terrace of débris gathered along the slope of the valley at a lower level than the former. The farther shrinking or bursting of the ice in like manner again lowered the lake, and so gave rise to the third and lowest shelf.

It is not necessary that I should enter here into all the arguments which can be brought against the marine theory. Those who desire to inquire more fully into the subject cannot do better than read the very excellent accounts of the parallel roads given by Chambers in his "Ancient Sea Margins," and by Mr. Jamieson in the "Geological Society's Journal." The bibliography of these remarkable terraces is a somewhat copious one, as will be seen from the note below, which is by no means exhaustive.* One very good proof of their non-marine origin is the fact that the highest terraces, although they go up to the summit-levels between two glens, yet do not reach a greater height nor cross these *cols* into the open valley beyond. This is just as we should expect, on the supposition that the terraces mark the margin of an old lake; for when the water reached the summit-level between two glens (such, for example, as that between Upper Glenroy and the valley of the Spey), it would immediately overflow, on the same principle as a cistern with its escape-pipe. But if the terraces were marine we should be at a loss to account for this connection between the highest terraces and the summit-levels. It would be very extraordinary, to say the least, if the sea had reached just to the cols between the glens and never any higher. Again, one may well ask why Glenroy and the little glens in its neighbourhood should be favoured above all other highland glens with sea-beaches. There is no reason in the world why such shelves (if they really be ancient sea-beaches) should not occur abundantly in the valleys, not only of the Highlands but of the Southern

* "On the Parallel Roads of Glenroy" (Macculloch), *Trans. Geol. Soc.*, First Series, vol. iv. p. 314; "On the Parallel Roads of Lochaber" (Dick Lauder), *Trans. Roy. Soc. Edin.*, vol. ix.; *Edin. Phil. Jour.*, vol. iv. p. 417; "Observations on the Parallel Roads," &c. (Darwin), *Phil. Trans.* 1839, p. 39; "Account of the Parallel Roads of Glenroy in Invernesshire," *Edin. New Phil. Jour.*, vol. xxvii. p. 315; "On the Parallel Roads of Lochaber," &c. (Milne-Home), *Trans. Roy. Soc. Edin.*, vol. xvi. p. 395; *Proc. Roy. Soc. Edin.*, vol. ii. pp. 124, 132; *Edin. New Phil. Jour.*, vol. xliii. p. 339; "On the Parallel Roads of Lochaber" (J. Thomson, M.A.), *Edin. New Phil. Jour.*, vol. xlv. pp. 49, 404.

Uplands also. A visit paid to Lochaber some years ago left me no alternative but to conclude with Agassiz, Jamieson, and others that a glacier did at one time fill Glen Spean, and that during the existence of this local glacier Glen Roy, owing to its limited drainage-area, could not support a glacier big enough to fill its own valley. Under such conditions this valley must in the very nature of things have filled with water, and so shelves and terraces would form, having precisely the same character as those that are found fringing its slopes.*

Having now rapidly reviewed the evidence supplied by our maritime brick-clays with their associated deposits, and by the moraine débris of the higher valleys, and the parallel shelves of the Lochaber district, I shall attempt in a few words to point out the succession of changes of which the phenomena we have been considering are the proofs.

It may be remembered that in Chapter XVII. we stated as the result of foregoing observations, that after the great confluent glaciers had shrunk up into the valleys, and the swollen rivers had thrown down massive deposits of gravel and sand, a period of submergence ensued. That submergence began and was continued until it approached its climax

* I have referred these terraces, according to the generally received opinion, to the period of local glaciers that succeeded the re-elevation of Scotland after the great submergence. We know, however, so little about the degree of depression experienced in the north of Scotland, that I shall not be surprised if more detailed examination should eventually compel us to refer the Glen Roy terraces to the period of great local glaciers that preceded submergence.

during a mild condition of climate. By-and-by, however, intense cold again prevailed, and ice-rafts and bergs sailed over the submerged parts of the country. The process of submergence was next reversed and converted into a movement of upheaval. It was during this depression of the land that the ancient river gravels were much denuded and rearranged, that kames were formed, and occasional erratic blocks were transported from distant localities.

The shelly brick-clays and associated marine deposits carry on the record. At the time these were being deposited the land still stood several hundred feet below its present level. The sea was cold, and tenanted by northern and arctic species of molluscs and other organisms. A long arm of water extended up the valley of the Clyde, as far as and even beyond Hamilton; all the low-grounds between the Paisley and Kilpatrick Hills were submerged, as also were extensive tracts in the maritime districts of Ayrshire, and the western counties. An arm of the sea connected the Clyde basin with the Forth on the one hand, and a second stretched up the valley of the Black Cart, and communicated with the open sea in that direction. At the same time wide areas in the Friths of Forth and Tay were drowned, and the ocean overflowed all the low-grounds along the east coast, and the maritime provinces of the Moray Frith and the north were likewise under water. Glaciers nestled in many of the higher mountain glens and valleys, and at the heads of not a few of the deep fiords of the west, they actually entered the sea and shed their ice-bergs. The rivers, flowing in greater volume than now, carried down immense quantities of fine glacial silt and mud, which gradually settled upon the quieter reaches of the sea-bottom. Coast-ice clogged the shores, and ever and anon breaking away took with it stones and blocks, which it sprinkled irregularly over the floor of the sea. In the mountain valleys the local glaciers renewed the polishing and striation of the rocks which had perhaps been altogether interrupted during the preceding mild period, while at levels which these local glaciers could not reach, the frosts were busy obliterating the traces of the massive ice-streams of earlier times. The wreck of the mountains overlooking the glaciers was slowly borne down the valleys and shot over the terminal front of the ice—here forming subaerial ridges and mounds, there accumulating as submarine banks. At the same time in some favourable localities glacial lakes were produced by the damming-up of lateral valleys.

Of the plants and animals that may have clothed and peopled the less desolate portions of the land at this time we know very little. Scotland was probably not a more cheerful residence then than North Greenland and Spitzbergen are now. The plants and animal remains which have been disinterred from some of the brick-clay pits, belong for the most part to the close of the cold period, when the climate was evidently becoming ameliorated.

Yet it is likely enough that Scotland may have been the favourite haunt in summer-time of sea-birds, just as arctic coasts are now. And to this date, therefore, may or may not belong the bird bones which have at rare intervals turned up in the brick-clays. It seems extremely improbable, however, that a country in our latitude, under such physical conditions as we have above attempted to describe, could present a less forbidding aspect than the dreary island of Georgia, in the Antarctic Ocean. And, perhaps, the extreme paucity in the arctic clay beds of land waifs, either plant or animal, may really point to the barren and sterile condition of the country at this time.

The process of re-elevation continued during the deposition of the shelly clays. Thus ere long the sand and mud beds which had gathered at the heads of sheltered bays and fiords gradually rose above the surface of the water. Hence it is that the clay deposits at the lower levels belong upon the whole to a later date than the others. For some unknown reason marine life would not appear to have been so plentiful during the deposition of the high-level stratified clays and gravels as it subsequently became. It may have been that the rivers at that time poured into the sea greater bodies of fresh water, and so rendered the sea less congenial to the taste of marine organisms. When the glaciers shrunk back and the rivers decreased in size, then the salt water in the sea-lochs would be less diluted, and marine life would begin to abound. But whether that be a reasonable explanation or not, the fact still remains, that the clay beds at and below a height of 30 ft. over the present sea-level, are much more fully charged with organic remains than the stratified beds above that level.

When the beds occupying the 30 ft. level began to be laid down the climate still continued frigid, but the area of land had considerably increased, and the sea in the Clyde basin no longer communicated with the estuary of the Forth. It is remarkable that the shell bed of Elie gives evidence of somewhat colder conditions having prevailed at this time than the equivalent fossiliferous deposits of the Clyde would seem to imply. From this it has been argued that the Elie shells belong to an older date than the Clyde beds—to a period when the cold was more intense. But for this conclusion there does not seem to be any need. The North Sea, filled with bergs and floating ice, derived from Scandinavia, and even from Scotland itself, would in all probability have a lower temperature than the freer ocean that washed the western sea-board; and, therefore, it is not surprising that this difference of temperature should have taken effect and influenced the distribution of the fauna.

We have now completed our rapid sketch of the Scottish glacial deposits; the accumulations which fall next to be described contain the record of the geographical and climatal changes that followed upon the close of the glacial epoch. But before we enter upon this latest chapter in the history of Scottish geology, we must for a little retrace our steps for the purpose of considering the origin of certain remarkable features in the contour of the land, of which as yet no special mention has been made.

CHAPTER XXI.

LAKES AND SEA-LOCHS OF SCOTLAND.

Different kinds of lakes.—Lakes occupying depressions in drift.—Lakes dammed by moraines and older drift deposits.—Lakes lying in rock-bound basins.—Origin of rock-basins discussed.—Ramsay's theory.—Sea-lochs.—Submarine rock-basins.—Their glacial origin.—Silted-up rock-basins.

WHEN we glance at a good map of Scotland,* one of the first appearances to catch the eye is the wonderful profusion of lakes. Moreover, it will not fail to strike us that these lakes are confined, for the most part, to the deep valleys of the Highlands. From the lowland tracts they are singularly absent. South of the Friths of Tay and Clyde comparatively few are seen, and these are nearly all restricted to the high-

* A good handy general map of Scotland has yet to be made. There is no scarcity of maps which can be relied upon for showing the proper direction of roads and county-boundaries, position of towns, &c., &c.; but one which shall by shading give an approximately correct impression of the outline of the ground does not exist. When will map-constructors cease to cover the area of the Highlands and Southern Uplands with a series of black caterpillars, which only serve to confuse and mislead? A plain unshaded sheet with a few heights indicated would be infinitely more instructive than a map elaborately covered with all those incomprehensible crawling masses of printer's ink, which do duty for mountains. This conventional manner of mountain-shading, which has been so long in vogue, is, I firmly believe, in no small measure answerable for the persistency with which crude notions as to the origin of hills and valleys are maintained. Had hill-shading really been made to represent truthfully the form of the ground, we should have heard a great deal less of mountains being proximately due to upheaval, and valleys owing their origin to gaping fractures.

grounds of Carrick, Galloway, and Peeblesshire. There would thus appear to be some connection between lakes and mountain valleys. We cannot but see this when it is pointed out; yet it is not more than some thirteen years ago that the attention of geologists was called to the fact. It was reserved for Professor Ramsay not only to indicate the very remarkable manner in which lakes are distributed in alpine and northern countries, but also to bring forward an explanation of the phenomena which has already commended itself to many physicists and geologists, and bids fair to become ere long one of the most generally accepted theories in geology.

The Scottish lakes may be grouped under three heads, viz.:—

1. Lakes occupying hollows in the till or other superficial deposits.
2. Morainic and drift-dammed lakes.
3. Lakes resting in basins of solid rock.

The lakes of the first group have no importance, and indeed are little more than shallow pools. They are developed chiefly in the Lowlands, and have at one time been much more numerous than they are now. Many have been silted up with alluvial matter, others have been converted into peat-bogs, and not a few have been drained. They rest sometimes in the hollows between banks of till, and not infrequently in cup-shaped depressions of sand and gravel. The most considerable assemblage of these lakes of which I know is in the island of Lewis—the low-lying tracts of which are literally peppered with lakelets. Not a few of these, however, belong to the drift-dammed series. But hundreds of them appear to rest in hollows of the till—their longer axis pointing north-west and south-east. They are, for the most part, very shallow, and have been much encroached upon by peat.

The lakes of the second group are also somewhat insignificant. Those which are dammed back by moraines are confined, as one would naturally expect, to mountain valleys. An excellent example of the kind is that of Loch Skene, in the south of Scotland.

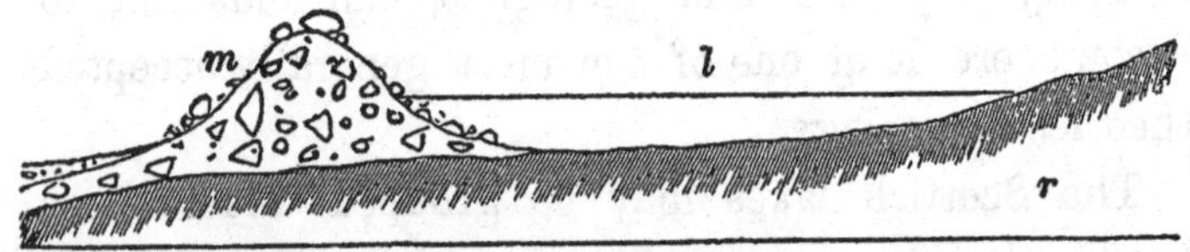

Fig. 46.—Morainic Lake.

In this instance a series of moraine mounds, left behind by one of the last local glaciers, extends from one side of a mountain-valley to the other, and has thus formed a barrier to the escape of the water. In not a few mountain valleys where such lakes have at one time existed, we may see how the waters have been drained off by the gradual cutting down of the moraines.

Besides the lakes which are confined behind barriers of morainic débris, we occasionally meet with small sheets of water which appear to rest partly in rock and partly in till or other superficial accumulation. Fine examples of this variety of lakelet occur in the low-grounds of Lewis. Unlike those lakelets of the first group, which occur in the same region, these drift-dammed lakes do not range from south-east to north-west, but exactly at right angles to these—namely, from south-west to north-east. They lie between parallel ridges of rock, and are dammed up at either end by accumulations of till. In Lewis * it is common to find both kinds of lake represented in one and the same sheet of water—one elongated portion of which will trend south-east or north-west, and another arm extend itself in a north-east or south-west direction, as the case may be.

In the Lowlands there are a few lakes such as Loch Leven (Kinross-shire), the original shape of whose bed it is difficult to determine. Probably most of them, however, are merely due to the unequal distribution of the till and other superficial deposits over the underlying rocks. Others, again, may really belong to the third group, and occupy rock-basins the nature of which is now concealed by the accumulation of alluvial matter along their margins. I am not sure but that Loch Leven is a case in point.

The third group embraces the largest and most important lakes in Scotland, and to it also belong a vast number of mountain-tarns which are neither large nor important. All these lakes and tarns rest in hollows of solid rock. In very many cases we may trace the rock all round their margins, and even when this is partially obscured by superficial deposits there are yet other circumstances which enable us to show that although these deposits were to be entirely removed

* The features referred to are beautifully delineated upon the Ordnance Survey map of Lewis. See paper by author, *Quart. Jour. Geol. Soc.*, 1873.

there would still be a lake completely surrounded by rock. These lake-hollows are what Professor Ramsay has termed *rock-basins;* and his ingenious theory of their origin I must now attempt to describe.

When we reflect for a moment we shall find that it is a very hard thing indeed to account for a rock-basin. The usual agents of erosion, those which we see at work in our own country, fail to afford any solution of the problem. We may, for example, dismiss the sea as utterly inadequate. The action of the sea upon the land is that of a huge horizontal saw; the cliffs are eaten into, and gradually undermined; masses of rock, loosened by rains and frosts, tumble down, and are pounded up by the breakers into shingle and sand. Thus in process of time a shelf or terrace of erosion is formed, and were the shore to be sufficiently elevated to-morrow, we should find that such a platform would extend all along our rocky coast-line—narrowing where the rocks were hard and durable, broadening out where the cliffs had yielded more easily to the ceaseless gnawing of the waves. But nowhere should we be able to detect anything approaching to the character of a rock basin; for it is self-evident that the sea "cannot make a hollow below its own average level." Its tendency, indeed, is quite in the opposite direction—much of the material derived from the denudation of the land being carried out and deposited in quiet depths.

If the sea does not help us to discover how rock-basins are formed, will rivers do so? Most assuredly they will not. Rivers will flow down any slope, but they cannot run up an inclined plane. What they do is to cut channels which carry them down persistently from higher to lower levels. The only approach to something like a rock-basin which can be excavated by river-action may be observed at the foot of a waterfall, where a more or less deep hollow always appears. This hollow is of course scooped out by the forcible impact of the falling water, aided by the filing action of the stones and pebbles which are kept in constant motion. But it would be idle to suppose that such has been the origin of the rock-basins. Yet this action of the falling water, as I shall try to show presently, may aid us in appreciating more fully Professor Ramsay's theory.

Since rock-basins owe their origin neither to rivers nor to the sea, may not they be simply due to disturbances of the strata? We know that the solid aqueous rocks of which our country chiefly consists have been elevated and depressed times without number since the date of their formation. We know further that they have been dislocated and displaced by movements of the earth's crust, and confused by the intrusion among them of melted volcanic materials; and this has happened over and over again. Strata which we have every reason to believe were laid down in horizontal or approximately horizontal planes have been puckered and thrown into innumerable folds, or here pushed up into ridges, and there carried down into troughs. May not the lakes then occupy such troughs, or rest in cracks and chasms or depressions caused by dislocation and displacement of the rocks?

To those who have made no special study of physical geology this may appear a ready and simple explanation. In sober truth, however, it is no explanation at all; for when we come to examine the rock-basins themselves, we do not find them occupying "hollows formed of strata bent upwards at the edges all round into the form of a great dish, the uppermost bed or beds of which are continuous and unbroken underneath the waters of the lake." No such synclinal troughs occur anywhere in Scotland; indeed, as Professor Ramsay remarks, they are the rarest things in nature. As a general rule, we find that synclinal troughs or geological hollows form hills, while conversely anticlinal ridges or geological hills give rise to valleys. And it not infrequently happens that the

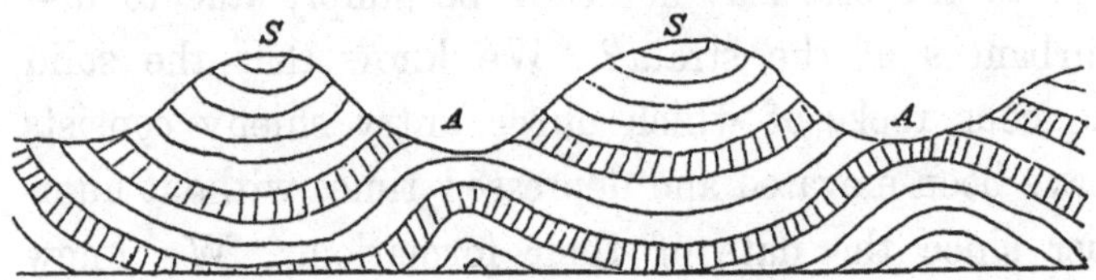

Fig. 47.—Diagrammatic view of synclines and anticlines. *AA*, anticlinal arches; *SS*, synclinal troughs.

hollow in which a lake lies, is geologically speaking, a hill or anticline. But still more frequently rock-basins occur in regions where the strata are "bent and contorted in a hundred curves all along and under the length of the lake," nor does the direction or slope of the basins bear any relation whatever to the prevailing inclination of the strata. We may conclude, then, that the bedding of the underlying rocks affords no clue to the solution of the problem.

Do the lakes lie in gaping fissures, or in chasms produced by dislocations of the solid rocks, or, as they are technically termed, *faults?* As a matter of fact, no single instance has yet been adduced, either at home or abroad, where a *fault* could be said to be the proximate cause of a lake-hollow. My duties in connection with the Geological Survey have afforded me exceptional opportunities for the study of dislocations and displacements. Some time ago I completed the mapping of the major portion of the coal-fields in the west of Scotland—perhaps one of the most abundantly faulted districts in Britain; and I found, as the result of a minute examination of all the carefully-prepared mining-plans and a close investigation of the ground, that none of the faults ever gave rise to any feature at the surface, save when a hard rock like basalt was brought into juxtaposition with a soft rock like sandstone. In such a case the basalt-rock almost invariably gave rise to a prominence. Now many of these faults are 20, 40, 60, and some even 100 fathoms in extent, and they frequently cross and shift each other; yet no yawning crack or irregular depression at the surface gives one any indication of their existence. In walking over a level turnip-field we may traverse a dozen in the space of a few hundred yards. My impression is that none of these dislocations ever showed at the surface. This is hardly the place to describe the observations which have led me to this conclusion; but I may state that, as a general rule, faults increase in extent downwards and diminish upwards, so that the upper seams are not dislocated to the same extent as the lower seams of the same coal-field. Some of the smaller faults, indeed, die out upwards altogether.

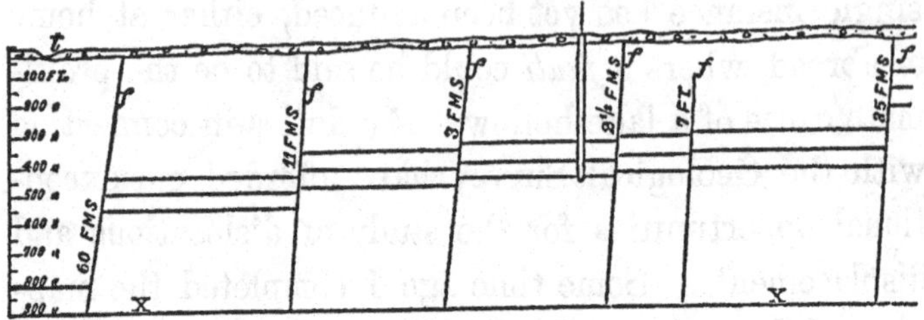

Fig. 48.—Section across coal-measures at Muirhouse Colliery, near Wishaw (horizontal and vertical scale the same). The horizontal lines represent coal-seams; *f*, faults, the amount of displacement shown in fathoms and feet; *X* line, 550 ft. below sea-level; *t*, till.

Now we know that hundreds of fathoms of solid strata have been planed off the central Lowlands of Scotland since the close of the Carboniferous epoch, and not a few of the dislocations that now come to the light of day may have died upwards long before the original surface of the coal-measures was reached. But even if these faults did actually traverse the whole thickness of the strata at the very time the movement of the earth's crust took place, yet they could not have produced any external feature—they could not have been actually visible at the surface. Had it been otherwise—had all the faults which we now see cutting and shifting the strata in every direction shown boldly at the surface—what a very remarkable landscape would have been the result!—straight cliffs intersecting at all angles; here a great parallelogram of strata forming a solid embankment, there a profound and extended chasm fantastically shifted in a hundred places; in one place a long descent, terminating abruptly against the blank wall of a frowning precipice; in another place an inclined plane, rising for a considerable distance, and all of a sudden stopping on the edge of a wall that dropped perpendicularly for 500 or 1,000 ft. Instead of all this wild vagary, we have a softly-outlined country, with its regular valley-system—the direction of the streams never being in any degree influenced by the rock-dislocations, which, as far as the trend of the valleys is concerned, might never have existed.

Judging, therefore, from the contour of the ground in this district, where the direction and amount of the faultings have been so well ascertained, it may be concluded that these dislocations have not been the proximate, nor even the remote cause * of the formation of the valleys. Many of the faults may have died out upwards without reaching the original surface, while others that did reach that surface may have displaced the strata so gently, by such gradual creeping, that atmospheric or other superficial denudation may well have kept pace with the movements, and so removed the inequalities as these arose.

* It may be well to remind the reader that I am speaking of the carboniferous areas of the west of Scotland. I am very far from affirming that faults have never in any case given the initial direction to a line of drainage. I could mention a number of instances where they have certainly done so. A good case in point is that of Glen App, in the south of Ayrshire, which coincides with a large fracture. Again, the great north-east and south-west fault that traverses Scotland from the shores of the Frith of Forth to the Irish Sea, gives rise in many places to a distinct feature, and streams occasionally follow it for some distance. The Great Glen would also appear to lie in a line of dislocation. I have never seen a gaping fault, and would travel a long way to see one.

Since, then, rock-basins do not occur in these highly faulted districts, it seems idle to speak of such hollows being due to dislocations, unless we are prepared to bring forward some well-proved case in point. But while no such case has been adduced, geologists have been referred by Professor Ramsay and others, both at home and abroad, to numerous examples of rock-basins which it can be shown have no connection whatever with gaping fissures and dislocations. Take, for example, that of Loch Doon, or that of Loch Trool in the wilds of Carrick. In neither of these cases do any shiftings and displacements or cracks and chasms occur, but the beds on the one side exactly correspond with those on the other. One may walk round the lower end of Loch Doon and never have his feet off solid rock all the way. The valley in which it lies cuts right across the strike of the strata, which, as the merest tyro may readily ascertain, are quite continuous.

Nay, more than this, rock-basins are of all sizes; many of them are no larger than an ordinary drawing-room; myriads are a great deal smaller—mere pools; and between this and basins that are square miles in extent we have every gradation. One may wade through or swim across the shallower ones, and satisfy himself that the rock is all solid underneath; and no one indeed has ever ventured to suggest that these smaller hollows are due to fractures. Yet it is just as difficult to account for a rock-basin fifty yards in length, as it is to explain the origin of one a hundred times larger. If big ones be caused by fractures, and little ones owe their existence to—something else, where are we to draw the line?—how many square yards are we to be allowed free of fracture?

Some of these objections apply with equal force to another explanation which has been advanced—to this, namely, that the lakes lie in special areas of depression; in other words, that the land has sunk down underneath each lake. It is simply incredible that such could have happened. No one will deny that special areas of depression do exist, but to have produced the innumerable lakes of all sizes that stud the surface of alpine countries and many northern regions, the rocky crust of the earth must needs have been in a condition nearly as unresisting as putty. If movements of depression were allowed to explain the existence of a sheet of water like Lake Superior, they would still leave utterly unaccounted for the vast number of smaller lakes that crowd the surface of the northern part of North America, as well as the innumerable lakes of Scotland, Cumberland, Wales, Ireland, Scandinavia, Finland, and Switzerland.

If, then, all the above "explanations" be rejected, we have only one agent left which can possibly account for the origin of rock-basins, and that is *ice.* Professor Ramsay points out in the first place that these remarkable lakes abound in every region which is known to have been subjected during the glacial epoch to the grinding action of glaciers; while conversely, in tropical areas or such countries as have not supported glaciers, true rock-basins do not occur. Thus, if we exclude the great African lakes, of which we know too little as yet to justify us in coming to any definite conclusion in regard to the mode of their formation, and if we also put aside crater-lakes and lagoons, we shall find that in our hemisphere rock-basins increase in number as we pass from south to north, and are always specially abundant in mountainous districts. And this peculiar distribution Professor Ramsay accounts for by the bold suggestion that the basins have been scooped out by the grinding power of glaciers.

He shows that the erosive action of a glacier must necessarily be less at the lower end, where the ice is comparatively thin, than further up the valley, where the ice attains a much greater thickness. Obviously, then, were a glacier to continue to flow for a sufficient length of time, this unequal pressure upon the underlying rock would produce some effect; there would be a great deal more tear and wear where the ice was thick than where it was thin, and thus a rock-basin would eventually be formed.

Take the case of a glacier creeping down an alpine valley, and spreading itself out upon the low-ground at the foot of the mountains. Let us suppose that in the upper part of its course, that is to say, within its mountain valley, the incline down which it moves is greater than the slope of the low-ground upon which it eventually deploys. When the ice reaches this latter point, it is evident that its flow must be retarded, and there will therefore be a tendency in the ice to accumulate or heap up. Now we know that the pressure of a body in motion upon any given surface varies with the degree at which that surface is inclined; as the inclination decreases the pressure

PLATE VII.

Loch Doon (upper reach): a Rock Basin, showing roches moutonnées. (By B. N. Peach.)

To face p. 295.

increases. It follows from this that when the glacier leaves the steeper part of its course, and begins to creep down the gentler slope beyond, it will press with greater force upon its rocky bed, and this increased pressure will be further intensified by the greater thickness of the accumulated ice. But as our glacier continues to flow on, it gradually loses in bulk, its rate of motion at the same time diminishes,* and thus its erosive power becomes weaker and weaker. The result of all this is the formation of a rock-basin, the deeper portion of which lies towards the upper end, just where the grinding force of the glacier is greatest.

Such is the effect we might naturally expect glacial action to produce. When we turn to the rock-basins in our own country, we find that these occupy precisely the very positions which theoretically might have been expected. And not only so, but they almost invariably reach their greatest depths towards the upper end, shallowing away gradually down the valley. Cases in point are Loch Doon, Loch Trool, and numerous other rock-basins amongst the Carrick and Galloway mountains. One of the best examples of a rock-basin is furnished by Loch Lomond, a map and sections of which, drawn on a true scale, will be found in the Appendix. No fault or dislocation of the solid strata is known to cross the area occupied by the Loch; but my colleague, Mr. R. L. Jack, has detected one that crosses below the foot of the loch, passing through Tullichewan Castle on

* *La Névé de Justedal et ses Glaciers*, by C. de Seue.

the west side and Haldane's Mill on the east; and the downthrow side of this fault is to the south—a clear proof, if such were needed, that faulting and unequal subsidences of the rock have had nothing whatever to do with the formation of the great hollow.

Another striking circumstance in connection with Scottish lakes is this—their dimensions are always proportionate to the extent of the drainage-system in which they occur. If the valley in which any particular rock-bound lake appears should be an important one, draining a wide tract of elevated ground, the lake is sure to be long and deep; if the valley be of inconsiderable extent, the rock-basin it happens to contain is certain to be proportionately unimportant. In other words, where large glaciers are known to have existed, we find large rock-basins, while in valleys which have been occupied by inconsiderable glaciers, the rock-basins are small.

These remarks on Scottish rock-basins would be incomplete if no reference were made to the fiords or sea-lochs so abundantly developed along the west coast. The hollows occupied by these arms of the sea are simply submarine continuations of land valleys, which, as everyone knows, stretch into the country for a less or greater distance from the head of every fiord.

If the reader will glance at the sketch-map of Scotland, he will observe that while fresh-water lakes are plentifully present on the eastern slopes of the great watershed that runs from the head of Loch Shin to the hills above Loch Linnhe, there are com-

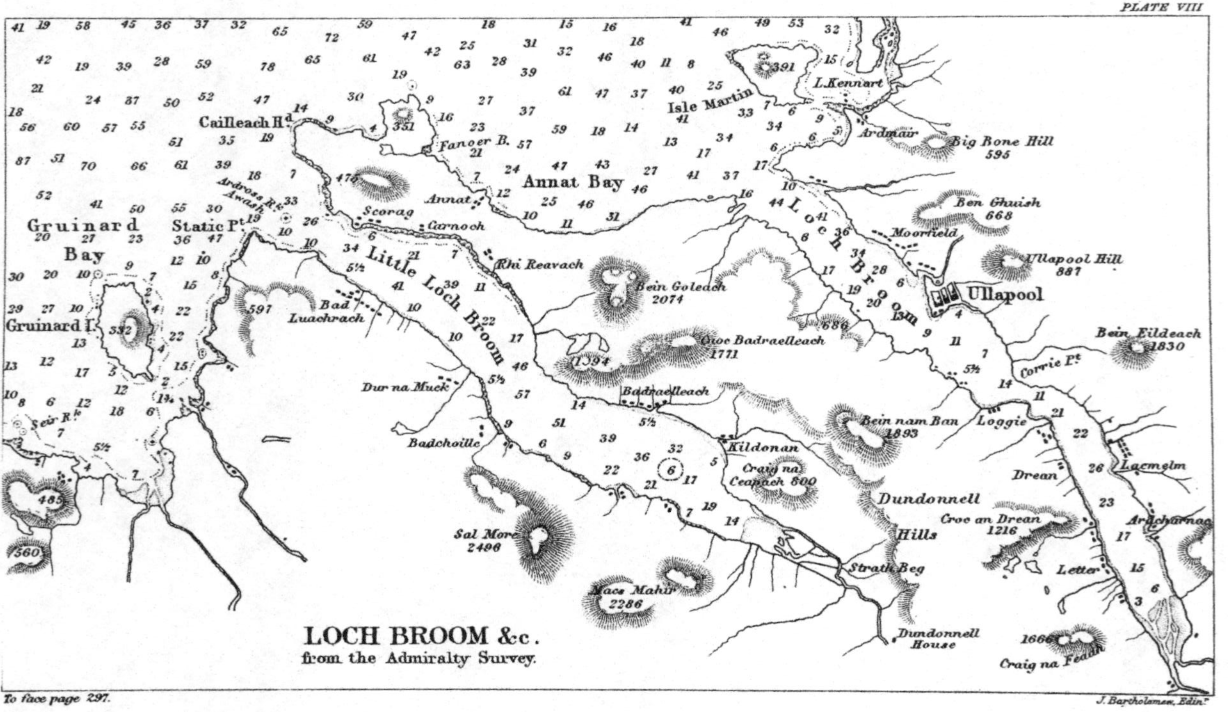

To face page 297.

J. Bartholomew, Edinr.

paratively few on the other side, but we have great sea-lochs instead. Now each of these submerged land-valleys contains at least one rock-basin, so that if the land were only to be elevated sufficiently, we should find in that region an exact counterpart of the appearances that present themselves on the eastern slopes of the watershed, namely, deep mountain glens with rock-bound lakes.

The Admiralty charts are excellent maps of the sea-bottom, and afford clear and definite ideas of its physical features. The tracings given are taken from the reduced Admiralty chart of the west coast of Scotland. One of these shows Loch Broom, and Little Loch Broom, with Gruinard Bay. This part of the coast has been selected for no other reason than simply because it happens to come nicely within the compass of one of these pages. Almost any other sea-loch would have served my purpose equally well—some of them, indeed, would have done better, as for example, Loch Long, Loch Fyne, or Loch Sunart.

It will be observed, upon glancing at the chart, that the upper reach of Loch Broom, between Corrie Point and the head of the loch, rests in a distinct basin, the lower lip of the basin opposite Corrie being reached at a depth of 11 fathoms, and its deepest part 26 fathoms, occurring near Lacmelm. There appears to be a second basin between Corrie Point and the mouth of the loch, but it is not so well marked. Little Loch Broom, however, is an admirable example. At the mouth of this loch the underlying rock comes actually to the surface in Ardross Rock, and between this point and the shores we have the maximum depths of 10 and 26 fathoms respectively. Yet the soundings half way up the loch show a depth of no less than 57 fathoms. Beyond the mouth of the loch the water deepens somewhat gradually until a depth of 119 fathoms is attained, beyond which it shallows again to 34 fathoms. Now it is worthy of remark that no part of the North Minch, into which Gruinard Bay opens, is anywhere deeper than some 50 or 60 fathoms, except immediately opposite the mouths of the great sea-lochs that open out from the mountain valleys of Harris and Lewis, and the mainland. The 100-fathom line is only reached when we get beyond the western Hebrides altogether. Yet it is no uncommon thing to get greater depths than this in many of the sea-lochs and sounds. In the Inner Sound of Raasay, for example, we find depths of 100, 128, and even in one place of 138 fathoms! (See Chart.) Thus, were Scotland to be lifted out of the sea for 100 fathoms, all the islands would be connected with the mainland, and numerous lakes would exist to mark the sites of the sea-lochs, one of which, lying between Raasay and Ross-shire, would reach as much as 528 ft. in depth.*

Here then is the singular fact that the deepest portions of the Scottish seas lie close in shore; nor in the vast majority of cases can there be any doubt whatever that these deep submarine hollows are true rock-basins. In many cases, indeed, the soundings actually prove this. Still some may think that since

* See Appendix: "Map showing Physiography of Western Scotland."

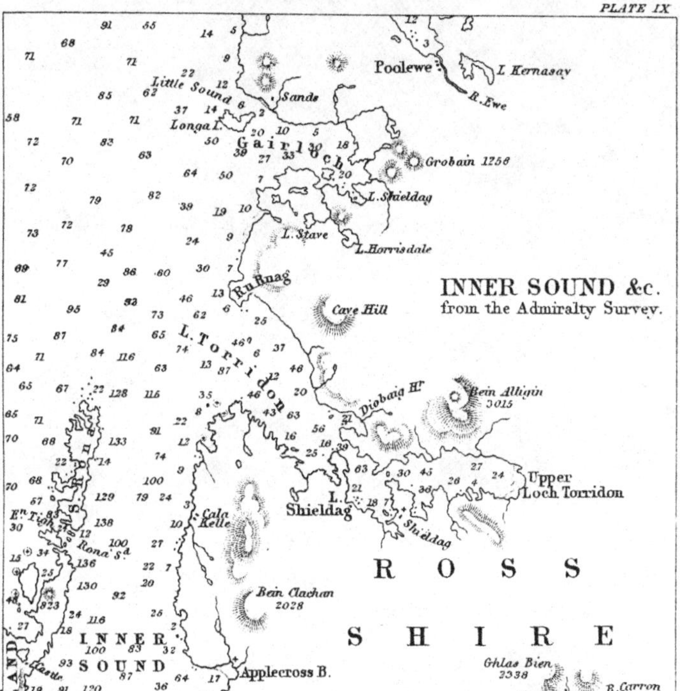

INNER SOUND &c.
from the Admiralty Survey.

Poolewe
L. Kernasay
R. Ewe
Little Sound
Sands
Longa I.
Gairloch
Grobain 1256
L. Shieldag
L. Stave
L. Horrisdale
Ru Rnag
Cave Hill
L. Torridon
Diobaig Hr
Bein Alligin 3015
Upper Loch Torridon
L. Shieldag
Shieldag
Cala Kelle
ROSS SHIRE
Bein Clachan 2028
Rona Sd
INNER SOUND
RAASAY ISLAND
Castle
Applecross B.
Ghlas Bien 2338
Tullich
R. Carron
Achinte
Scorr n caorachan 2397
Jeantown
Attadale
L. Kishorn
Ardrisaig
The Airds 1283
Ardnerr
Duncan
Cow I.
Tosgach
Caol Mor
Croulin Beg
Croulin More
L. Carron
New Harb.
Craig More 1245
Scalpa I.
En. Longa
1469
Balmacara
Dornuidh
Paba
Kyle Akin
L. Alsh
Sguir an Airgoid 2730
I. OF SKYE
Bradford B.
Bein na Cailleach 2387
Kyle Rhea
Bein Chourn 1912
L. Duich
Bein na Cailleach 2380
Glenelg
L. Beag
L. Slapin
Scour na Gour 1983
Sgur mhic Bharraich 2346
L. Eishart
Bein Scrial 3196
L. Ornsay

J. Bartholomew, Edinr.

we do not actually see the lower lips of these basins, it is at least doubtful whether they really consist of rock. Might not the lower lips be formed of mud or sand? We know that the sea has the power of heaping up banks of these materials across the mouths of estuaries; might not this be the explanation of the basins in our sea-lochs? Now there are many reasons which could be given to show why it is in the highest degree improbable that banks of the kind required would always accumulate in such places. But it is not necessary to go into this question, for it is quite evident, upon a little reflection, that even if the sea had the power to heap up, and were now actually engaged in heaping up banks across the mouth of every sea-loch on the coast, still this would not help us to explain the matter any better. For, assuming this to be the case, we should be compelled to admit, first, that the beds of all the sea-lochs did at one time fall away gradually from their present extreme depth up to and even far beyond the 100-fathom line; in other words, that the floor of the sea sloped persistently outwards from the greatest depth reached by the fiords. So that if the land had then been elevated above the waves there would have appeared mountain valleys containing no lakes, but showing rivers that flowed on uninterruptedly up to and even far beyond the edge of what is now termed the submarine Scottish plateau. But this plateau is not known to be breached by any such profound submarine valleys. Then, in the second place, we should be forced to hold the extravagant belief that the sea had filled up all these deep hollows with accumulations of sediment varying in thickness from 30 or 40 to considerably more than 100 fathoms, but had left the shoreward portions of its bed pretty much as they were before submergence ensued; that, in short, only the lower reaches of the submarine valleys had been silted up, the platform of sand and mud stopping abruptly opposite the mouths of the sea-lochs, and not a few of the sounds! Into such absurdities do we land ourselves if the rock-bound nature of the basins in the sea-lochs be denied.

Fortunately, however, there is at least one sea-loch of the rock-bound character of which we have ocular demonstration. I refer to Loch Etive. This loch, as the Admiralty chart will show, contains two basins, the lower one of which extends from Connel Ferry to near Taynuilt, the upper one from this place to the head of the loch. At Connel Ferry the passage is very narrow, and the rock so near the surface that at half-tide the water rushes over the reef with the roar of a cataract. The sight of the seething white water will, I should think, be enough to convince even the most sceptical that, above the ferry, Loch Etive is a true rock-basin. It will be observed that the greatest depth attained in the lower basin is 35 fathoms; in the upper basin we have a depth of not less than 76 fathoms. Here then we have a double rock-basin, the bottom of which is rather more than 400 ft. deeper than its lower lip.

Now while such deep submarine rock-basins can be traced in all or nearly all the sea-lochs, it is remarkable that no such basins occur opposite a low flat

PLATE X

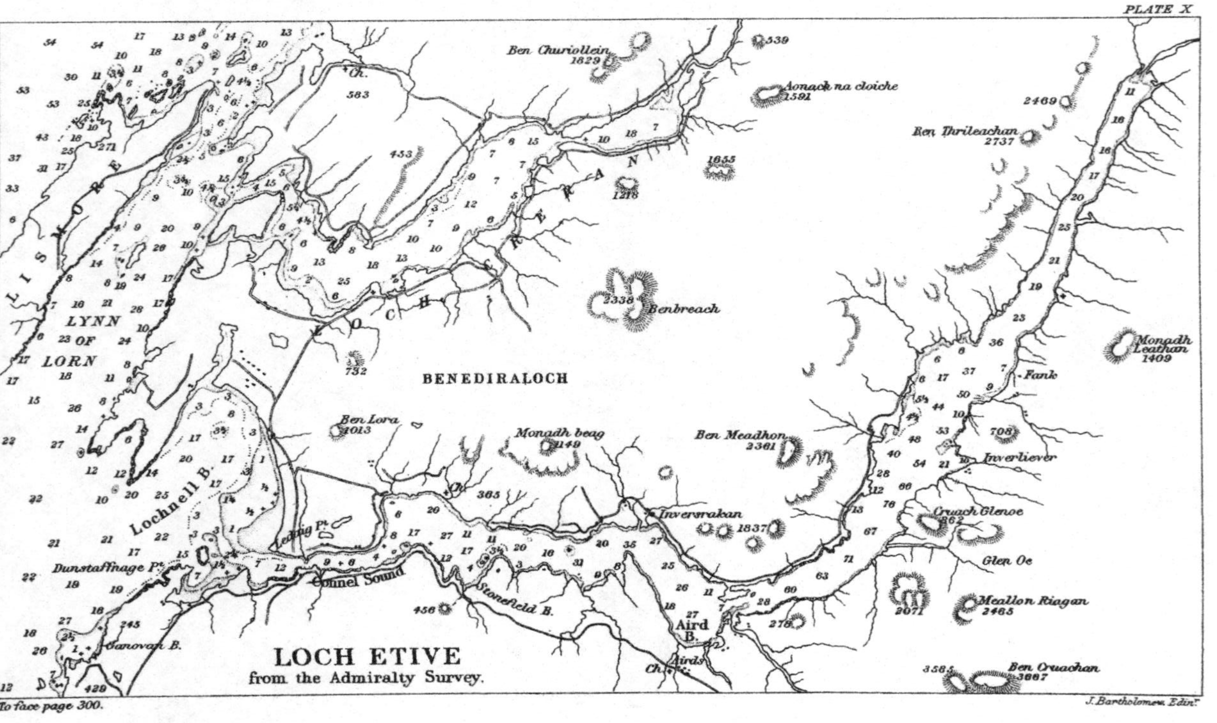

To face page 300.

J. Bartholomew, Edinr.

stretch of country. For example, along the western shores of Lewis, in which part of the island there are no deep valleys, we find that the sea-bottom inclines very gradually outward; and the same is the case off the low-lying districts on the eastern sea-board of Scotland. There seems to be an obvious connection, therefore, between submarine rock-basins and sea-lochs, just as there certainly is between sea-lochs and mountain-valleys.* Nor, if we admit that rock-basins, filled with fresh water, have been scooped out by glaciers, can we, as I think, escape the conclusion that the submerged rock-basins have had precisely the same origin; for the presence of the sea is a mere accident. We know for a fact that all those sea-lochs were once filled to overflowing with ice, and rock-basins occur just in those places where from theoretical considerations they ought to appear. An attentive examination of the physical features of the fiords, and a careful scrutiny of the Admiralty charts, will show that whenever the opposite shores† of a fiord approach each other so as well-nigh to separate the water into two separate sheets, two distinct rock-basins are almost invariably the result. This appearance is well explained by the erosion theory, but is inexplicable otherwise. When glacier ice filled such a fiord, it would be strangled in the narrow pass, and the motion of the ice advancing from behind would be impeded. Hence there would be a heaping up of the glacier, and

* For further details regarding submarine rock-basins, see Appendix, "Physiography of Western Scotland."

† I refer, of course to promontories of solid rock. Where a lateral stream enters a fiord, a spit of low land frequently projects for some distance into the water, but that is a very different matter.

intensified pressure upon the rocky bed would produce its natural effect—increased erosion.

At the same time, not a few double rock-basins are to be explained in another way. Each basin has probably been scooped out at a different period. During the advance of the glaciers there may have been—there most likely were—long pauses; each pause, if of sufficient duration, being marked by the formation of rock-basins. When, again, the glaciers began to retire, the same set of circumstances might well recur—pauses might take place, and basins be again deepened. Moreover, owing to the nature of the ground and the character of the drainage-system, one may be quite sure that a glacier would tarry much longer in some reaches of its valley than in others. Hence we might naturally expect the valleys often to contain more rock-basins than one. If the larger fresh-water lakes were drained, we should in all probability find that some of them would show separate basins, analogous in a measure to the upper and lower basins of Loch Torridon, Loch Cairn-Bahn, and others. Even as it is, however, many valleys do contain more than one lake. Good examples occur in the region of the Trosachs. Again, in the valley of the Tummel we have Loch Lydoch and Loch Rannoch. Further north, in the valley of the Conan, we come upon the wild Loch Fannich and, lower down, Loch Luichart. The valley of the Doon, in Ayrshire, contains the loch of the same name and Bogton Loch, which last, although now much silted up and of little extent, is yet proved to be a rock-basin, from the fact that the soft watery mud of its alluvium broke into some coal-workings in the neighbourhood, at a depth considerably below that of the rock, which comes to the surface, and crosses the valley a mile or so further down.

So long a time, indeed, has passed since the glaciers vacated the valleys, that streams and rivers, by carrying down and depositing gravel, sand, and silt, have often obscured the original rock-bound character of the lake-basins, and in several cases have even entirely obliterated them, so that we find broad flat meadows where lakes formerly existed. As an instance of a partially obliterated rock-basin, I may refer to St. Mary's Loch, in the south of Scotland, the depth of which has been ascertained to be at least 120 ft. At its outlet the whole valley is paved with gravel, but a series of borings made across the bottom of the valley at this place struck the rock at depths varying from 24 to 53 ft. The present bed of the lake is therefore 67 ft. at least lower than the lip of the buried rock at the point of outlet; what thickness of superficial materials may be lying upon the rocky bed of the lake has not been ascertained.

Of rock-basins completely obliterated, many examples might be given. There is one at the head of the Manor Water, in Peeblesshire. Another fine instance occurs in the valley of the Talla (same county), where an ancient lake once occupied the whole bottom of the valley from Talla Linnfoots for nearly two miles down. But, as might have been anticipated, the rock-basins that are so silted up are usually of small extent.

CHAPTER XXII.

POST-GLACIAL AND RECENT DEPOSITS OF SCOTLAND.

Raised beaches.—Faintly-marked terraces, &c., at high levels.—Platforms of gravel, &c.—Platforms and notches cut in rock.—Raised beaches at mouth of Stinchar, at Newport, and Tayport.—Highest-level beaches belong to glacial series.—Evidence of cold climatal conditions during formation of some post-glacial beaches.—Oscillations of level.—Raised beaches at the lower levels.—Human relics in raised beaches.—Proofs of oscillations of level.—Submarine peat-mosses, &c.—Blown sand.

THE accumulations which must next engage our attention carry on the story of the past from the close of the glacial epoch to the present—they are, therefore, termed by geologists the post-glacial and recent formations—and comprise raised beaches, blown sand, peat-mosses, and alluvium. These will be described in the order they are here mentioned, not because that order is strictly chronological, but simply for convenience' sake. When the conditions of their accumulation have been ascertained, we shall be in a better position to discuss the question of their relative age.

In Chapter XX. we considered the subject of certain beds containing marine organisms, and were enabled to come to these conclusions, namely, that in late glacial times Scotland was submerged to a great depth below the waves, and that in the later stages of this submergence the temperature of our seas was somewhat arctic. We found reasons for concluding that the depression reached probably as much as 1,100, or even 1,258, ft. below the present level of the sea. When the submergence had reached its climax the land again rose. During this upheaval, which may not have been, and probably was not continuous, but interrupted by occasional pauses, the waves were enabled to cut out certain horizontal shelves in the solid rocks, and to spread out here and there terraces or flats of shingle, gravel, sand, and silt; these elevated shelves and terraces are what we term *raised beaches*. They occur at many different levels, from 1,050 or 1,100 ft., down to a few yards above the present mean tide. At the higher elevations, however, they are seldom or never well-marked, and as far as known have not yielded any organic remains. Indeed, with the exception of the gravel flats of the Moorfoots, and the successive terraces at Eaglesham, I have never met with any raised beach in the interior of the country that could be certainly recognised as such. Dr. Chambers in his "Ancient Sea Margins" has given a number of instances of what he took to be old sea-beaches at heights approaching, and even exceeding 1,000 ft. But none of these I fear can be relied upon as evidence of wave-work. The shelves on the Eildon Hills, and the West Lomond Hill are cases of atmospheric erosion, and similar markings occur on every hill-side where the rocky strata are horizontal or approximately so. The examples quoted by the same author from the upper reaches of the valleys of the Clyde and Tweed, are, some of them, flutings formed by the action of ancient glaciers, others are merely river deposits, while very many of the terraces at much lower levels, as for example those at Peebles, Kelso, &c., are also of fluviatile origin. Again, the Parallel Roads of Glen Roy which have formed so fruitful a source of controversy are no longer recognised as ancient sea-margins but as lake-terraces. Indeed when we consider the conditions under which the great glaciers disappeared, we cannot but be chary of ascribing any deposits met with in the interior of the country to the action of the sea. Beds of gravel, sand, and silt, occur in such extraordinary positions in the Alps and other highly glaciated regions that we may well pause before deciding upon the marine origin of any such unfossiliferous deposits occupying similar positions in our own country. Yet, as we have seen, there are strong reasons for believing that certain of these deposits in Scotland have received much of their present configuration from the action of the sea, while others appear to be exclusively of marine formation. For the absence of fossils is not sufficient to prove the non-marine origin of a deposit —the physical evidence cannot be ignored, but on the contrary deserves very careful consideration. If we take the highest level at which fossiliferous beds have been detected as the limits of the submergence reached in Scotland, during late glacial times, this would indicate a depression of not more than 350 ft. or so; while, on the other hand, if we have regard to physical evidence we are seemingly compelled to admit a depression of at least 1,100 ft. or 1,258 ft. But whether the whole country was ever submerged to this extent we cannot yet say.

If we have reasonable doubts as to the marine origin of those shelves and terraces which may sometimes be detected in the interior of the country, we can have little hesitation in ascribing to the action of the sea those platforms of rock, and more or less broad terraces of silt, sand, and gravel which in so many places skirt the coasts. The terraced deposits are often plentifully stocked with the shells of the common littoral molluscs, and the whole aspect of the accumulations leaves one in no doubt as to their being ancient sea-margins. When such a terrace is followed inland from the coast it is usually found to abut more or less abruptly against a steeply-sloping bank or a well-marked cliff of hard rock, at the base of which we may often observe caves and hollows which are evidently the work of the sea.

As might naturally be expected, the raised beaches at the lowest levels are the best preserved—those at the higher levels occurring often in mere patches, and being hardly observable except by a trained eye. These have been so long exposed to the action of the atmospheric forces—to rain and frost—that their sharper outlines have frequently been smoothed off, and the terraced appearance well-nigh obliterated. This is especially the case where the shores are abrupt and rocky, and present bold cliffs to the sweep of the waves. Yet even on such rugged promontories we may occasionally detect shelves hewn into the seaward slopes of the hills. It is difficult indeed to perceive these when we are actually upon the ground, but when we retire to some little distance so as to catch the profile of the land, they are then seen to form prominent platforms and notches. Some admirable examples of the kind occur along the bold coast-line in the south of Ayrshire, particularly on the high-grounds that face to the sea at the mouth of the river Stinchar, others are noticeable on the shores of Fife, as near Elie. They may also be seen again and again on the rugged coasts of the western islands. The most continuous succession of shelves and terraces which I have observed at any one part of the coast, occurs between Newport and Ferry-port-on-craig (Frith of Tay). The lower terraces at those places consist of gravel and sand—the higher ones being excavated in the solid rock. They are seven and perhaps eight in number—their levels being respectively 25, 50–54, 75–78, 100, 145–150, 250, 290, and 350 ft. above mean water—the last mentioned one, however, is very faintly marked. Similar shelves and terraces occur at the same or approximately the same levels, and also at intermediate levels at many places both on the west and east shores of central and southern Scotland. The higher ones all belong to the glacial series, but even in those at the lowest levels we have indications of a somewhat colder climate than the country now experiences. And probably, when all the raised beaches have been examined in sufficient detail, it will be found that the movement of elevation that succeeded the great depression was not continuous, but interrupted by occasional pauses, and perhaps by movements in the opposite direction. There are indications of this even in some of the upper terraces. Thus, near Trinity (Frith of Forth), at a height of 70–80 ft. above the sea, Dr. Chambers observed a bed of peat under ten feet of sea-sand—the peat containing the remains of trees that were rooted in an underlying stratum of blue clay.* Facts of a similar kind have been noted not far from the same place;† and the inference is that after the land had been pushed up so far and the sea had retired, and vegetation had covered the country, there was a relapse to the former downward movement, when the sea again advanced for some distance upon the land, and a bed of sand was deposited above the peat and prostrate trees. A much more striking example of this oscillatory movement is furnished by the facts connected with the latest raised beaches, as I shall now proceed to point out.

When we get down to levels below 60 ft. we find that both on the east and west coasts of central and southern Scotland there are two well-marked raised beaches, the upper of which occurs at a height varying from 45 ft. to 55 ft. or thereabout above the present sea-level. This old beach often extends continuously, and with some breadth, for considerable distances, but as a rule it appears for

* *Ancient Sea Margins*, p. 17.

† *Op. et loc. cit.*

the most part only here and there in what seem to have been once sheltered bays. Nevertheless, on projecting promontories we not infrequently find it represented by a shelf cut in the solid rock. The terraced deposits consist principally of gravel and sand, and frequently contain shells belonging to existing British species, associated with which there have occurred certain shells that seem to be now restricted to more northern latitudes.* It is in this same terrace that the earliest traces of man appear to have been recognised. In the parish of Dundonald, Ayrshire, a rude ornament made of cannel coal was found at a height of 50 ft. above the sea-level, resting upon boulder-clay, and covered with gravel containing marine shells.† This evidence is perhaps slight, but it seems doubtful whether the human relics which have been disinterred from marine deposits so abundantly at lower levels than 50 ft. may not in reality sometimes belong to the age of the 45–55 ft. beach, or even to earlier times still: of this, however, more anon.

But by far the best preserved and most interesting of the old raised beaches is that which occurs at a height of 25–30 ft. above the present mean-tide mark in the friths of Tay, Forth, Clyde, and Solway, and generally both on the east and west coasts of central and southern Scotland. When traced north-

* Mr. Gwynn Jeffreys, *British Association Reports*, 1862; A. Geikie, "On the Glacial Drift of Scotland," *Trans. Geol. Soc. Glasg.*, vol. i. part ii.; the latter author suggests that possibly these more northern forms may have been washed out of deposits of glacial age.

† *Antiquity of Man*, p. 61.

PLATE XI.

THE CARSE OF STIRLING: a raised beach, or estuarine flat. (By B. N. Peach.)

To face p. 311.

east to Aberdeen, it appears according to Mr. Jamieson * gradually to lose in elevation until it sinks to a height of only eight or ten feet above high-water. Along the steep and rocky sections of the coast this raised beach exists only as a narrow shelf hewn into the solid rocks, and in not a few exposed places it seems to be entirely absent. But wherever the land falls away to the sea in a long gentle slope the 25–30 ft. beach is usually well-marked, forming a broad terrace that may sometimes extend inland for a distance of several miles. Of this nature are the great carse-lands of the estuaries—the Carse of Gowrie, the Carse of Falkirk and Stirling, and the broad flats through which the Clyde flows in its lower reaches, are all raised beaches or terraces. So constantly present is this strip of flat land round the coasts that there are very few sea-port towns and villages in Scotland which are not built upon it.

The silts, clays, and sands, &c., of which this beach is composed, have, in many places, yielded abundance of shells—all of which, without a single exception, are still living round the coasts. But, besides shells, there have also been found the skeletons of seals and whales. The latter were got in the Carse of Stirling at a height of twenty or twenty-five feet above the present tide, and at a distance of several miles from the sea. Along with one of the skeletons (discoverd in 1819, during the famous improvements on Blair-Drummond Moss), a perforated lance of deer's horn

* *Quart. Jour. Geol. Soc.*, 1865, p. 191.

was obtained.* Again, on a later occasion (1824), a similar implement, evidently intended for a harpoon and still retaining part of the handle, was found associated in the same way with another skeleton of a whale.† But these are not the only traces of man which the alluvial deposits of the low-lying carse-lands have yielded. A canoe, hollowed out of a single oak-tree, and measuring 36 ft. in length by 4 ft. in breadth, was exposed to view in the Carse of Falkirk by the undermining of its banks by the river Carron. It lay at a depth from the surface of fifteen feet, and was overlaid with successive layers of clay, shells, moss, sand, and gravel.‡ Another canoe was dug up near Falkirk at a depth of 30 ft. from the surface.§ In the 25–30 raised beach of the Clyde, at and near Glasgow,|| at least twenty canoes have been discovered at various times, while digging foundations for houses and in cutting drains. Most of these are formed of single trees; others indicate greater skill in construction—one being built of several pieces of oak, but without ribs—another having its base and keel formed of a single oak, to which were attached ribs, planks, and a prow with a high cut-water. In another specimen of these ancient craft a circular hole in the bottom was

* *Edin. Phil. Jour.*, vol. i. p. 395; *Ibid.* vol. xi. pp. 220, 415; *Memoirs Wernerian Society*, vol. iii. p. 327.

† *Mem. Wern. Soc.*, vol. v. p. 440.

‡ *Bibliotheca Topog. Brit.*, No. 2, part iii. p. 242.

§ *Beauties of Scotland*, vol. iii. p. 419.

|| See Chapman's *Picture of Glasgow*, p. 152; Chambers's *Ancient Sea-Margins*, p. 203; J. Buchanan's *Glasgow, Past and Present*; *Trans. Geol. Soc. Glasg.*, vol. iii. p. 370.

plugged with *cork*—pointing, as some antiquaries have suggested, to intercourse with Spain or southern France—but cork is light, and may have floated to the shores of the Clyde. A beautifully polished stone implement was found in the bottom of an old canoe, discovered so far back as 1780, while digging the foundation of Old St. Enoch's Church, Glasgow. The only recorded instance of the occurrence of human remains at any considerable depth in the old carse lands is that of a skull, which was dug up in 1843 21 ft. below the surface at Grangemouth.

No decided traces of ice-action have been detected in the deposits belonging to the 25–30 ft. raised beach. It is true that, in the clay-pits on both sides of the Forth at this elevation above the sea, contorted bedding is a common occurrence, but these contorted clay-beds belong unquestionably to the glacial series —the 25–30 ft. beach is merely cut into them, and here and there the gravel and sand of the old beach may be seen lying upon the denuded clay-beds, which have evidently suffered considerable erosion. Here and there large boulders rest upon the raised beach, but they have plainly been derived from the denudation of the till and boulder-clay. Boulders are scattered along the present beach in exactly the same way. The observer has, therefore, to be on his guard lest he should include among the deposits belonging to the 25–30 ft. beach every superficial accumulation that may chance to occur at that elevation. He must always remember that the beach is not entirely a terrace of deposition—but over wide districts is merely a shelf, broad or narrow as the case may be, cut out of pre-existing deposits of clay, sand and gravel, and sometimes out of solid rock.

Before leaving the subject of these raised beaches, it must be mentioned that beds of peat containing trunks and roots of trees frequently occur underneath the deposits of the carse-lands, and submerged peat-mosses and ancient forests appear at many places round the coasts, as will be more particularly described in the next chapter. There can be no doubt that these indicate an old land-surface that existed prior to the deposition of the carse-clays. They therefore prove that the see-saw movement of which there are some traces in the higher raised beaches, was repeated at a later stage. The land was pushed up until all the area now covered by the carse-clays was converted into dry ground. Trees then flourished and decayed, and were eventually submerged by the sinking-down of the land, and buried underneath great deposits of silt and sand. Then, finally, the land was re-elevated to its present level.

Now, since the peat and trees occur at the bottom of the carse-clays, and are usually found overlying glacial deposits, it is clear that, so far as direct evidence goes, they must be older than the 25–30 ft. raised beach, and they may be, and probably are, older even than the 45–55 ft. beach. When the sea stood at a height of 50 ft. above its present level—or, if the reader prefers it, when the land stood just so many feet below that datum-line—clay and silt must have accumulated in our estuaries and bays just as it did when the land had risen some 25 ft. higher. Indeed, if it be true, as some evidence would lead one to infer, that glaciers may still have lingered on at the heads of some of the higher valleys even after the 45–50 ft. beach began to be formed, then the rivers would bear seawards a much greater quantity of fine silt than they could at a later date, when the glaciers had dwindled down to utter insignificance, or even disappeared altogether. And it does seem reasonable, therefore, to hold that some part at least of the carse-clays was laid down upon the sea-bottom at a time when the waves beat along the 50 ft. level. From which it would follow that the bulk of the buried and submerged peat-mosses belongs to a period anterior in date to either of those two raised beaches: that, in short, the mosses indicate a depression and subsequent elevation of 50 ft. and more since the time that they and the trees which they enclose grew green upon the land. It seems also probable that some of the primitive canoes may have sailed in the waters of the Clyde when the sea reached not 25 or 30, but 45 or 55 ft. higher than now.

Of the *Blown Sands*, which are met with here and there along the coasts, it hardly falls within the scope of this work to say more than a few words. They occur generally on low-lying shore-lands, and very often at or near where a large river enters the sea. They frequently form long chains of dunes that extend parallel to themselves and the coast-line. The

Tent-Moor, between Tayport and the mouth of the Eden, is the best example of the kind which I have seen. The ridges there are wonderfully persistent and well-defined. In other places, as at Stevenston on the Ayrshire coast, they form irregular hummocks and banks. Some of the most extensive accumulations are found along the shores of the Moray Frith, and Mr. Jamieson thinks that these have probably some connection with the rivers entering the sea in their neighbourhood—the sands of Culbin, for example, having been derived in great measure from the sand brought down by the Spey, the Findhorn, and the Nairn.* But in many instances they appear rather to be derived from the denudation of pre-existing drift deposits. Such is certainly the case with those on the shores of Ayrshire, and the Frith of Forth. So much is this so, that it is often impossible to draw a line upon the map to indicate what is true recent blown sand and what is drift belonging to the glacial series. Blown sands are certainly very often connected with the great river valleys, but this appears in most cases that I have seen to be due to the fact that those river valleys contain abundant gravel and sand deposits of glacial age which, upon the low coast-lands at their mouths, become exposed to the combined action of the sea and the wind.

* *Quart. Jour. Geol. Soc.* 1865, p. 192.

CHAPTER XXIII.

POST-GLACIAL AND RECENT DEPOSITS OF SCOTLAND—*Continued.*

Peat-mosses, their composition.—Trees under peat, their distribution.—Submerged peat-mosses of Scotland, England, Ireland, Channel Islands, France, Holland.—Loss of land.—Continental Britain.—Climatal conditions.—Causes of decay, and overthrow of the ancient forests.—Growth of the peat-mosses.—Climatal change.—Decay of peat-mosses.

EVERYONE is familiar with the fact that very large areas in Scotland are covered with more or less thick coatings of peat. These are not confined to any particular region, but they certainly occur in greatest abundance in upland and highland districts, where the soil is frequently obscured over many square miles in extent.

That a peat-moss is entirely composed of vegetable matter I need hardly say—that it has been formed by the growth and decay of successive generations of plants no one doubts: and these plants, moreover, as is well known, are still indigenous to the country. Such being the case, there may seem to be no difficulty in understanding how it comes to pass that we have peat-mosses. The plants which go to form these turbaries are still growing, and if we only allow them sufficient time, no doubt they will give rise to more peat-mosses. But when we begin to look a little more closely into the matter, we find that we cannot settle the question quite so easily. It is well known that underneath the bogs and peat-mosses, roots, trunks and branches of forest-trees and shrubs occur in great profusion. Here, then, is a difficulty. This buried timber assuredly marks the sites of ancient forests. How, then, did the peat come to overwhelm them? To discover this, it is obviously necessary that we should first endeavour to ascertain what kind of trees are buried under the peat, and how they are distributed through the country.

The Scottish "mosses" have yielded oak, pine, birch, hazel, alder, willow, juniper, &c.—all of them species which are even now indigenous to the country. There would appear, however, to be an interesting exception to the rule, for it is said, on good botanical authority, that the cones of *Abies picea* (silver fir) have been dug out of the peat in Orkney—a tree which is common in Norway, but not now indigenous to Scotland. From the position occupied by the buried trees, it is positively certain that they actually grew in place. The stools are rooted in the old soil; the trunks, branches, twigs, and even the leaves, lie all about. Nay, more than this, each species is found rooted upon that particular kind of soil which it is known to prefer: thus pines occupy the lighter gravelly soils, and oaks the heavier clays. Again, we find that the pine predominates in high-level mosses, while the oak abounds most in the bogs of the lower grounds.

Then, as regards the distribution of the ancient forests, it is no overstatement to say that they occur everywhere. I know of but few areas of peat-moss in which they have not been detected; and this is not peculiar to the mainland, but even characterizes the little outlying islands. The visitor to Lewis is startled to find amidst the blank desolation and sterility of its extensive moorlands, the trunks of full-grown trees, consisting of oak, alder, birch, and especially Scotch fir. Now, the only trees in the island are those which Sir James Mathieson has, at great expense, reared around his residence. Yet, in some of the islands of the Outer Hebrides, a few stunted stems of hazel, birch, and mountain-ash may occasionally be seen clinging to the rocks, in places which are beyond the reach of sheep and cattle. The bare islands of Orkney and Shetland have also at one time supported large trees, while of the mainland itself it is difficult to say what district has not waved with greenery. The bare flats of Caithness, the storm-swept valleys of the Western Highlands, the dreary moorland tracts of Perthshire and the north-eastern counties, the peaty uplands of Peeblesshire and the Borders, and the wilds of Carrick and Galloway have each treasured up abundant relics of a bygone age of forests.

It would seem also that some of our trees had a greater vertical and horizontal range in old times than they have now. Mr. Watson gives 600 yards and upwards as the height reached by the Scotch pine at present. But he "has seen also small scattered examples at 800 or even 850 yards of elevation." These last, however, he thinks had probably been planted. "But that the pine," he continues, "has grown naturally on the Grampians at an equal elevation in former ages, is rendered certain by the roots still remaining in the peat-mosses of the high table-lands of Forfar and Aberdeen at 800 yards and upwards."* Again, in Glenavon, Banffshire, there are peat-mosses nearly 3,000 ft. above the sea, with abundant roots of the pine;† and in the north of England they have been met with at a similar height.‡ The Scotch pine now ranges from Perthshire into Sutherland within lat. 56°–59°, but in ancient times it must have grown indiscriminately throughout the length and breadth of Britain, since we meet with it in many of the English mosses, as well those of southern as of northern districts.

The common oak has a similar wide diffusion in the peat-mosses, and the same remark applies to other species. Nothing indeed is more common than to meet with buried trunks of very large dimensions, occupying levels and positions which are now in the highest degree unfavourable to the growth of timber; and this not only in the interior parts of the country, for great trees are frequently dug out of peat close to the sea-shore.§

But one of the most noteworthy points in con-

* *Cybele Britannica*, vol. ii. p. 409.

† Sinclair's *Statistical Account of Scotland*, vol. xii. p. 451.

‡ Mr. Wynch, quoted in *Cybele Britannica, loc. cit.*

§ *Edinburgh Philosophical Journal*, vol. xvii. p. 53; see also *Philosophical Transactions*, vol. xxii. p. 980; and the *Statistical Account of Scotland* (Old and New), *passim*.

nection with the peat-mosses remains to be mentioned. They are frequently found to pass below the level of the sea. This peculiarity has been observed in many places all round the coasts. It is needless to describe these submerged forests in detail, but I may note a few localities where they have been seen. On the coast of the Bay of Skaill (Orkney) an acre of peat-moss containing roots and trunks of fir-trees was exposed during a storm by the washing away of the superincumbent sand.* Again, on the north coast of the island of Sanday (one of the Orkneys) decayed roots of trees are seen at ebb tide upon the beach at Otterwick Bay,† and the like occurs in the bay of Sandwick, another of the same group of islands.‡

In the sea at Lybster, and under the sands of Riess in Caithness, my colleague, Mr. B. N. Peach, tells me he has seen sunk peat with large trees.

A number of years ago, while some improvements were being made in the harbour of Aberdeen, a good many trunks of oak of large size were dug up, and their position showed that they had not been brought down by the river, but had grown where they were found.§ In the parish of Belhelvie, in the same county, peat-moss occurs under the sea-level, and is covered to a depth of 10 or 12 ft. with sand. Oak-remains appear in this peat, and from the fact that during storms large cubical blocks of peat are

* *Edin. Phil. Journal*, vol. iii. p. 100.

† Sinclair's *Statistical Account*, vol. vii. p. 451; Barry's *Orkney Islands.*

‡ *New Statistical Account.* § *Ibid.*

often cast on shore, it seems probable that the peat-moss and its buried forest extend for some distance out into the bay.*

In the Carse of Gowrie it is well known that trunks of oak, willow, fir, hazel, and other trees lie buried at depths varying from 20 to 27 and even to 40 ft. All these are really at or below the sea-level, for the surface of the carse does not rise more than 20–30 ft. above that datum-line. At the Braes of Monorgan and Polgavie, the river Tay has cut down through the carse-land and exposed a bed of peat four ft. thick, containing trunks of oak, fir, alder, and birch, the roots of which penetrate an old soil. This peat, which now forms the bed of the river, is buried below some seventeen feet of alluvial matter, throughout which a good deal of vegetable débris occurs; towards the top of the section, cockles, mussels, and other sea-shells make their appearance. It is said also that in sinking wells in the carse-land, "deers' horns, skulls, and other bones" have frequently been found, along with the remains of the trees mentioned above.†

At Flisk, on the south side of the Frith of Tay, submerged peat-moss has been traced along the beach in one place for a distance of three miles, and in another for no less than seven miles. It contains hazel and hazel-nuts, and what appears to be alder, and the roots are said to occur in places at 10 ft.

* *New Statistical Account.*

† Sinclair's *Statistical Account*, vol. xvi. p. 556.

below the limits of the full tide.* A thin bed of peat, also, may be seen at low water on the shores of the Frith of Forth, a little to the east of Largo.

In the islands of the Hebrides the same phenomena may be studied. At Pabbay for example, moss with large trees is exposed at ebb of the spring tides, and on the coasts of Harris, wherever a high bank has been undermined and cut back by the sea, a rich loam or black moss is discovered. And this is by no means peculiar to Harris, but characterizes all the low-lying sandy shores of the Long Island,† as in North Uist and Vallay, at both of which places submarine peat and trees occur.‡ Again, on the north-west coast of Tiree, and here and there in Coll, the same appearances recur.§

Nor are similar phenomena wanting along the western shores of the mainland, for trees and peat have been found under low-water mark at Loch Alsh; ‖ while at Oban the bottom of the harbour, which is not less than twenty fathoms deep, is said to be covered with peat in some places.

The phenomena of drowned trees and peat-mosses are by no means confined to Scotland. They appear quite as constantly in England ** and Ireland,†† nor

* *Trans. Roy. Soc. Edin.*, vol. ix. p. 419.

† Sinclair's *Stat. Acc.*, vol. x. p. 373.

‡ *Ibid.* vol. xiii. p. 321.

§ *Edin. Phil. Jour.*, vol. vii. p. 125.

‖ *Op. et loc. cit.*

¶ Dr. Anderson's *Practical Treatise on Peat-moss*, p. 150.

** *Philosophical Transactions*, vol. xxii. p. 980; vol. l. p. 51; vol. lxxxix. p. 145; *Philosophical Journal*, April 1828; *Quart. Jour. Geol. Soc.*, vol. vi. p. 96.

†† Jukes' *Manual of Geology*, p. 740.

are they absent on the further side of the English Channel. Sunk forests abound along the coasts of Brittany, Normandy, and the Channel Islands, where trees have been observed at a depth which could not have been less than 60 ft. below high-water. Similarly, off the shores of Holland submarine peat-mosses occur.

Now these facts enable us to conclude that the sea has within geologically recent times made considerable encroachments upon the land. It is quite clear that our shores and the opposite coasts of the Continent have at some period in the past extended much farther seaward. We infer this not only from the facts connected with the submerged mosses, but also from the large size of the buried trees of the maritime districts. In most of the islands it is impossible for trees now to attain to respectable dimensions, unless they are carefully protected and looked after. When the buried forests of the Hebrides and Orkney and Shetland Islands were growing, no doubt the land stood at a relatively higher level and the sea at a greater ditsance.

But more than all this, peat has been dredged far out in the German Ocean and the English Channel; and fresh-water and littoral shells have been dredged in deep water at distances of fourteen and forty miles from the nearest land.* Add to these cases the frequent occurrence of mammalian remains—bones and tusks of the mammoth—which have been obtained often at a considerable distance from the English

* *Quart. Jour. Geol. Soc.*, vol. vii. p. 134; *Ibid.* vol. xxii. p. 160.

coast, both in the North Sea and the English Channel, and we shall hardly escape from agreeing with Mr. Godwin-Austen, who infers that at no distant date, geologically speaking, the bed of the North Sea was a great undulating plain traversed from south to north by a mighty river, which carried the tribute of the Thames, the Rhine and other streams, and poured in one magnificent flood into the Northern Ocean.

At this period the British Islands were in all probability joined to each other, stretching farther westwards than now into the Atlantic; even reaching perhaps to the edge of the 100-fathom plateau. We cannot of course positively assert that this last was the case, but a full consideration of the evidence in all its bearings, leaves one hardly any room for doubt. Botany, zoology, and geology alike seem to require a continental Britain to explain the facts. The plants and animals that now clothe and people our islands have been in large measure derived from the Continent, and some land-passage was needed in recent times to convoy them hither. The geological evidence proves that a much larger area of land must certainly have existed; and if the bed of the sea were only to be elevated so far as to expose to the light of day all the peat which has been reached in dredging and sounding, Britain would immediately become part and parcel of the Continent; for our seas are very shallow, no part of the German Ocean, the English Channel, or the Irish Sea, being anywhere so deep as Loch Lomond. An elevation of from 20 to 30 fathoms would drain nearly all the German Ocean between England and the Continent, and 20 fathoms more would lay dry the same sea between Scotland and Denmark. The average depth of the Irish Sea is not more than 60 fathoms, and that of the Minch does not exceed the same depth. By an elevation of only some 400 ft., every little islet off our coasts would unite with the mainland and the mainland with the Continent. Such a degree of upheaval, however, is not absolutely required to explain the phenomena of the submarine forests: 250 or 350 ft. would perhaps suffice.

The character of the buried timber indicates pretty clearly the kind of climate experienced at that time in Britain. It is impossible to look at the thick bark and the tough resinous wood of the bog-pines, without feeling sure that the winters then were much colder than they are now. A rigorous climate is required to produce the best pine-wood, and it has frequently been observed in the north of Scotland that trees grown in the most exposed situations show the reddest wood and the thickest bark. Yet pines dug out of mosses in the south of England compare favourably with the best timber in the old pine forests of Rothiemurchus. This of itself is enough to show that our climate has greatly changed within comparatively recent times.

But again, the presence of large oaks buried in peat, at heights and in situations that would now be considered most unfavourable to their growth, bespeaks the former prevalence of somewhat warmer summers. In short, we are led to conclude that the seasons during the growth of the ancient forests, were more strongly contrasted than they are now. Such conditions would naturally follow upon the union of Britain with the Continent. With broad wooded plains substituted for the German Ocean and the Irish Sea, and with a wider spread of land along our western sea-board, it can hardly be doubted that, other things being equal, our climate would be greatly affected, so much so as to approach in character to that of Germany.

We have now to inquire what it was that caused the decay and overthrow of the ancient forests, and induced the growth of peat-mosses instead.

It has been surmised by some writers that much of the old timber may have been blown down, and they have referred to the fact that all the trees in some peat-mosses often lie in one direction, as if they had met their fate at one and the same time.* The woods, as we know, grew thick and close, the trunks rose tall and straight, and their roots were fewer and not so widely spreading as they would have been had the trees grown in isolated positions. Hence, when a breach was once effected by the overthrow of the sturdier trunks that guarded the outskirts of the forest, the destruction of the less firmly-rooted timber, it has

* *Highland Society's Prize Essays*, vol. ii. p. 19 (Old Series); Rennie's *Essays on Peat-moss*, p. 31; Sinclair's *Stat. Acc.*, vols. iv. p. 214; v. p. 131; and xv. p. 484; *New Stat. Acc.*. *Paisley* and *Carluke*. *Vide* also, for similar phenomena in English and foreign peat-mosses, *Phil. Trans.*, vol. xxii. p. 980; Rennie's *Essays*; Degner, *De Turfis*, p. 81.

been thought, would speedily follow. Doubtless many acres may have been desolated in this way. But it seems extravagant to infer that all the buried timber met its fate after this fashion. We cannot suppose the peculiar position of the trees to be in every instance the result of violent storms. Trees are usually bent over in some particular direction by prevailing winds, and when any cause leads to their overthrow, whether it be natural decay or otherwise, they naturally fall as they lean.

Other causes of destruction have been suggested, such as lightning, and what are known in America as ice-storms. Neither of these causes can be quite ignored, yet they can hardly have been other than partial in their operation.

But when we come to ask what share man has had in the work, we find that he has been a far more potent agent of destruction than any of the causes yet referred to. Besides the evidence of his hand afforded by the charred wood under peat, we sometimes come upon marks of adze and hatchet.

The earliest historical accounts of North Britain have afforded abundant food for controversy to antiquarians, but when the geologist has gleaned together the few descriptive remarks which occur here and there, in the pages of Tacitus, Dion Cassius, Herodian, and others, he will find that his knowledge of the physical aspect of Scotland does not amount to much that is very definite. He will learn, however, that Caledonia was notorious on account of its impenetrable forests and impassable morasses. But the precise extent of ground covered by these woods and marshes must always be matter of conjecture. The forest-land known as *Sylva Caledoniæ* appears to have stretched north of the wall of Severus, but south of that boundary large forests must have existed; indeed, down to much more recent times, many wide districts of southern Scotland could still boast of their woodlands. Of the nature of those waste plains, described by the ancients as full of pools and marshes, we can have little doubt, although we cannot of course pretend to point out their particular site. Those who have traversed the central counties of Scotland must have been struck with the numberless sheets of alluvium which everywhere meet the eye, betokening the presence, in former days, of so many little lakes. In Bleau's Atlas, many lochans appear in spots that have long ago come under the dominion of the plough. These, however, must form but a small proportion of the lakes which have been drained since the time of the Romans. Such inconsiderable peaty lochans were not likely to merit particular mention by those who had gazed on the Alpine lakes, save as they became vexatious interruptions to their progress through the country, and surrounded, as many of them in all probability were, with treacherous morasses, the words of the old historians appear to have been descriptive enough of certain ample areas in the Scottish Lowlands.

It seems to have been the common practice of the Romans to cut down the trees for some distance on either side a "way," to prevent surprise by the enemy.

Several old "ways" have been discovered on the clearing away of mosses, and in their neighbourhood lie many trunks of trees, bearing evidence of having fallen by the hand of man. The presence of Roman axes and coins leaves us in no doubt as to who the destroyers were. But it is quite evident that such embedded relics do not enable us to fix the age of a peat-moss. They merely tell us that the origin of the peat cannot date back beyond a certain period, but may be ascribed to any subsequent time. Hence it is impossible to say what amount of waste we are to set down to the credit of the Romans. Some authors have, perhaps, been too ready to exaggerate the damage done by the legions. The buried forests which can be proved to have fallen before Roman axe and firebrand are not many after all; but we may reasonably suppose that these form only a portion of the woods which were cleared at that time.

We have, however, what appears to be direct evidence to show that some regions had been divested of their growing timber before the Roman period; for, if Solinus may be trusted, the Orkneys were, in the days of the Romans, bare and bleak as they have been ever since. He says, "They are three in number, and contain neither inhabitants nor woods; here and there they bristle with shaggy copse and herbs, but, for the most part, all they show is bare sand and rock." A patriotic Orcadian might insist that the statement "three in number" renders what follows untrustworthy; and perhaps he might prefer the testimony of Ossian, who, in his poem of "Carric-thura," says of some island in the group, "a rock bends along the coast with all its echoing wood." According to Torfaeus (historiographer to the King of Denmark),* the condition of the Orkneys in the year 890 agreed with the description given by Solinus.† For at that time Einar conferred a great boon upon his countrymen by teaching them the use of peat for fuel—there being then, as Torfaeus says, no woods in the islands. Yet it is well known that the peat-mosses of the Orkneys, and even those of the Shetlands, contain the remains of considerable trees.

My limits will not permit me to consider in detail accounts of the condition of the Scottish forests in times subsequent to the Roman period. Any reference by the chroniclers to the state of the woodlands is only incidental, and perhaps not always to be relied upon. It is interesting, however, to learn from Boethius that the *horrida Sylva Caledoniæ* had in his time become mere matter of history.‡ He further tells us that Fifeshire had formerly been well wooded (in the times of some of his early Scottish kings); but "it is now," says his old translator, "bair of woddis; for the thevis were sometime sa frequent in the samin that they micht na way be dantit, quhill the woddis war bet down."§ Again, Boethius describes the island of Isla (whose peat-mosses contain roots and trunks of trees)

* Torfaeus wrote about 1690. He was a native of Iceland, and died in 1720.

† Solinus is supposed to have written about A.D. 240.

‡ If this had not been the case, he would surely have quoted a less ancient authority than Ptolemy for the site of the ancient forest. See *Cosmographie and Description of Albion.*

§ *Croniklis of Scotland*, chap. xi.

to be an island rich in metals, which could not be wrought on account of the want of wood.*

After the period to which Boece refers, any allusions to the aspect of the country are best sought for in cartularies and such records. For the rights acquired by monasteries over various forests throughout the country, these cartularies afford abundant evidence. Chalmers † has enumerated many instances of special grants by kings and barons "of particular forests in pasturage and panage, and for cutting wood for building, burning, and all other purposes;" and Mr. Tytler ‡ has added to the list. It need hardly be remarked that the greater part of these woodlands has long disappeared. And yet the old cartularies "abound with notices of forests in every shire during the Scoto-Saxon period." I have not hesitated to quote the authority of those records, and the opinions of two such learned and correct writers as Chalmers and Tytler. No one can deny that the evidence of the cartularies is in favour of a better wooded condition for the country than now obtains. But we must guard against the mistake of supposing that all the area embraced under the designation of a "forest" was covered with forest-trees. And there can be little doubt that both Chalmers and Tytler read the cartularies in the light of the facts which are disclosed by the peat-mosses. The trunks of pine, oak, ash, and

* Bellenden's version of the passage is characteristic. He says, Isla is "full of metallis, gif thair wer ony crafty and industrius peple to win the samin;" but he quietly drops all allusion to the want of wood in the island.

† *Caledonia*, vol. i. p. 792, &c.

‡ *Hist. of Scot.*, vol. ii. chap. ii. third edit., and the authorities there quoted and referred to.

other hard timber dug out of the mosses, were regarded as proofs that the regions indicated by the cartularies were in reality the sites of great forests at the time to which those records refer. But it is probable, nay in many cases quite certain, that much of this buried timber belongs to a more remote period. But even with this reservation, Scotland, down to the fourteenth century, would appear hardly to have merited the description given by Æneas Silvius at a later date. During the civil commotions of the country,* and the long wars with England,† much wood seems to have been destroyed, and the gradual progress of cultivation also began to encroach upon the forest-lands. The great number of salt-pans that were early established in Scotland, and the right which the proprietors usually obtained to cut the requisite firewood from the forests of the country, was another cause of destruction, and much timber disappeared in this way from the maritime districts. But although wood appears to have been the fuel commonly employed in the manufacture of salt, yet it is not unlikely that peat may also have been burned in some cases. It is certain, at least, that peat was a common enough fuel in David I.'s reign, and that‡ "petaries became frequent objects of grant to the abbots and convents

* Foredun relates that Robert the Bruce defeated the Earl of Buchan near Inverury, and ravaged the district with fire. The marks of fire are said to be visible on the trees in the peat-mosses of that neighbourhood. Sinclair's *Statistical Account*, vol. xx. p. 144.

† Knighton mentions that in the reign of Richard II., the English, under the Duke of Lancaster, besides firing the forests, employed eighty thousand hatchets in the work of their destruction.

‡ *Caledonia*, vol. i. p. 793.

during the Scoto-Saxon period." This fact ought perhaps to be looked upon as a farther proof of the increasing decay of the forests.

But by far the most remarkable testimony to the bare condition of the country is furnished by the Acts of the Scottish Parliament. From the times of the first James, stringent acts were adopted by successive parliaments,* having for their object the preservation of the woods. Æneas Silvius (afterwards Pope Pius II.), who visited this country about the middle of the fifteenth century, relates: "We have seen the poor people almost naked, who came to beg at the doors of the temples, receive for alms pieces of a stone, with which they went away contented. This kind of stone, being impregnated either with sulphur or some other combustible material, they burn instead of wood, of which their country is destitute."† Such a statement regarding the bare condition of the country might have been thought somewhat exaggerated, for it is the testimony of a visitor from more favoured climates; but its truth is curiously illustrated by an Act of

* See *Acts of Scottish Parliament.* The more interesting acts referring to the state of the woods were passed as follows:—James I., Second Parliament, A.D. 1424; James II., Fourteenth Parliament, A.D. 1457; James IV., Sext Parliament, A.D. 1503; James V., Fourth Parliament, A.D. 1535; Mary, Sext Parliament, 1555; James VI., First Parliament, 1567, Sixth Parliament, 1579, Eleventh Parliament, 1587. It is curious to notice how, from the time of James I., the penalties imposed upon the destroyers of wood increase in severity. Pecuniary fines are succeeded in time by stocks, prison, or irons; the culprit is to be fed on bread and water during confinement, and to be scourged before parting from his jailers. The climax is reached in the following act, which became law in 1587:—"Whatsoever persone or persones wilfully destroyis and cuttis growand trees and cornes, sall be called therefore before the Justice or his deputes, at Justice Airs, or particular diettes, and punished therefore to the death, as thieves."

† *De Europa*, cap. 46.

Scottish Parliament, passed in the reign of James IV. —"Anent the artikle of greenewood, because that the Wood of Scotland *is utterly destroyed*, the unlaw theirof beand sa little: Therefore," &c.*

There are, of course, numerous traditions regarding the former wooded condition of various districts from which the trees have long since been stripped. Many of these refer to some of those woods which I have already mentioned, as being frequently named in the cartularies and similar records.

Another line of evidence is supplied by local names; but into this subject I cannot enter here.

The short outline of historical facts now given seems to prove:—

1st. That when the Romans entered Britain they found the surface of the country to some extent covered with forests, but diversified in many places with bogs and marshes.

2nd. That to this period we must refer the destruction of some portions of the ancient forests, whose remains are dug out of the peat-mosses; but what amount of damage the woods then sustained we have no means of ascertaining.

3rd. That from the time which elapsed after the departure of the Romans down to the eleventh century we have no certain records referring to the state of the preservation of any part of the Scottish woods, unless we except the statement of Boethius, who tells us that Fife had in great measure been divested of its forests by some of his early Scottish kings.

* Sext Parliament, A.D. 1503.

4th. That from the eleventh to the thirteenth century, and down even to later times, there appear to have been still considerable areas of forest-land, the rights to which were frequently granted to ecclesiastical communities and others.

5th. That during these centuries much forest was thus cleared and brought under cultivation; that at the same time woods were exhausted by building and burning, more especially as fuel for the salt-works; while extensive tracts were displenished and laid waste during times of war and civil strife.

6th. That from the time of James I. there appears to have been a progressive decay of the remainder of the Scottish woods.

There can be no doubt, then, that man must be credited with no small share in the work of destruction. It may be questioned, however, whether he was, after all, the chief agent. Certain considerations would seem to show that too much has been reckoned up against him.

We have seen that the general character and distribution of much of the timber in our peat-bogs points to the former prevalence of a somewhat excessive climate, and I have sought to connect this period with that continental condition of Britain which geologists are generally agreed did obtain within comparatively recent times. Now, when our area once more became insular, it is almost needless to say that this change would react upon the climate—the winters would become milder, and the summers cooler. The trees in what are now the maritime districts, would soon succumb to the influence of the sea air. Thus, wide areas along the coasts, and in the islands of Orkney, Shetland, and the Hebrides, would be displenished. The trees, falling to the ground, would obstruct the surface-drainage, and thus give rise to marshes in which the bog-mosses would speedily multiply, until, by-and-by, they overwhelmed the prostrate timber, and covered the whole with a mantle of growing peat. Nor can it be doubted, that, on the moist hill-tops, and in many places in the interior of the country, similar changes would transpire.

It is a mistake, however, to suppose that peat-moss always overlies a prostrate forest. There are cases where no trace of wood can be detected. Peat of this description is not uncommon in the upland districts of the south of Scotland, where it frequently clothes the tops and slopes of considerable hills to a depth of from 6 to 12, and even 16 ft. Here, then, there are no trees to account for its presence. Again, in the mosses of the higher hill-tops, when trees do occur, they are of small size—mere brushwood, in fact, the overthrow of which, we can hardly think, would have done much in the way of collecting moisture for the support of the bog-mosses.

From a consideration of these and other points, it seems not unreasonable to suppose that the decay of the woods and the growth of the peat were both alike to some extent induced by a change of climate. Insulation would render the country less capable of supporting an exuberant forest-growth, and would at the same time increase the moisture of the atmosphere, and thus favour the spreading of the bog-moss and its allies.

As the beginning of these changes carries us back far beyond the earliest dawn of history, it follows that much of the peat and buried timber of our country may be of great antiquity. And, indeed, in the case of many mosses, we seem more likely to err in ascribing too recent than too early a date to the period of their formation. We cannot estimate the time which has gone by since the western islands supported those timber-trees, the remains of which are dug out of the mosses. It is highly probable, that, at the period in question, those islands were joined with the mainland, and shared a continental climate. To the same date we may refer much of the buried timber of the Orkney and Shetland Islands. Again, the more elevated peat-mosses of the country must have been among the first to be formed; for, any change from a continental to an insular state would tell first upon the trees that grew along the sea-board, and at the higher elevations of the land. It seems reasonable, therefore, to conclude that, long before the Romans set foot in Britain, the overthrow of timber and the growth of peat-moss had made considerable progress; that, in short, the Caledonian Wood was but the relics of that great forest which in former ages had spread over all the area of these islands and the German Ocean.

Farming operations have encroached, and every year are continuing to encroach, upon the moss-lands.

But a very long time indeed will elapse before the great "flow-mosses" or quaking bogs, some of which exceed 40 ft. in depth, shall be improved out of existence. Draining, however, has done much to stop the growth of peat in many places, and is destined to do still more. I feel sure, from what I have myself seen, that the general decay of the peat-mosses (I refer more especially to those on flat hill-tops and sloping ground) far exceeds the rate of growth. Frost and rain are breaking down the peat and washing it rapidly away, and in many cases only a few shreds have been left scattered here and there over the tops and slopes of the hills. Under present climatal conditions, the eventual clearing of all these high-grounds is only a question of time. This change must, no doubt, be attributed, in large measure, to improved systems of drainage, but it seems not improbable that it may also in some degree be due to a lessened rainfall—the mosses not receiving so much moisture as they did in former times. But, for obvious reasons, this would be a very difficult matter to prove.*

* For general information on Scottish peat-mosses consult Rennie's *Essays on Peat;* Aiton's *Treatise on the Origin, Qualities, and Cultivation of Moss-earth;* and Steele's *History of Peat-moss.* Dr. Anderson's *Practical Treatise on Peat* is a whimsical attempt to prove that peat-moss is a plant *sui generis!*

CHAPTER XXIV.

POST-GLACIAL AND RECENT DEPOSITS OF SCOTLAND—*Continued.*

Action of the weather on rocks.—Erosion by running water.—Post-glacial erosion insignificant in amount as compared with denudation during last inter-glacial period.—Recent river-terraces.—Silted-up lakes.—Marl-beds.—Organic remains in fresh-water alluvia.—Human relics.—Conclusions regarding succession of events in post-glacial times.—General summary of glacial, inter-glacial, and post-glacial changes.

THE great erosion caused by the continuous grinding of gigantic masses of ice—the rounded outline given to crags and hills—the scooping-out of deep basins in solid rocks—and the accumulation of thick and wide-spread deposits of till—seem, by comparison, so immensely more important than the results effected by the action of rain and frost, as at first sight to render these somewhat insignificant. A little reflection, however, will serve to dispel this notion, and show us that, even in the absence of snow-fields and glaciers, the waste of the land by the other atmospheric forces must be enormous.

Let us recall the appearance presented by the Scottish mountains—bold hummocky masses of rock, for the most part, but often bristling with splintered crags and shattered precipices. See how frequently the hill-tops are buried in their own ruin, and how the flanks are in many places curtained with long sweeps of loose angular blocks and rubbish, that shoot down from the base of cliff and scaur to the dark glens below. All this is the work of rain and frost. When the whole country was swathed in snow and ice, the hills and valleys no doubt experienced considerable degradation from the grinding action of the confluent glaciers; but then, on the other hand, frost was debarred in large measure from carrying on its usual work of destruction—for only a very few of the highest mountains then lifted their heads above the level of the far-extended mer de glace. Glacial erosion during these conditions was, no doubt, excessive; but we must remember that the continuous grinding of a glacier upon its bed, produces less effect than the simple action of the frost upon such rocks as are exposed above its surface. Consequently, under extreme climatal conditions, the greater the area of bare rock, the more considerable will the waste be; the moraines of small glaciers being proportionately larger than those of more important ice-flows. Thus, long after the glaciers of Scotland had dwindled down to comparative insignificance, the waste amongst the mountains must still have been prodigious. The extreme cold of those old times has long since passed away, but even now, when no persistent snow-fields exist, the havoc effected by frost is yet very considerable.

Then, again, we have to bear in mind that the whole surface of the country is being subjected to the abrading action of running water. Under the influence of rain, soil is continually travelling down from higher to lower levels; rills and brooklets are gouging out deep trenches in the subsoils and solid rocks; streams and rivers are constantly wearing away their banks and transporting sediment to the sea. The gravel and sand and silt that pave the numerous water-courses, are but the wreck and ruin of the land, and it is easy to see, that, since the close of the glacial epoch, immense quantities of material have been thus abstracted from the country. The streams and rivers have been working deeper and deeper into the bottoms of the valleys, and leaving behind them terrace after terrace of alluvial detritus to mark the different levels at which they formerly flowed. And if we tried to estimate the amount of material which has been thus cut out of the valleys and carried seawards, we should no longer feel inclined to undervalue the erosive power exerted by running water. But we should not have proceeded far in our investigations before we became aware of the fact that, great as this erosion has been since glacial times, it is yet insignificant as compared with the vast denudation which was effected during the last inter-glacial period, that is, prior to the submergence of the country. It was then that the till and other glacial deposits were swept out of the valleys, and the solid rocks themselves deeply incised. During the succeeding period of submergence, the sea no doubt eroded the superficial materials here and there, and scarped out shelves on the slopes of the hills, but its work was chiefly that of re-arranging and re-assorting

—the clearing-out of the valleys, many of which had been choked up with débris by the glaciers, was well-nigh exclusively effected by the great rivers of the last inter-glacial period.

In the annexed diagram (Fig. 49) I have indicated the relative positions occupied by the modern or recent river deposits, and those that belong to the era that preceded submergence. The solid rock is shown at *r r*; the till at *t t*; the inter-glacial river gravels at *g g*; and the recent deposits at *a a*. These appearances compel us to infer, first, that till not only covered the side-slopes, but also cumbered the bottom of the valley to a considerable depth. Then came the last inter-glacial period, with its big rivers, which ploughed out the till and, as they deepened their channels, clothed the sides of the valleys with gravel and sand. The hummocky outline of the gravel may, in many cases, be due to mere subsequent atmospheric erosion, but frequently it would appear that marine action may have had something to do with it. At all events, we know that subsequent to the denudation of the till and the formation of the river gravels

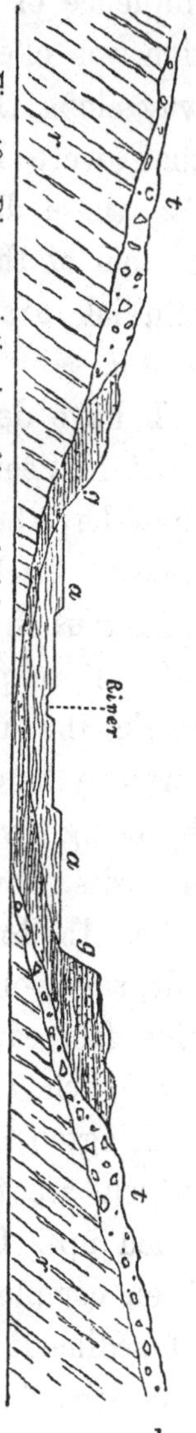

Fig. 49.—Diagrammatic view of inter-glacial and post-glacial river gravels: *t*, till; *g*, interglacial gravels; *a*, recent alluvium.

submergence ensued, and these fresh-water deposits must therefore have come under the modifying influence of waves and currents. Finally, re-elevation was effected, and rivers once more found their way along the valleys, and re-excavated the gravels that were laid down during the previous period. As these later streams cut their way into the bottom of the valley, they left successive terraces of alluvial gravel, sand and silt, which are shown at *a a*.

I have described the various kinds of lakes that exist in Scotland, and have shown how not a few of these have been silted up by the streams since glacial times. Such filled-up lakes are probably far more numerous than we have any idea of—for it is always difficult to prove that a wide flat of alluvial ground marks the site of an ancient lake. The barriers that formerly held in the water become obliterated, either by being swept away, or buried deeply under recent deposits. Such is the case with not a few rock-basins, where the lower lip of rock is often concealed below silt, sand, or gravel. And it is only by boring that this fact can be demonstrated. I have referred to two such cases, namely, Bogton Loch near Dalmellington, and St. Mary's Loch in Peeblesshire, which are two rock-basins in the course of being filled up. Others, again, as I have remarked, have been completely obliterated by the pouring into them of sediment. Very few of the fresh-water lakes and sea-lochs do not show more or less extensive flats opposite the mouths of the streams and rivers that flow into them; and even where no such flat ground appears, the soundings yet show that a delta is gradually increasing below the surface of the water. This is specially the case, of course, when the lakes are deep.

In the lowland districts there are numberless little sheets of alluvium and peat that mark the sites of shallow pools and lakelets. Many of these, especially in Ayrshire, rest in hollows of the till, and some in superficial depressions of drift sand and gravel. They have not been silted up by streams, but by the gradual washing down of the banks and slopes that surround them, under the long-continued action of rain and frost. And, indeed, I know hardly anything more calculated to impress one with the importance of these apparently insignificant agents of waste, than the phenomena connected with the obliterated lakelets referred to. The deposits that fill them up are often several yards in thickness, twelve and fifteen feet being no uncommon depth. Loam and loamy clay, which is sometimes used for brick-making, are the prevailing ingredients, but these often contain intercalated beds of peat and decayed trees, which have evidently grown *in situ*—principally willows and alders, but at the bottom oaks are frequently found rooted in the subjacent glacial clays.

Not a few old lake-beds are found to be in large measure filled up with marl, consisting of the remains of innumerable fresh-water shells. Many examples of this occur in Forfarshire, where they were long ago studied by Sir Charles Lyell and others.*

Now in these alluvia, both lacustrine and fluviatile, mammalian remains have frequently been discovered. Among other forms we get *Bos primigenius*, *B. longifrons*, wild boar, red deer, fallow-deer, roebuck, elk or moose-deer, Irish elk, reindeer, goat, wolf, wild-cat, fox, beaver. The *Bos primigenius*, or urus, is now extinct, but is believed to be represented by the white cattle of Chillingworth and Hamilton, and our present domestic breeds. *Bos longifrons* is also extinct, but from it some of our domestic cattle may have descended.

The true elk (*Cervus alces*) which ranges over the northern regions of America and Asia, from latitude 45° northwards to the shores of the Arctic Sea, but in Europe does not come farther south than the 64th degree of latitude, wandered in post-glacial times over all Scotland, its remains having been found in recent deposits as well in northern and central as in southern counties.† But there appears to be only one instance on record where remains of the Irish elk have occurred in recent deposits, namely, in a marl-bed at Maybole in Aryshire.‡ The reindeer is also rarely met with, but there can be no doubt that it also was a native of Scotland§ in post-glacial times; some writers, indeed, relying on a statement of Torfaeus, think that it may have survived in the extreme north

* *Transactions of the Geological Society*, vol. iv. p. 305. *Ibid.* Second Series, vol. ii. p. 73.

† *Proceedings of the Society of Antiquaries of Scotland*, vol. ix.

‡ *New Statistical Account of Scotland*, vol. v. p. 353.

§ *Proc. Soc. Antiq. Scot.*, vol. viii. p. 186.

down to the twelfth century. Wolves and beavers, though no longer natives of Scotland, were certainly so within historical times. Of the other animals mentioned nothing need be said here; they are all forms eminently characteristic of a temperate climate.

It is remarkable that nowhere in the recent deposits of Scotland have we any trace of the great pachyderms, so frequently met with in certain river gravels of England and the Continent. All the remains of the mammoth yet detected have appeared either in beds that underlie or are intercalated with the till, or else they have occurred in the actual till itself. There is a vague mention of the horn of a rhinoceros having been dug up in marl at the Loch of Forfar,* but I fear no reliance can be placed upon it.

The oldest relics of man in Scotland consist of stone implements. They turn up everywhere, and to give merely a list of the places where they have been detected would be to enumerate every district in the country, not even excepting the outlying islands. They occur either at the surface of the ground or embedded below peat, and in recent alluvial deposits, and they frequently appear associated with the remains of some of the animals referred to above. We cannot assign them to an older date than the Newer Stone period of archæologists, not a single relic of the Older Stone period having yet been met with north of the Tweed.

* *Edin. Phil. Journal*, vol. viii. p. 387; *Memoirs Wernerian Society*, vol. iv. p. 582, vol. v. p. 573.

Having now finished our very rapid survey of Scottish post-glacial deposits, I shall endeavour, in a few words, to summarise the results obtained.

During the deposition of the clays and sands with arctic shells, the land, as we have seen, stood relatively at a lower level than now. But the period of submergence was passing away; slowly and gradually the sea retired, leaving behind it, as the land emerged, certain shelves and terraces to mark the height at which its waves had formerly rolled. This recession of the sea appears to have been occasionally interrupted; nay, it even happened that sometimes the water regained a part of the territory it had lost, and spread its sediment over the site of prostrate forests. A good deal of ice seems to have been floating about in Scottish seas, even after the re-elevation of the land had made great progress. Rafts and, perhaps, bergs stranded in the shallow waters of the estuaries, and threw into confusion the fine silts and clays that were there accumulating. Often, too, they dropped stones and boulders as they floated along, and these became embedded upon the floor of the sea. In this way pieces of chalk, which may have come from Denmark, were covered up in the clay at Portobello. At this time glaciers still lingered in our mountain-valleys, and the rivers, turbid with silt, carried down great quantities of shingle and sand to the low-grounds, and poured their muddy waters into our sea-lochs and estuaries.

The sea continuing to retire, the British Islands became at last united to the Continent. We have no positive evidence as to the character of the climate which then prevailed in our area, but it could hardly have been other than excessive and quasi-continental. Edward Forbes considered* that our country at that time was in the condition of the barren grounds of North America—bare and treeless—and inhabited by reindeer, Irish elk, uri, bears,† foxes, wolves, hares, cats, and beavers. It is almost certain, indeed, that a bare and treeless state must have preceded the age of forests, but it would be hazardous to assert that trees did not exist in Britain before the German Ocean became for the last time converted into dry land. The remains of trees occur in Scottish post-glacial beds that are most probably of older date than the latest continental condition of Britain. It would appear, therefore, that trees existed in Scotland before the bed of the German Ocean was upheaved. Whence these were derived we can only conjecture; most likely, however, they came from England, and if so, this would indicate that England was joined to the Continent at an earlier period than Scotland. But, however this may have been, we know that the whole of the British area became at last one magnificent forest-land, continuous across the bed of the German Ocean with the great forests of northern Europe. Scotland at this time was inhabited by numerous races of wild animals, many of them now locally and some wholly extinct.

* *Memoirs of the Geological Survey*, vol. i. p. 397.

† Species of bear appear to have been got in true post-glacial deposits in Ireland, but nowhere in Scotland.

The reindeer, elk, wolf, beaver, &c., have disappeared from the country; the ancient wild bovidæ now only live in our domestic breeds, and the Irish elk is extinct. It was, probably, during this last continental condition of the country that the men who used the polished stone implements entered Scotland. They were a race of hunters and fishers, and seem to have frequented the neighbourhood of rivers and lakes, living a life not unlike that of the Indians in what were once the woody wildernesses of Canada. The climate would appear to have been somewhat excessive, the winters being considerably colder, and the summers, perhaps, a little warmer than at present—at least, on what is now the eastern side of our island.

How long this continental condition lasted we cannot even conjecture. We only know that, after it had endured for a considerable time, the sea once more began to encroach upon the land, and at last Britain became an island. The sea advanced until the water-line reached some 50 or 60 ft. above its present level, and then a considerable pause supervened. Man still inhabited the country, casting his nets in the lakes, and sailing about the sea-coast in his canoes hollowed out of single oaks. Again the sea retired, and once more paused when it had fallen to 25 or 30 ft. above its present level. The arctic mollusca that once lived around the shores had now vanished, and the present fauna occupied their place. But the seas were, perhaps, even then a little colder than they are now; as would seem to be indicated by the presence in the carse-clays of the stranded whales, which belong to the large Greenland species.

The final movement that carried the land up to its present level brings us down to recent times, after which it falls to the archæologist to continue the strange story of the past.

So many points of evidence have now been adduced, that it will be well, before we address ourselves to the study of the glacial deposits of other countries, to reckon up here in a few paragraphs what appears to have been the general succession of geological changes in Scotland from the advent of the Great Ice Age down to recent times.

1. The till, with its intercalated beds, indicates a vast lapse of time, during which there were several revolutions of climate—how many, we do not know.

2. The beds of till point to intense arctic conditions having prevailed at the time of their formation.

3. During the climax of glacial cold, glaciers not only filled the mountain-valleys to overflowing, but coalesced upon the Lowlands, and creeping outwards occupied all the shallow seas around the shores. Only the tips of the highest hills peered above the mer de glace, which became confluent with the ice-fields of Ireland, England, and Scandinavia, and terminated as a great wall of ice, beyond the shores of the Outer Hebrides, in the deep waters of the Atlantic.

4. The deposits of silt, clay, sand and gravel, with land-plants and mammalian remains, and occasionally with marine shells, all of which beds are intercalated in the till, clearly show that the intense arctic cold

which covered the country with an ice-sheet was interrupted, not once only, but several times, by long continuous ages of milder conditions. Some of these periods may have been warmer than others, just as some of the glacial periods may have been colder. The sea-shells, got in one place at a height of 512 ft. in an intercalated bed, indicate that there was at least one period of considerable depression during the accumulation of the Lower Drift.

5. So far as direct evidence goes, we cannot say that any one of the inter-glacial periods was characterized by a warmer climate than is now enjoyed in the forest regions of high latitudes in America.

6. Considering the sorely wasted appearance of the inter-glacial deposits, and keeping in view the nature of the conditions under which they have been preserved, it would be rash to conclude that they contain a complete record of the changes which ensued during inter-glacial times, or that we are entitled to argue, from the few fossils yielded by them, that the climate of the inter-glacial periods was never positively warm.

7. The boulder-clay, which occurs only at low levels, and in maritime districts, was laid down during the final recession of the great ice-sheets. It indicates a time when the ice had greatly diminished in thickness and extent, but was still broad enough to cover all the Lowlands. As the ice melted away, the sea occupied its place, following its retreating footsteps. But after a submergence of 200 or 260 ft. had been attained the glaciers ceased to reach the sea and began to deposit their moraines upon the land.

Perched blocks and erratics were stranded on the mountain-slopes and on the lowland hills at ever-decreasing levels as the ice drew back.

8. The sand-and-gravel drift overlying the till and boulder-clay was carried down from the melting glaciers, and deposited upon the low-grounds. These deposits, therefore, indicate the farther retreat of the ice, and the amelioration of the climate. It was during this period that the greatest erosion of the older glacial deposits ensued; the large rivers then occupying the valleys not only swept down the morainic matter which marked the retreat of the local glaciers, but ploughed into the till and boulder-clay, and scoured these deposits out of the valleys.

9. The country at length became submerged to a depth in the south-east and west midland districts of probably as much as 1,100 or 1,280 ft. Whether this great submergence extended over all Scotland we cannot yet say. The river deposits of the Lowlands were now partially re-assorted, or top-dressed, as it were, by the action of the sea. As the submergence approached its climax the temperature became colder; ice-rafts floated about, and dropped boulders over the sea-bottom: these are now found lying on the slopes of the re-assorted gravels, and enclosed in stratified clays, the character of the shells in which prove the climate to have been severe.

10. The sands and clays, with arctic shells, &c., were deposited while the land was being gradually upheaved.

11. Considerable local glaciers existed at that time,

and heavy rivers flowed down some of the highland glens, sweeping along sand and gravel to the sea, where they eventually settled down upon the clay that contained arctic shells.

12. The land continued to be upheaved, and several pauses in the movement of elevation were marked by the formation of what are termed raised beaches. With the exception of those at the lower levels, all the raised beaches belong to the glacial epoch.

13. Britain became continental by the conversion of the German Ocean into dry land. At first, probably, bare and treeless, it eventually passed into the condition of a great forest-land. The climate was continental, and the fauna temperate and cold-temperate. Men who used polished stone implements then lived in Scotland.

14. Submergence once more ensued. The destruction of the forest-lands and the increase of peat-mosses dated its commencement from this period. Climate insular, but colder than at present.

15. Final re-elevation and formation of the low-level raised beaches.

16. The present.

CHAPTER XXV.

GLACIAL DEPOSITS OF ENGLAND AND IRELAND.

Necessity of comparing deposits of different countries.—Glaciation of mountain districts of England and Wales.—Till, its character.—Direction of ice-flow in the north-west districts.—Lower boulder-clay of Lancashire, Cheshire, &c.—Middle sand and gravel.—Upper boulder-clay.—River gravels.—Morainic débris.—Succession of changes.—Lower, middle, and upper glacial series of East Anglia.—Irish glacial deposits.—Till.—Morainic débris.—Lower boulder-clay, middle gravels, &c.—Upper boulder-clay.—Eskers.—Erratics.—Sand-hills.—Marine shells at high levels.

NO one who shall endeavour to trace the origin and history of the drift accumulations within any particular area need hope to do so satisfactorily without continual reference to the superficial phenomena of contiguous regions. Individual sections, however clear and apparently consecutive they may be, yet do not contain all the truth; and it would be ridiculous to suppose that the drifts of any limited locality tell one everything that he might hope to learn of the physical history of the Great Ice Age. Two observers who should restrict their examinations, the one to a mountainous district, the other to low-lying tracts at a distance from the hills, would be sure to form very different ideas of what the glacial epoch really was. One might be all for glaciers—

the other all for icebergs. The earlier students of the Scottish glacial deposits held that the whole island had been swept from north-west to south-east by an ice-laden ocean current, and they pointed triumphantly to the direction followed by the rock-striations and the till in the great midland valley as conclusive proof of their theory. But had they been as well acquainted with the Southern Uplands and the Northern Highlands as they were with the low-grounds of the Lothians, this iceberg theory would probably never have been advanced. It is only when the geologist has gone over a sufficiently large tract of country and studied its superficial accumulations at all levels, in lowlands and mountains alike, that he can safely generalise. He will not, however, fully appreciate the results obtained if he ventures to ignore what other workers are doing elsewhere in the same field of inquiry. When he compares his own conclusions with those of others, he will often find reason to hesitate and proceed with caution, where, previously, he may not have perceived any difficulty. On the other hand he will not infrequently have his own inferences strengthened, and here and there catch a hint that may enable him to see newer and deeper meanings in his facts than he had any idea of before. Certain it is that we shall never acquire a proper knowledge of the physical changes that supervened during the glacial age, until the records of that age have all been correlated and compared. But what a vast amount of work remains to be done before this can be adequately accomplished! Nevertheless some approximation towards it can be, ought to be, indeed, attempted. A great deal has already been learned. The general succession of events that marked the progress of the Great Ice Age in widely-separated countries has been more or less clearly made out, and it becomes, therefore, a matter of importance to inquire how far the conclusions so arrived at harmonize with each other. If they shall be found to tally as closely as could have been expected, we shall have so far a guarantee of their accuracy, and be enabled, as we shall afterwards see, to form some definite opinion as to the succession of changes that followed upon the close of the glacial epoch. In this and succeeding chapters, then, I shall attempt, with what success I may, to compare the results obtained in Scotland with the conclusions arrived at by English, Irish, and foreign geologists.

So long as the observer confines his attention to the mountainous districts of England, he will experience no difficulty in detecting the traces of former extensive glacial action. He will find both in Wales and the Lake district, that the mountains frequently show that peculiar flowing and rounded contour which is so characteristic a feature of ground over which land-ice has passed. In the valleys he will see polished and striated surfaces of rock, and heaps of morainic deposits; and the presence of numerous true rock-basins will farther conspire to assure him that the influences under which Scotland assumed much of its most characteristic scenery, have also had no small share in designing, or at all events in adding some of the latest and finest touches to, that beautiful picture of hill and dale and lake that so charms one in Cumbria and Wales.

But when we leave the hilly districts and begin to traverse the broadly undulating low-grounds, the evidences of old ice-action become obscure and hard to read. And our difficulties increase the farther we recede from the mountains. The tough and firm rocks of Cumbria and Wales are replaced as we travel outwards from these centres by rocks less capable of retaining any ice-markings which may at one time have been graved upon them. Add to this the great thickness of superficial materials, underneath which the strata in the low-grounds are frequently buried, and the confused and intricate appearance often presented by these drifts, and it will be admitted that the geologist who sets for himself the task of unravelling the evidence, so as to educe a clear and consecutive story, has no easy work to accomplish. Not the least perplexing part of his task will consist in attempting to discover the meaning of the terminology employed by different observers. Similar deposits he will find are known under different names; while under one and the same designation accumulations are described which are certainly not the same, but in some cases as wide apart as they could well be.

The English glacial deposits are typically developed in two regions: 1st, in the region of Wales and the north-western counties; and 2nd, in East Anglia.

It will be most convenient to treat of these two districts separately. My limits, however, will not allow me to do more than trace the general succession established or in process of being established by English geologists. For convenience' sake I take first the north-western district.

The oldest deposit which has yet been recognised in this part of England, is a more or less tough stony clay that answers precisely to the Scottish till. It is quite unstratified, save here and there where the included stones show a rude kind of arrangement, similar to that which I have described as occasionally visible in the till of Scotland. Like the latter it also contains in places thin irregular seams and amorphous patches of earthy sand and gravel, while its colour varies according to the district in which it is found. Thus it may be yellowish brown, grey, dark blue, or red—the colour evidently depending upon that of the rocks from the degradation of which it has been derived. So far as yet known it would appear to be unfossiliferous. The stones are angular, blunted, striated, smoothed and polished—the more compact and finest-grained rocks receiving the best dressing. Moreover they are scratched most markedly in the line of their longest diameter—irregular-shaped stones not being striated in any one direction more than another. In these and other items this till tallies precisely with that of Scotland.

It rests usually upon a smoothed and striated pavement of rock, but sometimes the strata are bent over, crushed, and broken underneath, and their fragments commingled with the till.*

An examination of the rock-striations proves conclusively that from all the valleys of Wales and the Lake district, there formerly issued great streams of ice, which coalesced to form one gigantic confluent glacier. Mr. Tiddeman has shown † that the general trend of this ice-sheet in North Lancashire and adjacent parts of Yorkshire and Westmoreland, was towards the south or south-south-east, across deep valleys and over hills of considerable elevation. And he justly infers from this fact that some barrier existed in the Irish Sea which prevented it following the natural slope of the ground towards the south-west. This barrier was the ice deriving from the Lake mountains. But if so, then some other barrier must have ponded back the latter also; for had not such a barrier existed, the glaciers of Westmoreland and Cumberland would have found for themselves a more direct route to the sea than they appear to have done. The cause of this deflection was undoubtedly the presence of the massive ice-sheet that

* While these pages are passing through the press Mr. J. C. Ward's paper on "The Glaciation of the Northern Part of the Lake District" (*Quart. Jour. Geol. Soc.*, 1873, p. 422), has come to hand. He describes the till of that district as a "stiff clay stuck full of smoothed and scratched stones and boulders, and unstratified. It occurs every here and there in small patches among the mountains, in rock-sheltered spots, and may frequently be seen in the valleys, either by itself or underlying a more gravelly deposit." This latter he describes as consisting of "subangular gravel (very rarely containing beds of sand) in a clayey matrix, with large boulders in and upon it. It sometimes passes down into the till, and either forms sloping plateaux running up the valleys (as the till alone sometimes does) or wide spreads of a more or less moundy appearance." This deposit appears to be degraded terminal moraine matter. See infra, p. 361.

† *Quart. Jour. Geol. Soc.*, vol. xxviii. p. 471.

streamed from the south of Scotland, and had sufficient power to deflect the ice creeping out from Ireland. Such being the case it is not surprising to learn that the Isle of Man and Anglesea afford evidence to show that the united glaciers or ice-sheet actually overflowed both these islands.

At low levels the till is overlaid by another stony clay which, however, differs markedly from the true till. According to Mr. De Rance* it is a stiff reddish-brown clay, containing a vast quantity of packed stones and boulders that vary in size from fragments a quarter of an inch or so up to blocks measuring four yards across. At least 70 per cent., he says, are striated, but many have been more or less water-worn between the periods of scratching and deposition. The deposit shows rude bedding, and contains such shells as *Tellina balthica*, *Cardium edule*, *C. aculeatum*, *Psammobia ferroensis*, *Turritella terebra*, &c., and Mr. De Rance informs me that he has also obtained spiculæ of sponges. He calculates that 94 per cent. of the stones and boulders do not belong to the drainage-area in which any given section of the deposit appears; and he has not met with the clay at a greater height than 60 or 70 ft. above the sea.

This boulder-clay, as it is termed, bears a strong resemblance to the Scottish accumulation of the same name. The percentage of "stranger" stones, however, is considerably greater than one finds in the latter. It was probably deposited, as Mr. De

* *Quart. Jour. Geol. Soc.*, vol. xxvi. p. 641.

Rance has suggested, along the seaward edge of the ice-sheet at a time when this was melting and slowly retreating inland.

Above this boulder-clay come considerable depths of sand and gravel—200 ft. in places. They are well-bedded, and have been traced from near the sea-level up to heights of 1,200 ft. (Macclesfield), and even of 1,390 ft. (Moel Tryfan). As the boulder-clay does not rise more than 70 ft. above the sea, it is evident that the sand and gravel beds must overlap that deposit. Here and there they have yielded marine shells, many of which are still living in British seas, but upon the whole the fossils indicate colder conditions than now obtain in the Irish Sea; conditions, however, which were very far from approaching those of which the Scottish shelly clays are memorials.

Resting upon the marine sands * we find yet another clay. This deposit which is known as the upper boulder-clay, hardly deserves the name. According to Mr. Mackintosh † its prevailing colour is "red, with grey or blue partings." It contains, he says, "rather few stones, and exceedingly few large boulders." In the places where I have myself seen it, it seemed to me to resemble some of the glacial brick-clays of Scotland. The stones were scattered sparsely through the deposit, and never appeared so closely aggregated as in true boulder-clay. A further similarity between it and the Clyde

* Mr. Hull was the first to point out this succession; see *Memoirs of the Literary and Philosophical Society of Manchester*, 1865, p. 449.

† *Geological Magazine*, vol. ix. p. 190.

beds consists in the presence of marine shells of northern and arctic types.

Here then we have clear and decisive evidence of a change of climate. During the deposition of the sands and gravels little or no ice floated about, but at the time the overlying brick-clay (upper boulder-clay) was being accumulated, coast-ice kept forming, and ever and anon breaking off from the shore with stones which it scattered over the sea-bottom as it floated away. To what extent the north of England was submerged at this time would appear to be somewhat uncertain. According to Mr. Mackintosh, the upper clay "seldom rises higher than 400 or 500 ft. above the sea."

Considerable accumulations of fluviatile gravel occur in the valleys, often at great heights above the present rivers. This gravel, when traced up stream, becomes coarser and earthier, and not a few of the stones even show faint traces of striæ. As we follow it still farther into the mountains, it appears to pass into, or at least it cannot be distinguished from, moraine débris. Opposite the mouths of some of the mountain valleys, great deposits of hummocky angular and sub-angular gravel make their appearance, but elongated ridges of gravel and sand, like the more marked kames of Scotland, are either absent or of uncommon occurrence.

Lastly, I would refer to the immense quantities of moraine matter, and the numerous *blocs perchés**

* These erratics have travelled outwards from the mountains in precisely the same direction as that followed by the great glaciers; nevertheless, many geologists consider them to have been carried by ice-rafts and not by glaciers.

which are found in almost every valley in the Lake district and in Wales. This angular earthy débris hangs on all the hill-slopes, and gathers on all the bottoms of the valleys. Towards their upper reaches it often takes the form of low mounds and ridges—the lateral and terminal moraines of small local glaciers. But the terminal moraines of the great glaciers that ground out the rock-basins of such lakes as Llanberis, Coniston, and Ulleswater, and which must, at one time, have cumbered the valleys just below the lakes, have disappeared.

I have shown that in Scotland much of the moraine matter that occupies the mountain valleys cannot be referred to the latest period of local glaciers. The terminal moraines of the last glaciers are restricted to the higher mountain valleys, and even in many cases to the upper reaches of these; but much further down the valleys, and at levels on the hill-slopes which the glaciers of the last cold period never attained, great masses of true moraine matter exist. To what age do these masses belong? Certainly not to the period of the till, nor yet to that of the boulder-clay, but, as I have tried to make clear, to the great recession of the ice-sheet that followed upon the deposition of the boulder-clay—to a time when the ice had shrunk back from the sea into the mountain-valleys where it existed as a series of gigantic local glaciers.

Such are the kind of glaciers which Ramsay describes as having occupied the valleys of Wales*

* *Quart. Jour. Geol. Soc.*, 1851. See also *The Old Glaciers of Switzerland and North Wales.*

before the great submergence, and glaciers of a like extent are required to explain the phenomena of the low-level morainic gravels of the Lake district, which appear to be the direct successors in time of the lower boulder-clay of the districts of Lancashire and North Wales. This, in other words, implies that the ice-sheet crept gradually back from the sea, until, at last, its moraines began to be deposited upon the land. It then broke up into separate glaciers which filled all the mountain valleys, and from which, especially in summer-time, great rivers carried down to the low-grounds vast heaps of gravel and sand. How far these local glaciers receded we cannot tell. For aught that one can say they may have vanished altogether. Neither is there any evidence from this part of England that would indicate even approximately the extent of the land at that time. All that we know for certain, is, that the sea did not follow the steps of the retreating ice to the mountain valleys. When submergence at last ensued, the climate had greatly ameliorated, and the neighbouring sea was tenanted by a fauna closely approximating to that which at present characterizes the northern coasts of Britain. Currents and waves now redistributed the sand and gravel which the glacial rivers had carried down, and as the land continued to sink, the sea entered the mountain valleys and began to sift and remodel the coarse moraine drift which the glaciers had left behind them on their retreat. But, as the submergence approached its climax, another change of climate ensued. Snow again gathered on the hills and deepened in the valleys. Slowly the land rose out of the water. Glaciers appeared in some of the larger valleys, and streams and rivers, laden with fine silt, swept downwards to the sea. Thus, in process of time, a deposit of mud and clay accumulated upon the sea-bottom, and covered up those sands and gravels which marked the prevalence of milder conditions in the preceding period. Coast-ice, too, floated about, and ever and anon showered angular débris and boulders over the bed of the sea where the fine silt was accumulating. Such, in few words, appears to be the succession of changes revealed by the glacial and inter-glacial deposits of Wales and the north-west of England.

When we pass to the other side of the country, we find, in like manner, several stony clays and intervening beds of sand and gravel. From this circumstance some geologists have surmised that the East Anglian boulder-clays and associated beds are the precise equivalents of those of Lancashire and North Wales. There are certain facts, however, which, as Mr. S. V. Wood, jun., has pointed out, render this extremely improbable, if, indeed, they do not disprove it altogether.

But, before quoting what Mr. Wood has to say upon this interesting subject, I may first give a short outline-sketch of the deposits met with in the eastern part of England, arranging them in their order as ascertained by Mr. Wood and his associates the Rev. J. L. Rome and Mr. F. W. Harmer.

Along the cliffs of the Norfolk coast at and near

Cromer, magnificent sections of drift deposits are exposed. These have been long famous among geologists, and must ever be looked upon with interest, inasmuch as we obtain from them the only reliable evidence as to the kinds of plants and animals that clothed and peopled Britain in the old pre-glacial ages.

The cliffs show the following succession, beginning at the top:—

7. Sand and rolled gravel.
6. Contorted drift, with masses of marl and chalk.
5. Boulder-clay with erratics.
4. Laminated blue clay.
3. Fluvio-marine sand and clay.
2. Forest-bed.
1. Sand, gravel and loam (Norwich Crag).
Chalk.

The beds marked 1 contain a mingling of land and fresh-water shells along with many marine species, and bones of extinct mammalia. They thus indicate fluvio-marine conditions, having been laid down at or near the place where a river entered the sea. It is remarkable that, while all the land and fresh-water shells belong to still living species, 18 out of 124 of the marine mollusca are extinct. Another point of interest consists in the fact that, among the fossils a number of northern forms occur. This Norwich Crag, as it is termed, thus points to the gradual approach of cold conditions, but it is in deposits of still earlier age that we are to look for the first hint of what was afterwards to transpire in Britain.

On top of the Norwich Crag comes what is called the forest-bed. This consists of a mass of vegetable matter, showing stumps of trees standing erect, with their roots penetrating an ancient soil. Associated with this old forest, for such undoubtedly it is, occur the remains of many extinct species of mammalia, commingled with others that are still living, and some of which are even now indigenous to Britain. Of these animals may be mentioned the hippopotamus, three species of elephant including the mammoth, two species of rhinoceros, bear, horse, Irish deer and several other species of deer, beaver, wolf, &c.

The forest-bed is overlaid with alternations of fresh-water and marine deposits, showing that the conditions which obtained before the trees began to grow had returned. The accumulations next in order point to a still further depression of the land; but to what extent we cannot say. We have every reason to believe, however, that the submergence was accompanied by increasing cold climatal conditions. Glaciers must at that time have occupied the high-grounds of Britain, and even covered large areas of the low country, pushing out to sea and covering the bottom with stones and mud. At the same time, large bergs and ice-rafts, launched from the shores of Scandinavia, sailed across the North Sea, and dropped their boulders over what is now dry land. The contorted drift points to the continuance of similar conditions. The presence of sea-shells assures us that this deposit must be of marine origin. Moreover, it contains occasional large masses of marl which have evidently been dropped into their present position by floating ice. One of these masses, Mr. Wood tells us, is no less than 800 ft. and upwards in length, and 60 in height, and it is obvious, as he observes, that, "to float a berg adequate to such a marl-freight as this, many hundred feet of water are necessary."* The contortions may well owe their origin to the running aground of such huge islands of ice.

The beds now referred to, form what Mr. Wood terms the Lower Glacial series. It will be observed that the contorted drift is covered with sand and gravel. These are part of a great series of sand and gravel beds of marine origin, which Mr. Wood designates the Middle Glacial. They are clearly of younger age than the deposits just described, as is proved by their position, and also, as it would seem, by their fossil contents. The shells which they have yielded exhibit a preponderating southern aspect, and a greatly milder condition of climate than obtained during the deposition of the underlying contorted drift.

According to the same geologist, the accumulation of the Middle Glacial sand and gravel was succeeded in time by the deposition of a great thickness of stony or boulder-clays, which he distinguishes as follows (beginning with the oldest):—

The great chalky boulder-clay,
The purple boulder-clay with chalk, which passes up into
The purple boulder-clay of Yorkshire, *without* chalk.

These boulder-clays constitute Mr. Wood's Upper Glacial series. He believes them to have been formed, for the most part, underneath a great sheet

* *Geological Magazine*, vol. viii. p. 409.

of ice, and gradually extruded upon the floor of the sea. That they were thus accumulated is sufficiently evident, indeed, from the fact that they have yielded sea-shells, which sometimes occur in a distinct layer or bed, as at Bridlington, underlaid and overlaid by boulder-clay. Hence the deposits would seem to be analogous in every respect to the lower boulder-clay of the north-west side of the country, and the boulder-clay of Scotland. The shells met with in this East Anglian Upper Glacial series, are of a decidedly northern type, and approximate in character to the shells obtained from the boulder-clays of Scotland and Lancashire.*

Now, if we are to place any reliance at all upon the evidence derived from fossils, we must admit that Mr. Wood's Middle and Lower Glacial series cannot possibly represent the middle sand and gravel and lower boulder-clay of Lancashire, nor can they be the equivalents of any portion of the shell-bearing clays of Scotland. The fossils belonging to the deposits in

* I am aware that some English geologists consider these Bridlington beds to be older than any glacial deposits in Scotland; the reason being that while five out of seventy species of shells obtained at Bridlington are not known as living, "every one of the long list of shells from the Scotch *till* belongs to living species, and even to varieties which are still extant as inhabitants of the arctic regions, while about 60 per cent. are also British forms" (Lyell's *Antiquity of Man*, p. 287). Here the mistake is made of confounding the till with the fossiliferous clays which overlie it. I would reiterate that nearly all the Scotch shell-bearing beds belong to the very close of the glacial epoch; only in one or two places have shells ever been obtained with certainty from a bed in the true till of Scotland. They occur here and there *in boulder-clay*, and *underneath boulder-clay* in maritime districts, but this clay, as I have shown, is more recent than the till—in fact, rests upon its eroded surface. Moreover, the five species of apparently extinct shells that occur at Bridlington may yet be dredged up, as so many shells hitherto supposed to be extinct have already been within these few years.

the north-west of England and in Scotland have a decidedly northern aspect—that is to say, the more characteristic and typical shells belong to species which are even now living in the adjacent seas or in more northern latitudes. On the other hand, the shells of the East Anglian Middle and Lower Glacial series have, as Mr. Wood has shown, a preponderating southern aspect—the great majority occur in the Norwich Crag, or in other words, were denizens of our seas during pre-glacial times, and of these a number are either unknown now as living species, or as living nearer than the Pacific. The results obtained during the late deep-sea dredgings, should, of course, make us cautious as to deciding whether a species is extinct or not. It may eventually turn out that many, or even all, the shells obtained from the East Anglian deposits are really living in some hitherto unapproached region of the seas. But even bearing this contingency in mind, it appears to me that we cannot but admit with Mr. Wood, that his Lower and Middle Glacial series are older than any of the shell-bearing and glacial deposits in other parts of Britain.

To what part of the glacial series of the north, then, do they belong? I think they may, with all safety, be correlated with the true till and those fresh-water beds which are intercalated with that deposit. It is quite clear that no true till occurs in East Anglia, or, if it does, it has not yet been discovered. In Scotland, as I have shown, the till which underlies the oldest of the shell-bearing boulder-clays and brick clays, contains fluviatile and lacustrine beds, with vegetable remains and bones of the extinct mammalia. These, then, I take to be the equivalents of the Lower and Middle Glacial series of East Anglia. Nor is it difficult to understand why the English should bulk more than the Scottish deposits. The former are farther removed from the great centres of glaciation, and have escaped much of that grinding action of the old ice-sheets, under the influence of which the Scottish beds have been so extensively denuded.

The beds which, according to Mr. Wood,* succeed in time his Upper Glacial series, consist, first, of marine gravels, mostly occurring on plateaux and on hills. These indicate a time when the land was gradually emerging after the deposition of the Upper Glacial series. Then follow beds of gravel containing fresh-water shells, which are "associated with a littoral marine fauna"—facts which show clearly enough that the elevation of the land continued. A still more recent deposit of clay (Hessle clay), containing a few scattered stones and boulders, indicates from its position that the sea once more gained upon the land, but only to a very limited extent—not exceeding, perhaps, "350 or 400 ft. anywhere in Yorkshire, if indeed, it amounted to so much, and dying off to nothing towards southern Lincolnshire."

I will now add a few words about the Irish deposits.

* See *Geological Magazine*, vol. vii. p. 19, where the reader will also find references to other papers by Mr. Wood on the same subject.

No part of the British Islands exhibits better than the wilder parts of Ireland the extreme effects of glaciation. In the rugged western districts of Galway and Mayo especially, rounded and well-rubbed rocks and heaps of glacial deposits occur everywhere. The striæ upon the rocks, and the direction in which the till has travelled, mark out clearly the path taken by the great sheet of ice which wrapped up Ireland even as it enveloped Scotland.

The oldest glacial deposit recognised by the Irish geologists is a tough stony clay similar in all respects to the Scottish till. This is the chief drift of the central plain of Ireland. It usually lies upon a smoothed and striated surface of rock, and the stones which it contains are more or less blunted and well glaciated. Occasionally it contains nests or lenticular beds of sand and gravel, and sometimes of fine laminated clay. Not infrequently the deposit is arranged in a series of broad parallel ridges and banks, the trend of which has been ascertained to coincide precisely with the direction taken by the old ice-flows.*

Moraine débris occurs in great abundance in the hilly tracts, covering the slopes and bottoms of the valleys, in the lower reaches of which it appears frequently to graduate into coarse gravels.

In the north-eastern districts of Ireland it is in-

* See this beautifully shown upon the map accompanying a paper *On the General Glaciation of Iar-Connaught and its neighbourhood*, &c., by Messrs. Kinahan and M. H. Close. For detailed information on the Irish glacial deposits, the reader must consult the publications of the Geological Survey, and papers by Messrs. Kinahan and Close in the *Dublin Quarterly Journal of Science*, in which he will find references to other authorities on the subject.

teresting to find a series of gravel and sand beds resting upon and covered by stony clay.* These intercalated beds contain a number of sea-shells belonging, for the most part, to species now living, but indicating somewhat colder conditions than obtain at present in the neighbouring seas. Professor Hull believes these beds to be the equivalents of the middle sand and gravel series of the north-west of England. He also correlates with the same beds certain gravels at Wexford, described by Professor Harkness as underlying a mass of boulder-clay. But an examination of the fossils by Mr. A. Bell,† seems to show that this correlation cannot be maintained, but that the Wexford beds belong most likely to some part of the older glacial series of East Anglia, as defined by Mr. Wood and his associates.

The great elongated ridges of gravel called eskers, and the widespread deposits of similar materials which are met with so abundantly, especially in the central parts of Ireland, have long been famous. They are remarkable for being frequently dotted over with large erratic blocks.

On the sea-coast and in the interior of the country, sand-hills are very frequently found heaped up at or near the mouths of valleys. In the neighbourhood of the sea they form undulating dunes, which are continually being influenced by the winds. Mr. Kinahan was, I believe, the first to point out their connection with the valleys. He tells us that they not only occur

* Professor Harkness, *Geological Magazine*, vol. vi. p. 542; Professor Hull, *op. cit.* vol. viii. p. 294.

† *Geological Magazine*, vol. x. p. 447.

at or near the mouths of the valleys, but the greater and more extensive the valley, the greater, he says, is the accumulation of sand. He considers that the sand of which these hummocks consist was originally brought down by rivers at a time when glaciers occupied the valleys *—an inference which is largely supported by the appearance of similar phenomena in Scotland. The same author has described the occurrence under the till in Ireland of what he calls "pre-glacial drift," which occasionally contains the remains of trees.†

Gravel beds with marine shells have been traced in Ireland up to a height of 1,235 ft. on Montpelier Hill.‡

The general succession of events during the glacial epoch would seem therefore to be closely analogous to those experienced in the north-west of England—viz. :—

1st. A period of intense arctic cold, during which the whole island was wrapped in ice that coalesced with the mers de glace of England and Scotland. Underneath this ice the land was highly glaciated, and the till was formed. The Wexford beds may indicate a genial inter-glacial period, like that of the middle glacial series of East Anglia.

2nd. The shrinking back of the ice-sheet, and the stranding of *blocs perchés* and deposition of coarse moraine-matter in the mountain-valleys. At this time great bodies of water escaping from the melting gla-

* *Geological Magazine*, vol. viii. p. 155.

† *Dublin Quarterly Journal of Science*, vol. vi. p. 249.

‡ *Explanation of Sheets* 102 *and* 112 (*Geological Survey Maps*), p. 67.

ciers threw down large accumulations of sand at the mouths of the mountain-valleys.

3rd. Submergence of the land; deposition of the middle sands and gravels of the north-east; heaping up of the eskers, and formation of the high-level marine terraces; climate temperate.

4th. Land rises, accompanied by arctic conditions of climate; local glaciers, and much floating ice; formation of so-called upper boulder-clay, and scattering of erratics over the low-grounds.

The post-glacial alluvial deposits of Ireland and the north of England * have yielded similar results with those of Scotland. All that need be remarked about them, therefore, is simply this, that they indicate a gradual amelioration of climate from cold-temperate to the present temperate conditions. The land animals obtained in or below peat-mosses, &c., comprise the brown bear, the grisly bear, the Irish elk, the bison, bovidæ of extinct species, horse, wolf, fox, beaver, hare, &c. And, just as in Scotland, the polished stone implements used by primeval tribes are everywhere scattered about both in England and Ireland—the latter country being especially rich in these interesting memorials of our lowly-civilised predecessors.

* I discuss the age of the ancient river gravels with human implements and mammalian remains met with in south of England in chapters xxix. *et sqq.* The deposits referred to above are those which can be clearly shown to be of post-glacial age, from their super-position upon accumulations belonging to the *close* of the glacial epoch. As the ancient river-gravels in the south do not occupy this position, their age cannot be ascertained in so simple a manner, and, therefore, it is more convenient to treat of them separately.

CHAPTER XXVI.

SUPERFICIAL DEPOSITS OF SCANDINAVIA.

Extensive glaciation of mountainous and northern regions throughout the northern hemisphere.—Glaciation of Norway, Sweden, and Finland.—Lower and upper till of Sweden.—Inter-glacial fresh water deposits.—Till of Norway.—Åsar.—Erratics on åsar.—Clays with marine arctic shells.—Contorted bedding, &c.—Post-glacial deposits with Baltic shells.—Moraines.—Succession of changes.—Southern limits of glaciation in northern Europe.—The great northern drift.—Theories of the origin of the åsar.—Summary.

IF the British Islands, which are now in the enjoyment of a mild-temperate climate, have beyond doubt experienced in ages that are past the utmost severity of arctic and glacial conditions, it would only be reasonable to infer that other regions of the northern hemisphere should give evidence of having likewise been at the same time characterized by a rigorous climate. It is, *à priori*, in the highest degree improbable that a great ice-sheet should have enveloped a large part of our country, while other areas situate in similar, or nearly similar, latitudes escaped. On the contrary, the observer who knew nothing whatever of the geological records of any country save his own would be justified in believing that the evidence gathered in Britain alone is enough to convince one that during the intense cold of the glacial epoch the temperature of the whole northern hemisphere must have been similarly affected. Geological investigations have clearly shown that such was really the case.

In Scandinavia and northern Europe generally the evidence in favour of arctic conditions having formerly prevailed is overwhelming. So likewise in Switzerland the proofs of this are everywhere patent. But not only were the snow-fields and glaciers much larger at one time than now—they even existed in some mountain districts which at present lie far below the limits of perpetual snow. In the Black Forest, the Vosges,* and the volcanic mountains of Auvergne, we find ancient moraines. The Pyrenees, whose puny glaciers are confined to the higher mountain-slopes, and do not now descend into the valleys, appear† at one time to have been covered with great snow-fields, from which large glaciers streamed outwards, and even in some cases debouched upon the plains. Old moraines now covered with vegetation, and roches moutonnées rising behind quiet villages, and occasionally crowned with some church or chapel, are familiar scenes in many of the valleys on the French side of the mountains. Along their southern slopes similar traces of ancient glacial action have been observed, especially in Galicia, by the late Casiano de

* An excellent account of the old glaciers of the Vosges is given in Hogard's *Coup d'Œil sur le Terrain erratiques des Vosges*, 1851; see also an interesting paper on the same subject by M. Ch. Grad, *Bulletin de la Société géologique de France*, Tom. i. (3 série) p. 88. Professor Ramsay was the first to detect traces of glacial action in the Black Forest.

† "Glaciers actuel et Période glaciaire" (Martins), *Revue de deux Mondes*, 1867.

Prado. Farther south, in the Sierra Guadarrama,* to the north of Madrid, the same geologist has detected evident marks of old glaciers, and described great deposits of loam, sand, and gravel which have been laid down by rivers escaping from masses of melting ice. Nay, even as far south as latitude 37° the former existence of glaciers in the Sierra Nevada has been proved by the researches of MM. Schimper and Collomb; and so on with most, or perhaps all, the hillier regions in Europe, great snow-fields existed where now there are none, and the present ones are merely the insignificant successors of mighty ice-sheets which have long since melted away.

As we pass out of Europe we are met with the evidence of similar changes. The valleys of the Caucasus once brimmed with ice;† now the glaciers are comparatively small. Traces of glacial action have been noticed among the Atlas mountains, and the cedars of Lebanon, as Dr. Hooker‡ tells us, grow upon old moraines. The great glaciers of the Himalaya have in past times attained gigantic proportions; and in parts of Asia where there are now no glaciers at all, one may yet readily trace the marks of their former action. Thus in North China huge boulders are found scattered over the valleys, often at considerable distances from the mountains;§ nor are

* *Descripcion fisica i geológica de la Provincia de Madrid,* por Don Casiano de Prado, p. 164.

† *Travels in the Central Caucasus and Bashan,* by Douglas W. Freshfield.

‡ *Natural History Review,* January 1862.

§ See Williamson's *Journeys in North China.* Describing the country between Si-gnan and Toong-kwan (Shensi), the author says:—"For many miles the country was like one continued splendid park, with knolls and rock-basins wanting, especially in that rough undulating country which is passed through on the route from the Great Wall to the Siberian frontier.*

In North America the evidence in favour of intense glacial conditions having formerly prevailed is overwhelming. From the shores of British Columbia† to the borders of the Atlantic, and from the coasts of the Arctic Ocean down to the latitude of New York, ice-marks are everywhere; while in mountain regions even farther south, as in the Sierra Nevada‡ of California, the morainic débris of great glaciers lies scattered over the lower reaches of the valleys.

Having assured ourselves that a glacial or arctic climate did within comparatively recent times characterize a very large area of the northern hemisphere, it becomes a matter of no little interest and importance to inquire whether we can detect in foreign glacial deposits any proofs of that remarkable succession of changes

lawns and winding paths, leading round some huge fantastic boulder, which had descended from the mountains which lay contiguous on the south," vol. i. p. 387. Again: "We found the country (north-east corner of Shang-Tung) strewn with huge boulders of various sizes, looking like so many cattle at rest with elephants around them," *op. cit.* p. 189. See also pp. 429, 430, where the author describes conical hills and ridges crowned with huge boulders. These were probably roches moutonnées, or moraines.

* "In all these instances the depressions are entirely in the solid rock, and vary in size from a few yards to several thousand feet across. They have the appearance of having been produced by erosion, and not by sinking." Richard Pumpelly, "Geological researches in China, Mongolia, and Japan," *Smithsonian Contributions*, 1866.

† *Quarterly Journal of the Geological Society* (Bauermann), 1860, p. 202; see also *American Journal of Science*, vol. c. (1870), p. 318, where Dr. R. Brown shows that British Columbia, Vancouver's Island, Washington Territory, and the Queen Charlotte Islands, all exhibit fine rock-striations and scattered erratic blocks. Some American geologists had previously held the opinion that the glacial formation disappeared west of the Rocky Mountains.

‡ For a lively account of that region, see Clarence King's *Mountaineering in the Nevada*.

which, as we have seen, may be traced in the drifts of our own country. With this view, I propose to give a short sketch of the glacial phenomena of the better-known regions—viz., Scandinavia, Switzerland, and North America.

I have already had occasion to mention the great ice-sheet that enveloped Scandinavia, and to point out that it in all probability coalesced with that of Scotland upon the floor of the German Ocean. No grander display of ice-action could one wish to see than that which the fiords and fiord-valleys of Norway present. The smoothed and mammillated mountain-slopes, the rounded islets that peer above the level of the sea like the backs of great whales, the glistening and highly-polished faces of rock that sweep right down into deep water, the great perched blocks, ranged like sentinels on jutting points and ledges, the huge mounds of morainic débris at the heads of the valleys, and the wild disorder of crags and boulders scattered over the former paths of the glaciers, combine to make a picture which no after amount of sight-seeing is likely to cause a geologist to forget. The whole country has been moulded and rubbed and polished by one immense sheet of ice, which could hardly have been less than 6,000 or 7,000 ft. thick.

The Gulf of Bothnia appears to have brimmed with ice, which pressed up against and even in some places overflowed the lofty Norwegian frontier, through the valleys of which it found its way into the North Sea.*

* See Hörbye's *Observations sur les Phénomènes d'Erosion en Norvège*, where the striæ are indicated as crossing the watershed between the two countries.

Mr. Törnebohm, of the Geological Survey of Sweden, informs me that the glacier-carried erratics of Jemtland clearly show that the ice has passed from east to west—that is, right against the slope of the land; and, according to Keilhau, similar blocks which could only have come from Sweden are now found in Trondhjemsfiord. The most remarkable circumstance in connection with some of these blocks consists in the fact that they occur at a considerably greater height than the rock from which they have been derived. Thus at Åreskutan, Törnebohm found blocks at a height of 4,500 ft. which could not possibly have come from any place higher than 1,800 ft.

How far south the Scandinavian ice-sheet extended we cannot tell. We know that it not only filled the Gulf of Bothnia, but occupied the whole area of the Baltic Sea, overflowing the Åland Isles, Gottland, Öland, Bornholm, and Denmark, and passing south-east over Finland into Russia, across Lake Onega, Lake Ladoga, and the Gulf of Finland.* [See map.] But its farthest limits have not been determined, and that for a very good reason, as we shall see presently. The direction of the glaciation in the extreme north of Scandinavia, the peninsula of Kola, and north-eastern Finland, demonstrates that the great mer de glace radiated outwards from the high-grounds of Norway and Sweden, flowing north and north-east into the Arctic Ocean and east into the White Sea, and thus

* Professor Nordenskjöld, *Beitrag zur Kentniss der Schrammen in Finland*, Helsingfors, 1863; for direction of Norwegian glaciation, see Hörbye's work, *op. cit.*; for that of Sweden, consult the maps of the Geological Survey of that country.

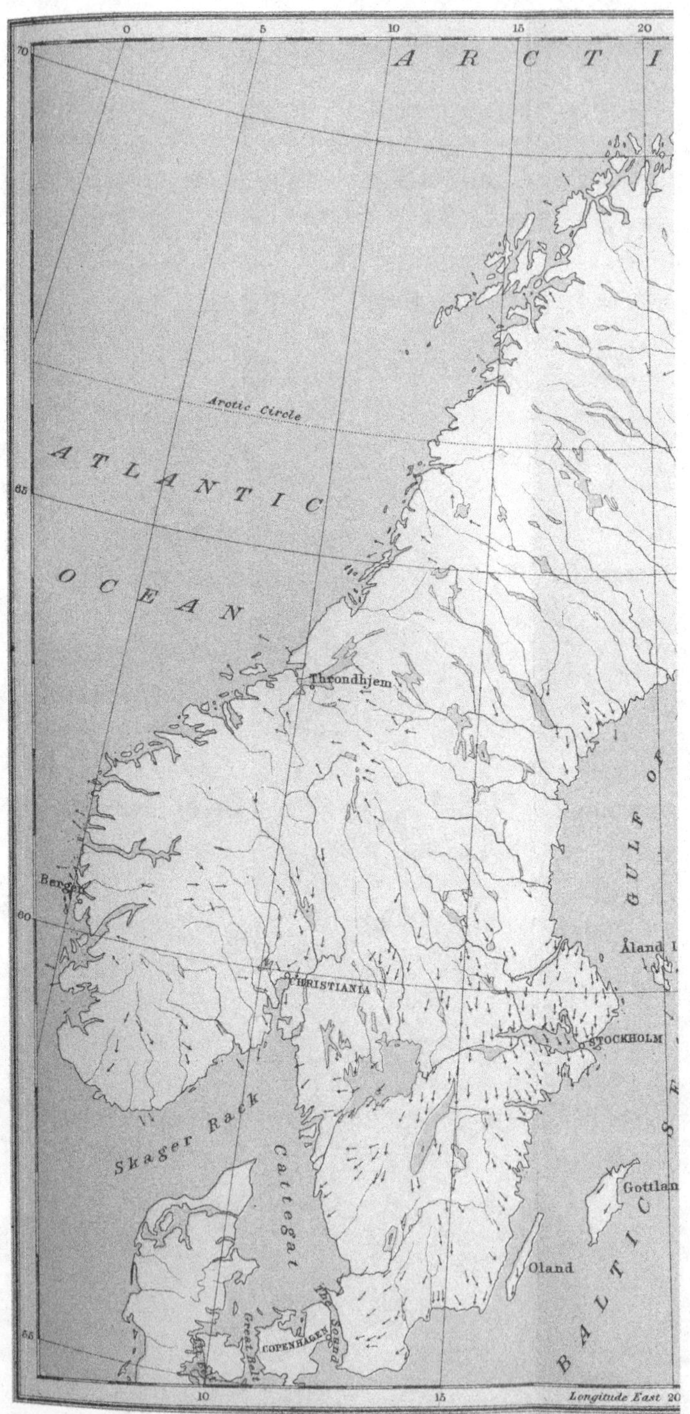

0
5
10
15
20
70
A R C T I
Arctic Circle
ATLANTIC
65
OCEAN
Throndhjem
GULF OF
Bergen
60
Åland I
CHRISTIANIA
STOCKHOLM
Skager Rack
Cattegat
Gottlan
Oland
55
Great Belt
COPENHAGEN
The Sound
BALTIC
10
15
Longitude East 20

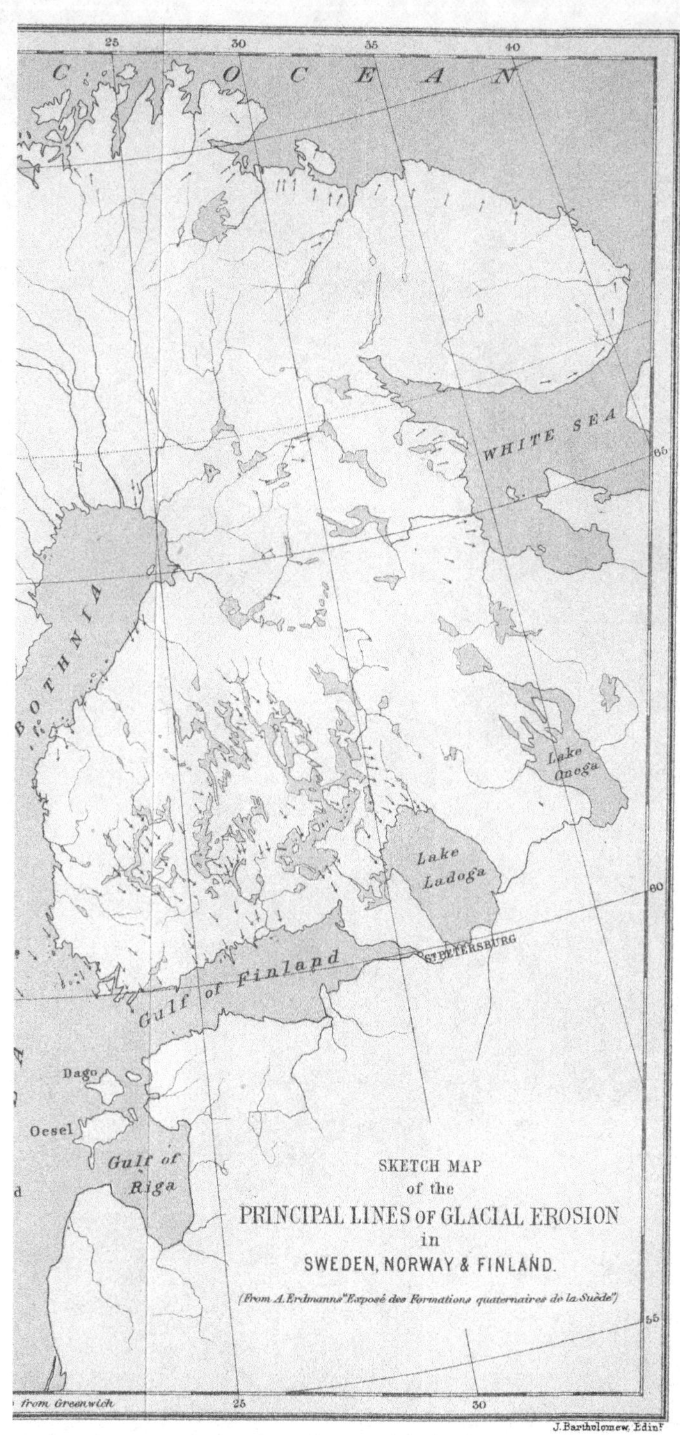
C O C E A N
WHITE SEA
BOTHNIA
Lake Onega
Lake Ladoga
ST PETERSBURG
Gulf of Finland
Dago
Oesel
Gulf of Riga
from Greenwich
SKETCH MAP
of the
PRINCIPAL LINES OF GLACIAL EROSION
in
SWEDEN, NORWAY & FINLAND.
(From A. Erdmann's "Exposé des Formations quaternaires de la Suède")
J. Bartholomew, Edinr.

clearly proving that northern Europe was not overflowed by a vast ice-cap creeping outwards from the North Pole, as some geologists have supposed.

The oldest of all the glacial deposits of Sweden is till, which, in a more or less continuous sheet, covers all the low-grounds. It usually lies upon a polished and striated surface of rock, but occasionally thick beds of sand come between it and the underlying rock. It "consists of two distinct layers, the lower of

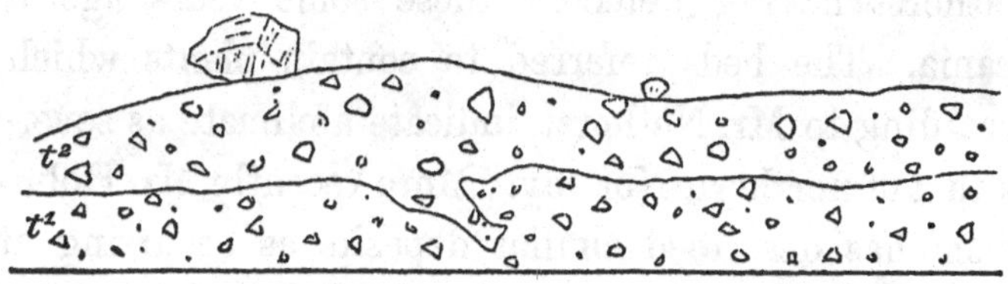

Fig. 50.—Upper and lower till in railway cutting in Wermland. (A. E. Törnebohm.)

which is generally darker in colour, and contains fewer big stones than the upper. Both, however, have evidently been formed in the same way, and may be considered true *moraines de fond.* There is usually a sharp line of demarcation between them, and in some places the lower till has been partly broken up and denuded before the upper till was deposited." * The Swedish geologists have no doubt whatever that these masses and sheets of till were formed and accumulated by land-ice, exactly in the same way as our own till. Moreover, the fact that their lower till exhibits such marked evidence of denudation underneath the

* Mr. Törnebohm, in a letter to the author. See *On Changes of Climate during the Glacial Epoch*, p. 24. He says: "The till is often so hard and packed, that it is easier to break the stones that are in it than to dislodge them."

upper and overlying mass, "seems to point out," says Törnebohm, "that during the glacial epoch there was a great interval of comparatively mild climate, during which the ice retreated to the mountains; the land, however, was not at this time submerged. When the ice-sheet once more overspread the country it would obliterate any fresh-water deposits that might have been laid down in the interval." Traces of glacial fresh-water beds, however, are not wanting—Mr. Nathorst having detected these some years ago in Scania. The beds referred to contain plants which, according to Mr. Nathorst, indicate a climate as severe as that of northern Norway. More recently Mr. Holmstrom has described similar deposits as occurring at Klågerup, in Scania, in the following descending series:—

Yellow till.
Brown sand and yellow glacial fresh-water clay.
Grey marly sand.
Blue glacial fresh-water clay.
Blue till.

The fresh-water clays have yielded—1st. (Shells) *Pisidium pulchellum* (Jen.); *P. obtusa* (Pfeiff); *P. Henslowianum* (Jen.); *Anadonta anatina*; *Limnœa lagotis* (Schr.); 2nd. (Plants) *Dryas octopetala*, *Betula nana*. From these facts Mr. Helmström considers that he has evidence of an inter-glacial period. He says the lower blue till is very thick, and extends almost continuously over the whole country. It points, as he thinks, to the former existence of an extensive mer de glace which covered the whole land, destroying all life. The shells and plants found in the fresh-water clays must therefore, according to him, have come in from the south when the ice retired. Then afterwards, at a later period, some local glaciers crept down from the great mass of ice that still lingered in the north, covering with morainic matter the fresh-water clays which, during the interval, had accumulated in pools upon the surface of the older till.*

In Norway a deposit of till also occurs; it does not, however, cover so wide an area as in Sweden, but appears often of considerable thickness in sheltered places, as on the lee-side of crags and rocky knolls.†

Resting sometimes on till, but oftener, perhaps, upon solid rock, appear certain great natural embankments, or long winding ridges, which are known in Sweden as åsar.‡ They generally rise abruptly to a height that may vary from 50 to 100 ft. above the average surface of the ground. Sometimes, however, they reach as much as 180 ft., while now and again they sink to 30 or 20 ft., or even disappear altogether below newer deposits. Their sides have an inclination of from 15° to 20°, but occasionally as much as 25° or even 30°, and the two declivities very rarely slope at the same angle.§

* Ofversigt af bildningar från och efter istiden vid Klågerup i Malmöhus län; af L. Holmström; *Ofv. af K. Vet.-Akad. Förh.* 1873, No. 1. I am indebted to Mr. Törnebohm for calling my attention to these interesting observations; the same obliging correspondent informs me that Mr. Nathorst, who was the first to detect traces of glacial fresh-water clays in Sweden, will shortly publish an account of further observations made by him. For an interesting account of stratified beds, underlying and intercalated with Swedish till, see a paper by E. Erdmann, *Geologiska Föreningens i Stockholm Förhandlingar*, Bd. i. p. 210.

† *Iagttagelser over den postpliocene eller glaciale Formation*, af Sars og Kjerulf.

‡ Ås singular, åsar plural; similar ridges in Norway are termed *Raer.*

§ *Exposé des Formations quaternaires de la Suède*, par A. Erdmann, pp. 40, 61.

Often beginning in the interior of the country, the åsar follow the valleys down to the low coast-land, across which they pass as well-defined ridges out to sea, after a course of not infrequently more than a hundred English miles.* In the mode of their distribution they show a striking resemblance to river

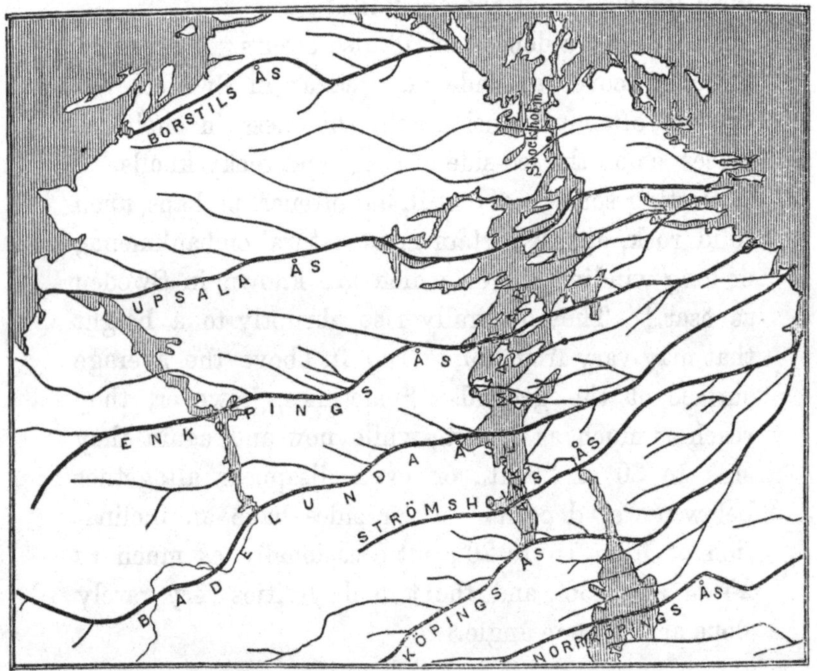

Fig. 51.—Map of Åsar in basin of Mälar Lake.

courses, as will be seen from the accompanying sketch-map, on which the black lines represent the åsar.†

* Erdmann mentions as examples, Upsala ås, which is about 200 kilometres in length; Köping ås, very nearly 240 kilometres; Enköping ås, 300 to 340 kilometres; Badelunda ås, about 300 kilometres. *Op. cit.* p. 44.

† The map is from a paper by A. E. Törnebohm in *Geologiska Föreningens i Stockholm Förh*, Band I. No. 4, and is meant to illustrate an ingenious theory advanced by that geologist in explanation of the åsar. See *postea*, p. 392.

At greater heights than 300 ft. above the sea these remarkable ridges are, as a general rule, confined to the valleys, but at lower levels they seem to be tolerably independent of the present configuration of the ground. They are met with at all levels up to and beyond 1,000 ft.* The materials of which they are composed may consist either of coarse shingle, or of pebbly gravel, or of sand, or it may be made up of all three. In some parts of a ridge shingle and gravel predominate, in others sand is the principal ingredient. In one place the stratification will be distinct, in other places obscure, and not infrequently false-bedding appears.

In many respects then the åsar seem analogous to the British eskers and kames. These deposits have yet another character in common—they are unfossiliferous. For, as we shall see immediately, the shells which have been described as occurring in the åsar, do not properly belong to them.

Erratic blocks are frequently found perched upon the top of an ås, or plentifully sprinkled along its sides, and sometimes also they occur in the interior, especially towards the top, or at the base.

A fine glacial clay containing arctic shells† often covers the slopes of the åsar. The same clay also forms deposits of considerable thickness, especially on the low-grounds upon the borders of the Baltic and

* The highest mentioned by Erdmann occurs at Herjeådal, between 1,300 and 1,400 ft. above the sea, but Mr. Törnebohm informs me that in the mountain-valleys of the north they go up to elevations of over 2,000 ft.

† The shells belong to such genera as *Yoldia*, *Saxicava*, *Leda*, *Cyprina*, *Arca*, *Natica*, *Astarte*, &c.

the North Sea. Traced inland it gradually gives place to sand and gravel. Here and there immense accumulations of shells of northern and arctic species* flank the hill-slopes, and have been dug for making lime and other purposes, just as we in this country quarry chalk or limestone. At the hill of Capellbacken, near Uddevalla (207 ft. above the sea), they were energetically digging the shells in Linnæus's time, and the same heap is being quarried now.

Frequently the shelly clays exhibit disturbed and contorted bedding, and they often contain large erratics. The greatest height at which shell-bearing beds have been obtained is some 600 ft. above the sea, but Erdmann seems to have no doubt that marine strata go up very much higher—to 700, 800, and even 1,000 ft. and more.

The beds next in succession consist of marl, clay, sand and gravel. They contain numerous shells of Baltic species, and large erratics often rest upon them.

In the interior of the country well-marked moraines are not very common. Some fine examples, however, are occasionally met with. Thus the southern part of Lake Wener is crossed by three parallel terminal moraines.† Moraines also occur in the mountain valleys of Sweden, and are well marked in those of Norway.‡

* The highest of these great shell-banks is that at Gustafsfors, near Westra Silen Lake, 500 ft. above the sea. (The Swedish foot is $\frac{1}{8}$th less than the English.)

† Mr. Törnebohm is my authority.

‡ See the paper by Sars and Kjerulf already cited.

The annexed diagram shows the mutual relations of the Swedish deposits, and will help to make the preceding notes more intelligible.

The succession of changes thus indicated would appear to be as follows:—

1. Period of intense arctic cold, with a vast mer de glace covering the country (lower till).
2. Intervening period of milder conditions, when the ice-sheet drew back to the mountains.
3. Return of the ice-sheet (upper till).
4. Retreat of the ice-sheet; stranding of blocs perchés; great rivers carrying down immense quantities of sand and gravel.
5. Submergence of land, and shaping-out of the åsar.
6. Deposition of fine clay; climate arctic; abundance of floating ice.
7. Gradual re-elevation of the land; climate changing from arctic to temperate.

Fig. 52.—Sweuish deposits: *t* till; *e* ås; *c* shelly clay, &c.; *p* post-glacial deposits; *b* erratics; *r*, underlying rocks.

How far south the Scandinavian ice-sheet extended

has not been determined. When we pass beyond the limits of Scandinavia and Finland into the low-grounds of Northern Russia and the great plains of Germany, we encounter vast accumulations of sand and gravel dotted over with blocks and boulders, all of which have come from the north. Erratics from Finland and Lapland are widely scattered over Poland and Russia, and fragments of rock broken from the mountains of Norway and Sweden strew the surface of Denmark, Holland, and Northern Germany. Underneath these deposits the solid rocks disappear, and we have, therefore, no means of ascertaining, with any approach to certainty, the extreme limits reached by the continental ice-sheet. It is by no means improbable, however, that it may have stretched nearly as far south as the great "Northern Drift" extends.

In its general character this Northern Drift very much resembles the remodelled drifts of the British Islands and Scandinavia—the eskers, kames, and åsar. It often forms undulating hills and mounds, which bear on their tops and slopes large boulders and angular erratics. Here and there in the neighbourhood of the Baltic it is said to have yielded shells, but certainly the great bulk of the deposit is quite unfossiliferous, and it is possible that these shells may belong to some more recent overlying formation; just as the shells which were formerly believed to occur in the åsar, are now known to be of later date, and to lie in beds that rest upon these peculiar ridges.

The general unfossiliferous character of the Northern Drift would be inexplicable if we could believe that the materials of which that drift is composed were deposited upon the bottom of what was once open sea. But for this supposition there are really no good grounds. We have seen that the materials of which the kames consist were laid down by the swollen rivers that issued from the retreating ice-sheet. And although subsequent submergence brought these materials under the influence of the sea, yet this only resulted in denuding and remodelling them here and there, while in many cases their original stratification remains undisturbed. In like manner the åsar of Sweden are merely the denuded and partially rearranged portions of old torrential gravel and sand, and morainic débris; and that the great bulk of the Northen Drift of Europe has had a similar origin may safely be inferred.

The remodelled aspect of this drift, however, and the presence of the shells and perched erratics, proves that after the ice-sheet had retreated and left the ground covered with sand, gravel, and moraine matter, a process of submergence ensued. Diluvial sand and gravel were now everywhere subjected to the action of the sea. Large rafts of ice set sail from the unsubmerged portions of Scandinavia, and drópped their burdens of stones and boulders over the sea-bottom. Step by step the land sank down, but to what extent the submergence reached is still a much-disputed question. Some Scandinavian geologists who maintain that the åsar are banks heaped up by the sea, insist that the land sank to a depth sufficient to allow of the formation of the highest of these ridges, which would give a depression of not less than 2,000 ft. Others, again, think the absence of shells from the åsar is proof that these ridges cannot be of marine origin, and since sea-shells are nowhere found in Scandinavia at a greater height than 600 ft. above the sea, they would limit the depression to that amount.

The whole question thus turns upon the origin of the åsar. How were they formed? It is difficult to see how otherwise than by the action of the sea, and yet at the same time it is hard to understand how sea-currents of any kind could form a series of long ramifying banks like those shown upon the sketch-map. Quite recently Mr. A. E. Törnebohm advanced a theory on the subject which I may here briefly describe.*

His belief is that the åsar are ancient river-courses, and he makes pointed reference to their peculiar river-like ramifications. In those valleys which contain the åsar detached patches of sand are sometimes found, perching high on the side slopes. These patches, according to him, are the wreck of a great deposit of sand, which at one time filled the valleys from side to side. While the valleys were still filled with this thick bed of sand, rivers began to flow just as they now do, and cut their way down in the sand. The running water carried along with it coarse sand and gravel, and deposited these on the beds of the rivers, which thus became paved with coarser materials. By-and-by this state of things changed—denudation set

* See foot-note, *ante*, p. 386.

to work upon the whole deposit, and removed the fine sand, but had not power to carry away the coarse gravel which had filled up the old river-courses. This gravel, therefore, remained behind, and very frequently has protected a considerable thickness of underlying sand. The annexed woodcuts, which are taken from Törnebohm's paper, will further illustrate his meaning.

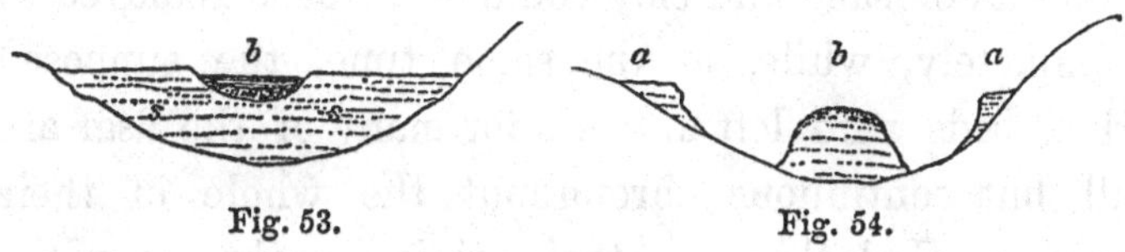

Fig. 53. Fig. 54.

Fig. 53 shows the section of a valley partly filled with sand, *s*, in which is cut the river-bed, paved with coarse sand and gravel, *b*. Fig. 54 represents the aspect of the valley after denudation has removed the greater portion of the sand, patches of which are seen at *a*, *a*. At the bottom of the valley the river gravel rests upon some depth of sand, forming together an ås, *b*.

The close connection between the åsar of the valleys and those that strike across the low country, clearly shows that in both districts they must have been formed in the same way. As an example, Mr. Törnebohm cites the åsar that occur in the basin of the Mälar Lake. (See the map, p. 386.) To apply his explanation to the åsar of that region, it is necessary to suppose the Mälar basin to have been filled up with sand and mud, through which the rivers, coming from the melting mer de glace, cut their way to the sea.

After the rivers had thus coursed across the broad deposits of sand which are inferred to have covered so large a part of Sweden, a movement of subsidence ensued, the land sank down, and the Mälar basin became converted into a shallow sea. During this depression the fine sand which was unprotected was washed away, and thus the åsar were formed. It is difficult, however, to comprehend how such immense deposits of sand and clay could have been removed so completely, while, at the same time, the supposed river-beds were left intact; for many of the åsar are all but continuous throughout the whole of their course. Had this been their origin, surely one might have expected to find these ridges more frequently interrupted than is the case, and many miles of them often swept away.* Mr. Törnebohm, who is one of those who consider that the land was not submerged more than 600 ft., is of course compelled to assume that the åsar above that level owe their origin to the action of rivers. And there can be no doubt, indeed, that streams flowing across a broad terrace will clear out hollows and leave intermediate banks or ridges of more or less prominence. There is great difficulty, however, in conceiving how stream action could so denude a terrace as to leave nothing but a great central ridge extending for many miles along the course of a valley.

Robert Chambers, who examined some of the åsar

* Most of them, indeed, are more or less interrupted with gaps, but these are not so wide as to make it a difficult matter to identify the different parts of the same ås. Badelunda ås is, one might say, all but continuous for a distance of more than 160 English miles.

many years ago,* at once recognised their strong similarity to our kames and eskers. And although it is quite true that a thing may be very like without actually being the same, still the points of resemblance between the Swedish and the Scottish and Irish ridges are too many to be the result of pure accident. Besides, they occupy the same geological position; upon the whole, therefore, I feel inclined to accept the marine theory first started by M. Ch. Martins,† and subsequently adopted by several Scandinavian geologists.‡ I am very far from denying, however, that we have still much to learn in regard to the mode in which the åsar have been heaped up or carved out. The whole secret has not yet been discovered.

But whatever may have been the precise origin of the åsar, of this there can be no doubt, that those at the lower level received their latest touches from the sea. The shell-beds that rest upon them, and the erratics with which they are so plentifully besprinkled, sufficiently attest the truth of this.

From the fact that erratics are abundantly scattered over the tops and slopes of the åsar in Sweden, and the diluvial sand of northern Europe generally, while they are comparatively seldom seen embedded in these deposits, we may perhaps infer that but little ice was floating about in European seas at the time these ridges and banks were being shaped

* *Tracings in the North of Europe* 1850, p. 238, *et passim; Edinburgh New Philosophical Journal*, 1853.

† *Bulletin de la Société géologique de France*, 1845, 1846.

‡ Erdmann, *op. cit.* p. 42.

out and heaped up. But after this process had been completed, or nearly so, a great dispersion of erratic blocks took place.* In further proof of such changed climatal conditions we get the glacial clays with their arctic shells, contorted bedding, and erratics. Nevertheless there is no instance yet on record of shells that would indicate a temperate climate occurring in the åsar underneath the glacial clays. Our inference rests upon physical evidence alone. For some cause or other the Swedish åsar and the German diluvial sand contain few or no erratics, while numerous blocks dot their surface. It is possible, therefore, that these deposits assumed much of their present form during a comparatively mild condition of things, and that when afterwards a colder climate ensued, coast-ice was formed in greater abundance, and erratics distributed in larger numbers over the submerged land.

The post-glacial clays, sands, and gravels, that overlie the arctic shell-beds, show that the glacial climate at last passed away, the land gradually rose, and the living Baltic fauna succeeded the arctic one. But from the presence of erratics resting upon these post-glacial beds, we gather that large ice-rafts still sailed about in the Baltic. Indeed, even

* At the same time, one cannot help surmising that some of these erratics, at least, may have been left behind by the retreating ice-sheet. In Switzerland we find large angular blocks perched on the tops of stratified sand and gravel, and yet no one imagines that these were dropped from floating ice. At present it is the prevailing belief amongst geologists in this country that the Northern Drift of the continent is a purely marine deposit, and that all the erratics associated with it have been dropped by floating ice. This is probably an extreme belief, and one that I cannot help thinking will yet be greatly modified.

in our own day floating ice continues to scatter stones over the bed of that sea, and sometimes immense rafts are driven ashore, causing great damage. During the winter of 1862–63 a vast pile of ice was cast ashore on the southern coast of the Gulf of Finland, overwhelming many dwellings and whole forests. When this ice had melted away stones and blocks were found piled in great quantities upon the ground. The change from arctic to temperate conditions has thus been less complete in Scandinavia than with us.*

* The gradual change from arctic to temperate conditions is well seen in the shell beds of Norway, as Sars has shown. He has also pointed out the curious fact that certain species of molluscs occur in these shell-beds which are now restricted to seas farther south, ranging from England to the Mediterranean; while others again now living off the Norwegian shores occur also in the Mediterranean, but are not met with in intermediate regions (*Iagttagelser over den postpliocene eller glaciale Formations*, p. 66). Mr. Gwyn Jeffreys, as is well known, noted similar facts at Uddevalla, *British Association Report*, 1863. Similar facts again have been observed in connection with the glacial and post-glacial deposits of Britain and Ireland. We have no doubt still much to learn concerning the natural history of glacial, inter-glacial, and post-glacial seas. But it already appears that there were several great changes in the distribution of the marine fauna of north-western Europe during the glacial epoch. I would ask palæontologists whether the occurrence of southern forms in late glacial times (arctic shelly clays) is not further proof that a mild climate of long continuance supervened after the retreat of the great ice-sheet which deposited the till and boulder-clay, and before the extreme submergence in Wales, Scotland, and Ireland. These southern forms I would take to be relics of the fauna that tenanted our seas during the climax of the last mild inter-glacial period. See the question of this last warm inter-glacial period discussed in chap. xxix. *et sqq.*

CHAPTER XXVII.

SUPERFICIAL DEPOSITS OF SWITZERLAND.

No trace of sea-action or of floating ice.—Erratics of the Jura.—Glaciers of the Rhine and its tributaries.—Glacier of the Rhone.—Moraine-profonde in Dauphiny.—Moraine-profonde of Swiss low-grounds.—Ancient alluvium or diluvium overlying moraine-profonde.—Inter-glacial deposits of Dürnten, &c.—Morainic débris, &c., overlaying inter-glacial deposits.—Reason why the moraines of the second great advance of glaciers are large and well-preserved.—Post-glacial deposits.

THERE is no region where the marks left by the gigantic glaciers of the Ice Age, have been more assiduously studied than in Switzerland. Besides the many eminent native geologists who have devoted themselves to this subject, hosts of enthusiastic visitors from a distance, some of them men of great distinction, have won for themselves scientific laurels amongst the glaciers of that beautiful country. For us especially the Swiss glaciers, ancient and modern alike, have many valuable lessons. Geologists in this country are frequently puzzled to decide what part of the glacial phenomena here ought to be ascribed to the action of land-ice, and what portion must be assigned to rafts and bergs. In Switzerland nowadays there is no such difficulty. Glacialists are unanimous in considering that all the marks of old ice action in that country have been produced entirely by glaciers. Rock-scratching on the grandest scale—striæ running across the tops of considerable hills—erratics which have crossed deep broad valleys, often for great distances, and stranded at last on steep mountain-slopes—till, formed and accumulated under ice (a process which some geologists even yet cannot be persuaded has ever taken place), may all be studied to the greatest advantage in Switzerland. It is not my purpose, however, to dwell much upon these matters. The object I have in view is simply to point out the succession of the older glacial deposits, for the purpose of instituting a comparison between these and glacial accumulations elsewhere.

The Jura Mountains, as every one knows, extend in a long series of parallel ridges from south-west to north-east, between the valleys of the Rhone and the Rhine. From the base of these mountains the low-grounds of Switzerland roll themselves out to east and south-east until they sweep up against the great barrier of the Alps. Now upon the southern flanks of the Jura we find numerous scattered blocks and boulders, all of which have been carried from the Alps across the intervening plains, and left where we now see them. Some of the blocks are of enormous dimensions: many contain thousands of cubic feet, and not a few are quite as big as cottages. Indeed one of them—the great granite-boulder of Steinhoff—might. be compared as Mr. Maclaren has remarked, to "a goodly-sized house of three storeys." Such blocks have been observed on the Jura at a height of no less than 2,015 ft. above the surface of Lake Neufchatel, or 3,450 ft. above the sea; and from this elevation downwards they are strewn in greater or smaller numbers along the whole mountain-slope that faces the Alps.

Towards the north-east, where the Jura begins to lose in height as it approaches the valley of the Rhine, we find the erratics scattered not only along the southern slopes, but even over the tops of the mountains. According to Swiss geologists these erratic blocks and boulders have been carried down from the Alps on the surface of a mighty mer de glace, underneath which the whole of the central low-grounds were at one time buried. This vast sheet of ice, not less than 3,000 ft. in thickness, stretched continuously outwards from the Rhone Valley, and abutted upon the Jura, the higher ridges of which rose above its level. Other gigantic glaciers descending the valleys of the Rhine, and its tributaries became confluent with the glacier of the Rhone, and each of these carried its quota of rubbish and boulders, which it stranded along the slopes of the mountains.

When the Rhone glacier advanced across the plains of Switzerland and abutted upon the Jura, it was of course compelled to flow south-west and north-east along the flanks of that range. That portion which crept towards the north-east coalesced with the glaciers of the Aar, and its tributaries, and these last again with the great glacier of the Rhine.

How far, then, to the north did this vast ice-flow extend? Its limits in that direction are at present unknown. The old moraine matter has been traced down to the Rhine at its confluence with the Aar, but beyond that, the track of the ice has not been followed. No one, however, can believe that the great glacier stopped abruptly on the banks of the Rhine. The mass of ice which overtopped the Jura and flowed down the Frickthal to the Rhine, must of necessity have continued its course much farther. Nor when we remember that at this very time the Black Forest also had its great glaciers, can we doubt that the ice of that region was confluent with the Swiss mer de glace.

That part of the Rhone glacier which flowed to the south-west, was not until recently thought to have extended beyond some twenty miles below Geneva. But quite lately the path of the old ice-stream was followed by MM. Falsan and Chantre,* from Geneva as far as Lyons, passing by Seillon, Châtillon, Ars, and Sattonay, and even farther south to Valence in the department of Drome—that is to say about 130 miles as the crow flies from Geneva. On its way it overflowed the rubbish heaps of limestone blocks brought down by the local glaciers that descended from the valleys of Savoy, and deposited above them its own moraines of crystalline rocks. What a picture does this give us of the old Ice Age.

* According to these geologists, all the plain of the Dombes and lower Dauphiné is covered with the moraine-profonde of the Rhone and its tributaries. *Bull. de la Soc. géol. de France*, tome xxvi. 2e Serié, 1868; *Bibliothèque Universelle* (1872), vol. xliv. p. 46; *Nature*, vol. viii. p. 468.

Truly if Scandinavia and Britain had then their great enveloping mer de glace, central Europe was not much behind with its colossal glaciers; a stone dropped upon the Rhone glacier at its source might in those days have been carried upwards of 270 miles before it reached the end of the ice-flow In comparison with this what are the present glaciers of the Himalaya and the Karakorum, the very largest of which attains a length of only thirty-six miles?*

But erratic blocks and loose rubbish strewed along the mountain-slopes are not the only deposits which these great glaciers accumulated. In the low-grounds of Switzerland† we get a dark tough clay packed with scratched and well-rubbed stones, and containing here and there some admixture of sand, and irregular beds and patches of earthy gravel. This clay is quite unstratified, and the strata upon which it rests frequently exhibit much confusion, being turned up on end and bent over, exactly in the same way as the rocks in this country are sometimes broken and disturbed below till. The whole deposit

* It would seem that during a comparatively recent period the desert of Sahara was submerged, recent marine shells having been found widely distributed over its surface, and embedded at some depth in the sand. It is highly probable that, as Escher von der Linth has suggested, this submerged condition of the Sahara obtained during the glacial epoch, and that much of the moisture which then fed the great snow-fields of the Alps was brought by the prevalent winds flowing from Africa across the Sahara Sea and the Mediterranean.

† For descriptions of the ancient moraine-profonde of the plains of Switzerland, see Necker's *Etudes géologiques dans les Alps;* and A. Favre's *Recherches géologiques dans les Parties de la Savoie,* &c., tome i. chap. iv. Some of the best sections of this stony clay or till which I have seen are exposed in the steep banks of the river Arve, a little above Carouge, where a thickness of 80 or 90 ft. is seen.

has experienced much denudation, but even yet it covers considerable areas, and attains a thickness varying from a few feet up to not less than thirty yards. At Durnten a shaft was sunk in it to a depth of thirty feet without reaching the bottom.

The Swiss glacialists believe this deposit to be the material that gathered underneath the great mer de glace, and hence they term it *grund-moräne* or moraine-profonde. That it has been subjected to great pressure, is evident from the exceedingly stiff and compact nature of the clay, in which, as in other particulars, it closely resembles, as indeed it is the counterpart of the Scotch till.

I have mentioned the fact that the grund-moräne has suffered much denudation. Resting upon its eroded surface we find sand and gravel, and the same deposits are spread far and wide over the low-grounds, attaining often a great thickness, and frequently rising to considerable heights. These constitute the ancient alluvium of Swiss geologists. When they are traced from the low-grounds up to the base of the Alps they generally become coarser, and show many angular stones, especially at or near the bottom.* Of course they do not cover the ground continuously; in many places, indeed, they are absent altogether.

The origin of the ancient alluvium is sufficiently obvious. When the great glaciers were retiring and leaving trains of blocks and boulders upon the moun-

* Carl Vogt's *Lehrbuch der Geologie und Petrefactenkunde*, third edition, vol. ii. p. 36.

tain-slopes at ever decreasing heights, they by-and-by came to shoot their rubbish upon the low-grounds. Streams of water issuing from the melting ice then washed these moraines down, rounded the angular stones, and spread sand and gravel far and wide. The ancient alluvium, therefore, is exactly the counterpart of those great accumulations of sand and gravel, which bulk so largely in and at the mouths of our own mountain valleys; accumulations which at a later date, as we have seen, were worked up into cones and ridges. In Switzerland, however, no subsequent submergence ensued as in northern Europe, and neither eskers, kames, nor asar occur anywhere in the low-grounds of that country.

The beds which I have now to refer to, carry the story still further on, and are of much interest and importance.

At Durnten and Wetzikon in the canton of Zurich, certain seams of lignite have long been worked, and in the canton of St. Gall, similar seams occur at Utznach and Mörschweil, on the Lake of Constance. The lignite at these different places is believed to be the same, or at all events to have been accumulated at approximately the same time. It varies from two to five feet, and even occasionally reaches a thickness of twelve feet. It appears to be made up chiefly of peat-forming plants. Like much more recent peat-mosses, it also contains numerous remains of trees, such as pines, oaks, birches, larches, &c. From an examination of these remains Professor Heer has concluded that the climate of Switzerland during the formation of the lignite was similar to what it is now.

Associated with this ancient peat-moss are found the bones of the Asiatic elephant, a species of rhinoceros, the urus or great ox, the stag (*Cervus elaphus*), and the cave-bear. In addition to these fossils a number of insects are found whose shining wing-cases speckle the upper surfaces of the beds. Some of the coleoptera described would seem to be now extinct.

Such in a few words is the general character of the lignite. At Dürnten it rests immediately upon a layer of fine yellow sand and clay, beneath which comes an unknown thickness of grund-moräne.* Overlying the lignite we find a considerable thickness of gravel and sand, in beds which are surmounted with several large alpine erratics.†

Now from these facts we gather that the great mer de glace eventually vanished from the low-grounds, and the glaciers shrunk back again into the deep mountain-valleys. The climate grew as mild as it is at present. Oaks, pines, and other trees overspread the ground, and many large animals became denizens of Switzerland. That this condition of things must

* *Die Urwelt der Schweiz*, p. 486. It was at Dürnten where they drove a shaft in the grund-moräne to a depth of 30 ft, and were prevented from going farther by an influx of water.

† Near the village of Hermance, on the borders of the Lake of Geneva, there is an interesting section showing a bed of turf (with trunks of trees and recent species of land and fresh-water shells) which overlies a blue tenacious clay with scratched stones, and is covered with reddish clay containing erratic blocks and a little gravel. None of the blocks in this upper clay are striated. A. Favre, *Recherches géologiques*, &c., tome i. p. 71.

have endured for a long time no one can doubt. Nor could the change from the intense glacial climate of the great mer de glace have been other than gradual. The glaciers would slowly retire, and many ages would elapse before the conditions became such as to induce the growth of oak-trees. After the genial climate that nourished these trees had lasted for untold centuries, the cold again increased. Slowly the glaciers crept down the valleys. Little by little, year by year, they continued to advance until at last, escaping from the mountain-valleys, they deployed upon the low-grounds. And now, encroaching upon, and eventually occupying the basins of the Alpine lakes, they crept out from these and piled up great end-moraines upon the low-grounds beyond.

But the erratic blocks that overlie the lignite beds are not the only evidence of this second advance of the glaciers. That the ice after retiring from the Jura to the mountain-valleys did again invade the low country, had been inferred before the interglacial character of the lignite beds was discovered. It had been known for years that the first ground-moraines and ancient alluvium were overlaid by newer ground-moraines, terminal moraines, and alluvium; the meaning of this having been pointed out by Morlot as far back as 1854.* It was extremely satisfactory, however, to get the further evidence supplied by the lignite beds, as it enabled geologists

* *Bulletin de la Société vaudoise des Sciences naturelle*, 1854, iv. pp. 39, 41, 53, 185; *Edinburgh New Philosophical Journal*, 1855, p. 14.

to appreciate more fully what the retreat and advance of the glaciers really meant.

A similar succession of deposits has been detected by Professor Hanns Höfer, as occurring in Carinthia. In the lower reaches of the valleys of that region ground-moraine is well developed, and perched blocks and erratics are found at great elevations, while the glaciated aspect of the mountains further shows that the valleys at one time must have brimmed with ice. Overlying the ground-moraine come massive deposits of river gravel, &c. (near Klagenfurt), which have yielded remains of the woolly rhinoceros, the steinbock (*Ibex cebennarum*), and *Bos taurus*. These fresh-water beds Professor Höfer correlates with the gravel beds that immediately overlie the Dürnten lignites (corresponding to the *Inter-glaciale Geröllbildung* of Heer). A younger series of large moraines met with in Carinthia near the Raiblersee, and in the Möll and Malnitzer Valley he considers to be the equivalents of the great moraines of the "second glacial period" in Switzerland.*

The glaciers of the second period, although of very much larger dimensions than their puny descendants of to-day, yet were themselves but pigmies as compared to the gigantic ice-flows of the first period. Nevertheless, strange to say, while the end-moraines left by the former are large and well defined, those of the latter do not exist, or exist only in the form of scattered blocks. For this apparent anomaly

* "Studien aus Kärnten": *Neues Jahrbuch für Mineralogie, Geologie, and Palæontologie*, 1873, p. 128.

several reasons may be given. In the first place we have to consider that during the melting of the great mer de glace that filled to overflowing the low-grounds of Switzerland, the rivers must have been enormously swollen, and excessive denudation of the moraines would necessarily follow. Next we have to take into account the greater antiquity of the deposits belonging to the first period. So long a time passed between the date of their formation and that of the second set of moraines, that we should naturally expect to find the latter in a better state of preservation than the former. Again, it is very doubtful whether the end-moraines of the first period were much larger or even as large as those of the second. The moraines of a small glacier, as Mühlberg * has well reminded us, bulk more in proportion than those of an ice-stream of greater pretensions. During the first period an immense frontage of rock disappeared below the ice, and was thus prevented from showering down its débris. In the second period, however, the glaciers did not reach nearly so high upon the sides of their valleys, while the snow-fields were less continuous, and thus a much greater area of rock was exposed to the action of the frost. Yet, further, we cannot tell whether or not the glaciers of the first period remained long stationary when they had reached their maximum size. But we know that those of the second period did. Finally, when we consider that the end-moraines of the former were dropped upon

* *Ueber die erratischen Bildungen im Aargau*, &c., p. 85.

the bottoms of two of the largest river-valleys—those of the Rhone and the Rhine—it does not appear at all improbable that much of the moraine débris may yet be lying concealed below those deep and wide-spread deposits of loam (löss), through which the present rivers make their way after escaping from Switzerland.

The change from the last glacial conditions to the present climate appears to have been gradual; the glaciers retiring slowly up the valleys, but perhaps occasionally advancing for a short distance, during some exceptionally severe year.

The mammalian remains met with in post-glacial deposits in Switzerland, belong for the most part to the same species and genera as those which characterize similar deposits in Ireland, Scotland, and North of England. Thus we get the bison, extinct bovidæ, several species of deer, &c., and also the mammoth and the woolly rhinoceros. Human relics likewise occur plentifully in the post-glacial beds, but none of these go back to an older date than the newer stone period of archæologists.

CHAPTER XXVIII.

SUPERFICIAL DEPOSITS OF NORTH AMERICA.

Glaciation of the northern regions.—The Barren Grounds.—Profusion of lakes.—Pre-glacial beds.—Oldest glacial deposits.—Unmodified drift, or till.—Scratched pavements in till.—Marine arctic shells in boulder-clay of maritime districts.—Inter-glacial deposits.—Iceberg drift.—Later glaciation.—Morainic débris.—Gravel and sand mounds, ridges, &c.—Erratics on gravel and sand ridges.—Limits of submergence uncertain.—Glacial lake-terraces in White Mountains.—Laminated clays with marine arctic shells.—Local moraines.—Tables showing succession of glacial deposits in Europe and America.

IT would be interesting to ascertain how far the results obtained from an examination of British, Scandinavian, and Alpine glacial deposits, harmonize with what may be learned from the records of the glacial epoch in other European districts, as in the Pyrenees and the Carpathians, but these districts have not yet been studied in sufficient detail. The day is no doubt coming when it will be possible to correlate the glacial deposits of such mountain areas with those of better-known regions. Nay, the student of glacial geology may even look forward to a time when it will be possible to compare in detail the relics of the Great Ice Age that are known to occur in northern Asia with the similar accumulations in Europe. But if we are as yet imperfectly acquainted with the superficial deposits in the northern latitudes of Asia, it is otherwise with those in the corresponding regions of North America. The literature of American glacial geology is already very extensive, and every year is adding to its bulk. Much light has been thrown upon the whole question of these researches, and the conclusions arrived at must be carefully studied by geologists on this side of the Atlantic if they would seek to gain an adequate conception of those great revolutions of climate that transpired in our hemisphere during the glacial epoch. In the following slight sketch, however, I will not attempt to do more than place before the reader what seems to have been the general succession of events.

It is no exaggeration to say, that, the whole surface of North America, from the shores of the Arctic Ocean to the latitude of New York, and from the Pacific to the Atlantic, has been scarped, scraped, furrowed and scoured by the action of ice. The ice-worn rocks of Labrador, Canada, the Northern States, and British Columbia, have been examined and described by many specialists; but, so striking are the appearances presented, that they have not failed to arrest the attention of other observers in those desolate regions of the far north which have been but seldom traversed. No geologist can read the accounts of the Barren Grounds to be found in the writings of Franklin, Richardson, Back, and others, without recognising everywhere the evidence of ancient glacier-action. If Dr. Richardson had been a professed glacialist, he could hardly have described in more expressive language the aspect of the ice-worn tracts traversed by him in company with the hapless Franklin.* Everywhere we meet with references to "round-backed ridges," "very obtuse conical hillocks," "bare, rounded masses of granite and gneiss," "land-locked sheets of water," &c. Many years afterwards, when this traveller once more threaded his way through the Barren Grounds, he did not fail to observe also the furrows and scratchings upon the harder rocks.† Another observant traveller, Captain Back, who followed the course of the Great Fish River (Back's River) down to the Arctic Sea, gives the following graphic sketch of a scene on the skirts of the Barrens; as a faithful picture of a highly ice-worn surface, it might have been drawn by a glacialist himself. "There was not the stern beauty of alpine scenery, and still less the fair variety of hill and dale, forest and glade, which makes the charm of a European landscape. There was nothing to catch or detain the lingering eye, which wandered on without a check over endless lines of round-backed rocks, whose sides were rent into indescribably eccentric forms. It was like a stormy ocean suddenly petrified. Except a few tawny and pale green lichens, there was nothing to relieve the horror of the scene."‡

But the rounding of rocks is not the only modification effected by the grinding of land-ice. Not only

* *Narrative of a Journey to the Shores of the Polar Sea in the years* 1819-22.

† *Journal of a Boat Voyage through Rupert's Land.*

‡ *Narrative of Arctic Land Expedition to the Mouth of the Great Fish River*, p. 178.

are sharp edges and projecting points smoothed away, but hollows are scooped out in the solid rocks. The lakes and sea-lochs of Scotland and Scandinavia, the innumerable lakes of Finland, and those of Switzerland and Italy, are both far outnumbered, and far surpassed in extent also, by the freshwater seas (for such not a few are) of North America. Lakes are profusely distributed over the whole vast tracts that drain into the St. Lawrence, Hudson's Bay, and the Arctic Ocean, and they are equally abundant in Labrador and Newfoundland. By far the great majority of these lakes must owe their origin directly or indirectly to the grinding power of ice—for they either rest in rock-bound hollows or are dammed back behind irregular ridges of glacial deposits.

But if the glaciated aspect of North America is only that of Europe on a larger scale, we shall find a no less close correspondence between the superficial accumulations of the two continents. The lowest glacial deposit recognised by Canadian and American geologists is "unstratified boulder-clay," or "unmodified drift." In some places this deposit is found to overlie beds of sand, gravel, and clay, and these beds have occasionally yielded vegetable remains. Dr. Dawson cites* the case of "a hardened peaty bed, which appears under the boulder-clay on the northwest arm of the River of Inhabitants in Cape Breton." "It contains many small roots and branches apparently of coniferous trees allied to the spruces." In an interesting paper by Mr. C. Whittlesey,† reference

* *Acadian Geology*, p. 63. † *Smithsonian Contributions to Knowledge*, vol. xv.

will also be found to the occurrence below and in the "unmodified drift" of decayed leaves and the remains of the mastodon and elephant. Generally speaking, however, the "unmodified drift" appears to rest directly upon the rocks, which are polished and striated below it. But over wide regions in the Barren Grounds and Labrador, the lowest member of the drift series appears to be entirely absent, or rests in patches among the hollows of ice-worn hummocks and hills. Considerable tracts of it, however, seem to occupy the low-grounds upon the western borders of Hudson's Bay.* In Labrador it appears not to be very plentiful. Dr. A. S. Packard says:—"Nowhere did I see on the coast of Labrador any deposits of the original glacial clay or unmodified drift. Upon the sea-shore it has been remodelled into a stratified clay; and the boulders it once contained now form terraced beaches."† Professor Hind, however, mentions its occurrence capped by sand, and forming banks "rising seventy feet above the level of the Moisie River, twenty miles from its mouth." It is well developed in Canada, where, according to Professor Dawson, it assumes the character of a "hard grey clay, filled with stones, and thickly packed with boulders." The stones and boulders are often scratched, and the whole deposit is usually devoid of stratification.‡ Thick masses of it are encountered in Maine, where it

* *Narrative of a Journey to the Shores of the Polar Sea* (Franklin and Richardson), p. 499.

† *On the Glacial Phenomena of Labrador and the Maine.*

‡ "Notes on the Post-pliocene Geology of Canada": *Canadian Naturalist*, New Series, vol. vi.

presents precisely the same character as in Scotland, —a tough, unstratified clay, crammed with angular and subangular, smoothed and striated stones. In the State of New York it is described as "sometimes loose, but frequently partially aggregated by argillaceous matter, that renders a pick necessary to dig it."* Mr. Whittlesey also makes frequent reference to its occurrence in Michigan and Ohio, where it is described as a firmly-compacted "mixture of clay, sand, and gravel, or fragments of rocks in a confused or imperfectly stratified condition, which is locally known as 'hardpan.'"† A similar account is given of the Illinois glacial drift by Professor E. Andrews,‡ and, scattered through the various admirable Reports of the State Geologists, numerous descriptions will be found that tally with the foregoing.

The somewhat rare phenomena of "scratched pavements" in the till have also been observed in North America. Professor O. N. Stoddard refers to an excellent example in the till of Miami. The upper surfaces of the embedded stones were all striated in one and the same direction, but when the boulders were picked out of the clay, their other sides showed scratches running in different directions.§

In the interior of the country the "unmodified drift" is quite unfossiliferous, but in certain maritime districts it appears to have yielded marine organic

* *Geology of New York,* part iv. p. 160 (Prof. W. Mather).

† *Smithsonian Contributions to Knowledge,* vol. xv.

‡ *American Journal of Science* (1867), vol. xciii. p. 75

§ *Op. cit.* (1859) vol. xxviii., p. 227.

remains. This, according to Professor Dawson,* is the case in the lower part of the St. Lawrence River. Further up in the vicinity of Montreal, however, it has not been observed to contain fossils. From Professor Dawson's description of the fossiliferous stony clay, that deposit would seem to be analogous to the boulder-clay of maritime districts in Scotland.

Another interesting feature in the American glacial deposits, is the occurrence of intercalated fossiliferous beds. These are passed through frequently in sinking wells and pits, and in digging foundations. A few examples will illustrate the general arrangement of the strata. The first is taken from a paper by Mr. Whittlesey:—†

ARTESIAN WELL, COLUMBIA, OHIO.

Surface 215 ft. above Lake Erie, and 780 ft. above tide.

	Ft.
1. Soil	4
2. Sand, gravel, and boulders	10
3. Coarse sand	2
4. Blue clay and boulders	4
5. Fine quicksand	2
6. Blue clay, inclosing a log	17
7. Hardpan	3
8. Quicksand	1
9. Hardpan to cliff limestone	37
	80

The same author gives several sketch-sections, which show that the beds are by no means of regular thickness, but thicken and thin out in a somewhat inconstant manner.

* "Notes on the Post-pliocene Geology of Canada," *Canadian Naturalist*, New Series, vol. vi.

† *Smithsonian Contributions to Knowledge*, vol. xv.

The following interesting section was obtained* during the sinking of a shaft at the city of Bloomington (Illinois):—

	Ft.
1. Surface soil and brown clay	10
2. Blue clay	40
3. Gravelly hardpan	60
4. Black mould, with pieces of wood, &c.	13
5. Hardpan and clay	89
6. Black mould, &c.	6
7. Blue clay .	34
8. Quicksand, buff and drab in colour, and containing fossil shells	2
Clay shale (coal-measures).	
	254

Again, Professor Newberry has described† the occurrence of a regular forest-bed intercalated among true glacial deposits, and bones of the elephant, mastodon, and great extinct beaver, have been found in the same position. According to Dr. Newberry and his associates ‡ on the State Survey (Ohio), the succession of deposits in the valley of the Maumee is as follows, beginning with the oldest:—

a Glacial drift.
b Erie clays.
c Forest bed.
d Iceberg drift.
e Alluvium.
f Peat, calcareous tufa, shell-marl.

The next section shows the nature of the deposits which overlie the "unmodified drift" or till:—

* *Geology of Illinois*, vol. iv. p. 179. The "shells" are of fresh-water species.

† *Nature*, 1871, p. 155. *Report of Progress* (Geol. Surv. of Ohio), for 1869.

‡ *Report of Progress* (Ohio Survey), 1870, p. 340 *et sqq.*

SECTION OF ARTESIAN WELL IN STATE HOUSE YARD, COLUMBUS.

	Ft.
Surface earth	1
Brown earth	2
Sand and gravel	11
Blue clay with boulders (Erie clay)	4
Sand	2
Quicksand	3
Leafy blue clay	1
Blue clay and sand (Erie clay)	18
Clay and gravel	3
Sand, clay, and gravel	9½
Cemented clay, sand, and gravel (unmodified drift)	68½
Lime rock at	123

The Erie clays are generally of a bluish or greyish colour, and always contain boulders and pebbles in greater or less abundance. These stones, moreover, are frequently scratched, and in some localities few of them are found without striæ. None of the clay beds has yielded any fossils.

Immediately overlying the Erie clay series occurs the forest-soil referred to above. The beds below are usually discoloured for some distance down with vegetable mould, and contain, mingled with their own substance, quantities of leaves, branches, roots, and tree-trunks. In some districts of Ohio, the old forest seems to be everywhere present, although it is more frequently met with on the high plateaux than in the valleys. The trees have been identified as sycamore, hickory, beech, and red cedar, the latter being by far the most abundant.

Heavy masses of gravel, sand, and boulders overlie the buried forest.

The succession of changes which appears to be

indicated by these phenomena, are, according to Professor Newberry, as follows:—

1st. A period of a great continental glacier or ice-sheet.

2nd. The retreat of the ice, and the appearance of a vast fresh-water lake (covering a large part of Ohio), in which were deposited the Erie clays, &c.

3rd. The silting up of the lake, and the advent of a luxuriant forest-growth.

4th. The submergence of the land and the deposition from floating ice of blocks and boulders.*

When the great continental glacier, which appears to have covered so large a portion of North America, began to retire, a newer set of glacial striæ was engraved upon the rocks, the direction of which does not coincide with that of the primary glaciation. In the district of Lake Superior, for example, there are traces of two distinct glaciations—the main or continental glacier having flowed from north-east to south-west, while the local and latest ice-flow was from north to south.† The thickness attained in Connecticut by the continental ice-mass, has been

* These details are taken chiefly from *Report of Progress* (Ohio Survey) for 1870; see also *American Journal of Science*, vol. c. 1870, p. 54, where a bed of peat, 10 to 15 ft. thick, is described as underlying 80 ft. of clay and gravel. Ash, hickory, and sycamore, together with grape-vines and beech-leaves, are said to occur in a similar position—a stratum of soil 2 ft. thick often underlying these vegetable remains. Two mastodon tusks were taken in 1870 "from the northern part of the same drift-bed to which the peat belongs, and at about the same level." Phosphate of iron (vivianite) occurs in small pockets in the beds associated with the peat, and is with some probability supposed to represent remains of bones.

† *Report on Lake Superior*, part i. p. 205 (Foster and Whitney); see also N. H. Winchell on "The Glacial features of Green Bay," &c. *American Journal of Science*, vol. cii. p. 15.

estimated by Dana to have reached 6,000 ft. or 8,000 ft.*

The withdrawal of the ice-sheet was marked by the stranding of perched blocks and the accumulation of moraine rubbish and diluvial gravel and sand, which, again, were partially rearranged during the submergence that carried down the Ohio forest lands below the sea. It is thus difficult sometimes to distinguish true moraine drift from reassorted marine drift; the one, in short, seems to shade into the other. This, however, does not mean that the deposition of the morainic débris and its reassortment by the sea took place at one and the same time. The ice may have melted away from all the low-grounds of New England, and shrunk back to the valleys among the White Mountains before subsidence of the land began; as, indeed, the evidence of the Ohio beds would seem conclusively to prove.

In reading descriptions of the mounds of gravel and sand which cover large tracts of country in the New England and the north-western States, and also in Canada, one cannot fail to notice how closely all the appearances coincide with those we are familiar with in this country. Professor Hitchcock, describing the drifts of New England, says they "form ridges and hills of almost every possible shape. It is not common to find straight ridges for a considerable distance. But the most common and most remarkable aspect assumed by these elevations is that of a collection of tortuous ridges, and rounded and even

* *American Journal of Science*, vol. cii. p. 240.

conical hills, with correspondent depressions between them."* This description would apply word for word to some of the larger areas of kames in Scotland. The American mounds and cones are almost invariably composed of well water-worn materials, usually gravel and sand; and they are, moreover, not infrequently false-bedded. Occasionally boulders are found inside these mounds; but this is certainly quite exceptional, and such included stones are usually more or less rounded. Now and again a mound appears to be composed of coarse shingle and rounded boulders. But when boulders occur in mounds of sand and fine gravel, they seem to be confined chiefly to the upper parts of the deposits.†

Immense numbers of large erratics cumber the surface of the ground in many parts of New England, the north-western States, Canada, and Labrador, and are scattered over the tops and slopes of the mounds and ridges of sand and gravel. Even much farther north the same phenomena are so striking as to arrest the attention of the traveller who is not strictly a geologist. I was much interested some years ago in reading the accounts given of the Barren Grounds of North America by various writers who had visited these inhospitable regions. Sand-hills and huge erratics appear to be as common there as in the countries farther south. Captain Back gives a very

* *Trans. of the Assoc. of Amer. Geol. and Natur.*, 1840–1842, p. 191.

† See *Report on the Geology of Lake Superior Land District*, p. 235; also *Geology of New York*, part iii. p. 121, where Lardner Vanuxem says: "With some exceptions, they (erratics) are generally found upon the surface, frequently upon the tops of hills or on their sides, appearing in almost all their localities as if but recently dropped," &c.

graphic account of the isolated cones and "chains of sand-hills," which he saw in several places stretching far away on either side from the valley of the Great Fish River. He tells us that "the ridges and cones of sand were not only of great height, but singularly crowned with immense boulders, grey with lichen, which assuredly would have been considered as having been placed by design, had not the impossibility of moving such enormous masses proved incontestably that it was Nature's work." This was in 66° N. lat. In another place "the country was formed of gently undulating hills, whose surfaces were covered with large fragments of rock and a coarse gravelly soil."* In the Barren Grounds to the west of the bleak country traversed by Back, sand-hills and huge erratics are equally abundant, the erratics being often abundantly dotted over the hillocks of drift.† The same appearances have been noted again and again by explorers. Between Lake Winnipeg and the Saskatchewan River, the number of boulders perched on the sides and summits of ridges and mounds seems to be specially remarkable.‡

The erratics sometimes show striæ which not infrequently are placed parallel to those of the bare rock upon which the erratics rest. A good example of this appearance is quoted by Mr. J. De Laske.§ He states

* *Narrative of Arctic Land Expedition to the Mouth of the Great Fish River*, &c., pp. 140, 346.

† See Franklin's *Journey to the Shores of the Polar Sea*, and his *Second Journey*; also Sir J. Richardson's *Journal of a Boat Voyage through Rupert's Land.*

‡ *Royal Geographical Journal*, vol. xxx. p. 275 (Palisser's Exploration).

§ *Second Ann. Rep. upon Nat. Hist. & Geol. of Maine*, 1862. These erratics, however, were most likely left upon the ground during the retreat of the ice-sheet.

that "around one of the quarries to the west of Carver's Harbour the ground is literally covered witl boulders, some of which are enormous. Many of these turned out of their beds exhibit the polishing and scratching of the common floor-rock of the island (Mount Desert Island). Furthermore, if carefully turned over, we find some of them left just where they had last been employed in scratching the ledges —the parallel scratches of the boulder being placed parallel to those of the rock beneath."

Thus in northern America, as in the northern latitudes of Europe, we find the ground covered throughout wide areas with groups of ridges, mounds, and cones of sand and gravel; and these peculiar hillocks are everywhere dotted over with large erratics, in such a way as to show that the sand and gravel must have been laid down and heaped up before the erratic blocks were dropped. And, from the rare occurrence of embedded boulders in the sand and gravel, it is only reasonable to infer that at the time the sand and gravel were heaped up there could not have been much ice floating about. It is true that piles and mounds of coarse unstratified débris and boulders are occasionally found associated with the re-assorted drift; but these, according to Professor Agassiz and several other American geologists, are moraines, and not the droppings of icebergs. The mounds of well water-worn sand and gravel are singularly free of boulders, except on the outside.

After the old diluvial deposits had been added to and re-assorted upon the bed of the sea, the climate of the northern hemisphere, which had been moderate during the period of subsidence, again became cold. Fleets of icebergs and icerafts set sail from every coast that remained above the sea, and dropped their burdens as they journeyed on. But to what extent North America was submerged we cannot tell. Professor Hitchcock describes many so-called *raised beaches*, one of which occurs at a height of 2,449 ft. above the sea in the White Mountains: but it seems most likely that these terraces are, as Dr. Packard and other writers have suggested, of fresh-water origin—being the relics of glacial lakes like the Parallel Roads of Glen Roy. Many of the terraces in question are strewn with huge boulders, as if these had been stranded by rafts of ice.

During the re-elevation of the land, beds of clay accumulated off the coast, and became gradually stocked with shells of an arctic type. These are the "Leda clays" of Labrador and Maine, so ably described by Dr. Dawson, Dr. Packard, and others. It can hardly be doubted that they are the equivalents of the Scottish and Scandinavian shelly clays. The fossils which they contain are very decidedly arctic in the lower beds, but in the upper beds they give evidence of a gradually ameliorating climate.

In the valleys of the White Mountains, and in those of the Rocky Mountains and the Sierra Nevada, a number of terminal moraines mark the sites of local glaciers, which gradually crept up the valleys, and vanished as the cold of the glacial epoch passed away.

We have now completed our review of the glacial

deposits, and have learned that the succession of changes during the Age of Ice was singularly uniform over large areas in the northern hemisphere. All this points unmistakably to the operation of cosmical influences. It is impossible to believe that mere local elevations and depressions of land were the causes that induced that remarkable rotation of cold and mild periods which we term the glacial epoch. On the contrary, we have seen that great oscillations of climate supervened independently of the varying distribution of land and sea. Thus, in Scotland and the north of England we have evidence to show that after the retreat of the ice-sheet a wide land surface appeared, and that the mild conditions of climate which then obtained continued to characterize our area until the land had sunk to a considerable depth in the sea, when at last another cold period ensued. Similar changes characterized the glacial epoch in America. But in order to bring these and other matters clearly before the mind, I shall here summarise the general results of our review in a series of short tabular statements. In each table I begin with the oldest deposits:—

SCOTTISH GLACIAL DEPOSITS.

1. Till, with intercalated and subjacent deposits.	Intense glacial conditions (general ice-sheet) with intervening periods marked by milder climates.
2. Boulder-clay in maritime regions; till in the interior; and perched blocks at high levels.	Ice-sheet melting back; gigantic local glaciers entering the sea.
3. Morainic débris; perched blocks; and ancient river-gravels or diluvium with animal remains.	Further retreat of the ice; local glaciers; large rivers; climate passing to temperate.

4. Kames	Depression of the land; climate temperate or cold-temperate.
5. Brick-clays, &c., with arctic and boreal shells; erratics.	Return of glacial conditions; period of floating ice; climate not so intense as during accumulation of till; re-elevation of the land.
6. Valley moraines . .	Final retreat of the glaciers.

English Glacial Deposits.

1. Till of north-west, &c.; lower and middle glacial of East Anglia.	Intense glacial conditions with intervening milder period.
2. Boulder-clay (lower) of Lancashire, &c.; great chalky boulder-clay, &c., of East Anglia; boulder-clay of Northumberland coast; perched blocks at high levels.	Ice-sheet melting back; gigantic glaciers entering the sea.
3. Older morainic débris; perched blocks; gravel and sand in and opposite mountain valleys.	Further retreat of the ice; local glaciers; large rivers.
4. Middle sands of Lancashire, Moel Tryfan, &c.	Depression of the land, climate temperate, or cold-temperate.
5. Upper boulder- or stony clay of Lancashire, &c.; Nar Valley beds; Hessle gravel and clay; erratics.	Return of glacial conditions; climate not so arctic as during accumulation of till and lower boulder-clay; land rising.
6. Valley moraines . .	Final retreat of the glaciers.

Irish Glacial Deposits.

1. Till	Intense glacial conditions (inter-glacial beds not certainly known to occur).
2. Boulder-clay (lower); till in the interior; perched blocks at high levels.	Ice-sheet melting back; gigantic local glaciers entering sea.
3. Morainic débris; perched blocks; sand, &c., opposite mouths of glacier valleys.	Further retreat of the ice; local glaciers; large rivers.
4. Eskers and middle sands, &c.	Depression of the land; climate temperate, or cold-temperate.
5. Upper boulder- or stony clay.	Return of glacial conditions; climate not so cold as during accumulation of lower boulder-clay and till.
6. Valley moraines. . .	Final retreat of the glaciers.

Scandinavian Glacial Deposits.

Deposit	Conditions
1. Till, lower and upper, inter-glacial fresh-water clays, &c.	Intense glacial conditions interrupted by intervening mild period or periods.
2. Perched blocks at high levels [no shelly boulder-clay corresponding to that of maritime districts in Britain].	Ice-sheet melting back.
3. Morainic débris and perched blocks of mountain regions; great deposits of sand and gravel.	Further retreat of the ice; large rivers from melting mer de glace.
4. Asar.	Deposits derived from the melting ice-sheet (see 3) denuded and remodelled during submergence.
5. Clays with arctic shells; erratics.	Climate arctic, but not so cold as 1; re-elevation of land.
6. Moraines . . .	Retreat of the glaciers.

Glacial Deposits of Central Europe.

Deposit	Conditions
1. Grund-moranen or till.	Intense glacial conditions; Swiss ice coalescent with that of the Black Forest; Rhone glacier descends as far, at least, as Valence, in the south-east of France.
2. Perched blocks at high levels, as on the Jura, Mont Salève, &c., in Switzerland; Ullrichsberg in Carinthia, &c.	Commencement of retreat of the glaciers.
3. Perched blocks; degraded morainic débris; great deposits of gravel, &c., or diluvium; lignite beds of Dürnten, &c.	Continued retreat of the glaciers; large rivers,, and erosion of the river-terraces; climate changing from cold to temperate and genial; elephant, rhinoceros, &c.
4. Gravel-beds overlying Swiss lignites; ancient river-gravels at Klagenfurt in Carinthia.	Climate becoming cold; woolly elephant and woolly rhinoceros, &c.
5. Moraines overlying the older glacial deposits; newer river-gravels and löss in part.	Return of glacial conditions; new advance of glaciers: climate not so intense as during formation of older grund-moränen.
6. Newer moraines; river gravels and löss in part.	Periodic retreat of the glaciers.

North American Glacial Deposits.

Deposits	Conditions
1. Till, or unmodified drift with subjacent and intercalated beds.	Intense glacial conditions (general ice-sheet) with intervening milder periods.
2. Boulder-clay of maritime regions in Canada.	Ice-sheet melting back, and followed by the sea.
3. Morainic débris and perched blocks at high levels; probably much of the sand and gravel in the interior of the country; lacustrine deposits and buried forests of Ohio, &c.	Continued retreat of the ice; great lakes appear; these partially silted up; climate becoming temperate, a great forest-growth overspreads the latitude of Ohio, &c.; mastodon, great extinct beaver, &c.
4. Mounds, ridges, &c., of gravel and sand.	Depression of the land; remodelling of pre-existing drifts and heaping up of kame-like ridges and mounds.
5. Leda clay, &c., erratics, &c.	Return of glacial cold; period of floating ice; land rising.
6. Valley moraines	Final retreat of the glaciers.

A glance at the foregoing tables will show that the oldest glacial deposits* have yielded evidence of inter-glacial mild conditions in the following countries:—viz., in Scotland, England, Scandinavia, and North America. That inter-glacial deposits should not be equally well developed in the older glacial deposits of other countries, ought not to surprise us. We must bear in mind that the preservation of such loose incoherent beds must always be exceptional under the conditions which are known to have obtained in early glacial times. But as I have already enlarged upon this point while treating of the Scotch glacial deposits, I need not weary the reader with a repetition of the remarks referred to.

There is one important point, however, to which I desire to call special attention. It is this: that in

* That is to say, the *till, unmodified drift, and grund-moranen.*

every country where the glacial deposits have been studied with any attention to details, we have clear and convincing proof of a mild inter-glacial period having supervened in what one may term the later stage of the glacial epoch. This—the last inter-glacial period—has left its mark alike in the British Islands, in Scandinavia and northern Europe, in the Alps and central Europe, and in northern America. In Switzerland the climate of the period referred to was mild and genial, the elephant and the rhinoceros being at that time denizens of the country. In North America, during the same period, a widespread forest covered Ohio and doubtless other regions, and the mastodon and elephant roamed over the land. Turning to the British deposits, we find that after the disappearance of the great ice-sheet underneath which the till was accumulated, a land-surface appeared, and large rivers eroded the older glacial deposits and laid down great beds of gravel and sand. In Scotland these ancient river-deposits have yielded only the remains of water-rats and frogs. But we have to remember that the fresh-water beds in question were at a later date submerged and remodelled by the action of the sea. During this submergence we know from the evidence supplied by the English middle sands (No. 4) that the climate was still temperate, although after the depression had considerably increased it gradually became colder.

The middle sands of Lancashire, &c. (No. 4) are generally believed to be the only representatives we have in England of this last inter-glacial period. I have endeavoured to show, however, that in England, just as in Scotland, dry land appeared after the great ice-sheet had crept back to the mountains, and that rivers flowed in our valleys and effected great erosion long before the land sank down below the sea to receive its covering of "middle sands." Have we, then, any remains of that old land-surface in England? are there any English river-gravels that can be correlated with the lignite beds of Switzerland? I believe that such river-gravels do exist, and that these have hitherto been erroneously referred to post-glacial times. The consideration of this important point brings us, as we shall presently see, face to face with the question of the antiquity of man; for it is in these so-called post-glacial deposits that the earliest traces of man have been detected.*

* In a former publication (*Changes of Climate during the Glacial Epoch*) I attempted a correlation of the Swiss and Italian glacial deposits, and Professor Gastaldi has recently done me the honour to criticise the views therein advanced. ("Appunti sulla Memoria del Sig. J. Geikie," &c.: *Atti della Reale Accademia delle Scienze di Torino*, vol. viii.) Desiring to occupy the text with as little controversial matter as possible, I have thought it better to reserve to the Appendix what I have to say in regard to the correlation referred to. (See Note F.)

CHAPTER XXIX.

CAVE-DEPOSITS AND ANCIENT RIVER-GRAVELS OF ENGLAND.

Pre-historic deposits.—Stone, bronze, and iron ages.—Palæolithic or Old Stone period.—Neolithic or New Stone period.—Universal distribution throughout the British Islands of neolithic implements.—Animal remains associated with neolithic relics.—Palæolithic implements.—Absence of intermediate types.—Break between palæolithic and neolithic periods.—Caves and cave-deposits.—Kent's cavern, Torquay.—Succession of deposits.—River-gravels, &c., with mammalian remains and palæolithic implements.—Geographical changes during palæolithic period.—Gap between neolithic and palæolithic deposits.

A FASCINATION attaches to the early history of every people. We long to penetrate that mystery which the lapse of ages has drawn like a thick curtain round the cradle of our race. How eagerly do we scan the oldest written records that have any reference to our country and its people; and how assiduously do we try to shape a coherent story out of those vague myths, legends, and traditions which have come down to us from the long-forgotten past. But there are memorials of man in this as in other countries, which date back to so remote a period that even the oldest traditions have nothing whatever to say about them. The English historian begins his narrative with the

Roman invasion, and the archæologist until recent years could hardly trace the story farther back; but now he can tell us of a time infinitely far beyond the first dim beginnings of history and tradition, when races of savage men and tribes of wild animals, some of which have long been extinct, were denizens of Britain. Hitherto we had been taught to look upon Stonehenge and the so-called Druid-circles as the oldest memorials of man in this country—mysterious monuments belonging to the shadowy past, about whose age and uses only vague conjectures could be offered. If older races than the builders of Stonehenge ever lived in Britain, we knew nothing, and could hardly hope to know anything about them. The past was apparently separated from us by a gulf which it was vain to think that any ingenuity would succeed in bridging over.

Now all this is changed. The massive monoliths of Stonehenge, however venerable their antiquity, seem but as structures of yesterday; the standing-stones of Avebury, of Callernish, and Stennis, the so-called vitrified forts, the round towers of Ireland, and all those remains of ancient camps, dwellings, and burial-places so abundantly met with throughout the British Islands, are of immeasurably more recent date than certain rude stone implements which our cave-deposits and ancient river-gravels have yielded. Since Stonehenge rose upon Salisbury plains no great change in the physical geography of Britain has taken place. The destruction of ancient forests and the cultivation of the soil have doubtless in some measure altered the aspect of the land, and influenced the character of the climate. Our hills and valleys, however, we are sure have remained the same, and even the coast-line has experienced probably little change. Changes undoubtedly there have been, yet none so considerable as to invalidate the truth of the statement, that since the days of the builders of Stonehenge no great geological revolution has transpired in Britain. But the rude stone implements to which I have referred date back to a period when the appearance presented by our country differed greatly from that which obtains now; and for so vast a time did the old tribes who used these rude implements occupy the British area, that the slowly-acting forces of Nature were enabled, during that time, to bring about many geological changes, each of which required long ages for its evolution.

What, then, is the nature of that evidence which has weighed with archæologists and geologists in assigning to man this great antiquity? It would lead me far beyond the limits I have set for myself were I here to attempt anything like a detailed account of the archæological evidence. My object will be sufficiently served if I give only a brief outline of the general results?*

All those monuments and memorials of man which belong to pre-historic times have been arranged by archæologists under three groups. The classification

* The reader who desires fuller information must consult Sir J. Lubbock's *Pre-historic Times*, and Mr. Evans's *Ancient Stone Implements of Great Britain.*

adopted is based chiefly upon the distinguishing features presented by such objects of workmanship as weapons, implements, and personal ornaments. The oldest group comprises implements of *stone;* next in order come *bronze* relics; and these are succeeded by tools and weapons of *iron.* The articles of stone indicate upon the whole a much lower grade of development than those of metal, and hence archæologists have inferred that the races who used stone-knives, hatchets, and hammers preceded in time those to whom the use of metal was known. In like manner they have argued that cutting-tools and weapons of bronze were supplanted by weapons of iron—the use of iron for such purposes evincing more knowledge of the metals and a greater advance in civilisation. Hence it is customary to speak of the Stone Age, the Bronze Age, and the Iron Age.

Geological investigations have strongly supported this classification—deposits containing the stone implements having been proved in many ways to be older than those which have yielded relics of bronze, just as these latter have been shown to belong as a whole to an older period than weapons and tools of iron. But while we speak of these three ages it must be distinctly understood that they are not separated from each other by any hard and fast line. The manufacture of stone tools and weapons did not die out all of a sudden, and the employment of bronze immediately succeed. On the contrary, stone hammers and implements continued in some use long after bronze had been introduced, just as this latter certainly found favour and was extensively employed, especially for ornamental purposes, after the advantages of iron for knives and swords had become recognised.

With the bronze and iron ages, however, the geologist proper has comparatively little to do: although, did space permit, it could be shown that even in regard to the history of these two ages, he might have something not uninteresting to say. But as I have already remarked, the physical changes which have supervened since the beginning of the bronze age, sink into insignificance before those which can be shown to have taken place during the preceding stone age. It is to the relics of this latter age, then, and the lessons which these seem to teach that I wish now to direct attention.

The stone age is subdivided into two periods which are termed respectively the Neolithic or New-stone period, and the Palæolithic or Old-stone period. To the former belong those implements and weapons which are often more or less polished and finely finished, and which in variety of form and frequent elegance of design evince no inconsiderable skill on the part of the old workmen.

These relics occur, it may be said, throughout the whole length and breadth of the land—from the extreme north of Scotland to the south of England, and they are equally abundant in the sister island. As a rule, they are met with either at or near the surface of the ground, and they very frequently appear associated with the remains of such animals as the dog, the horse, the sheep, the pig, and certain species of oxen which are believed to have been the precursors of the present breeds.

The weapons and implements belonging to the older or palæolithic period are altogether of ruder form and finish. They are merely chipped into the requisite shape of adze, hatchet, scraper, or whatever the implement may chance to be. Although considerable dexterity is shown in the fashioning of these rude implements, yet they certainly evince much less skill on the part of the tool-maker than the relics of the newer or neolithic period. It is somewhat noteworthy also that while the implements of the neolithic period are made of various kinds of stones, those of the palæolithic period consist almost exclusively of flint; and so characteristic are the shape and fashion of the latter, that an experienced archæologist has no difficulty in recognising and distinguishing them at once from relics of neolithic age.

We find no tools or weapons of intermediate forms which might indicate a gradual improvement and progress from the rude types characteristic of palæolithic times to the more finished implements used by neolithic man. The one set of relics is sharply marked off from the other. Even a casual observer cannot fail to notice the marked difference between a collection of neolithic implements and one of palæolithic flint hatchets. Nor can we help thinking that the people who used the latter were far less advanced than, and decidedly inferior in mechanical skill and contrivance to the race or races by whom the polished implements were fashioned.

A distinct passage can be traced from the new-stone period into the age of bronze. We can see plainly how neolithic man continued to improve the shape and fashion of stone implements, until at last he acquired a knowledge of the use of metals. But between the disappearance of palæolithic man and the advent of his neolithic successor occurs a blank which hitherto the ingenuity of archæologists has failed to bridge over. Why is it that while a passage can be traced from the new-stone period into the bronze age, the two periods of stone should yet be separated by a clear line of demarcation? The social conditions and degree of advance shadowed forth by the appearance of the rude palæolithic implements belong to a lower grade of civilisation than that which the polished stone implements represent. It is not probable, but in the highest degree unlikely, that men would suddenly and completely abandon the use of rudely-chipped flint-hatchets, and all at once, as by a kind of inspiration, acquire the skill to fashion elegantly-shaped and even finely-polished implements of various kinds of stone. Long years were surely required for such a measure of progress. Had the palæolithic period merged continuously into neolithic times, it seems certain that the older rude forms would have been gradually improved upon, both as regards shape and finish. Moreover, if one may judge from modern analogies, this improvement would be a very slow process. We know that many savage tribes in building their huts continue repeating the same kind of structure from generation to generation, just as the birds do with their nests. We know also that in the fashioning of their implements and weapons they often show extremely little or even no advance whatever upon their forefathers. But, as just stated, we find no trace of any intermediate age of gradual progress linking on, as it were, the palæolithic to the neolithic period. If such an intermediate stage had obtained in Britain we should long ere this have discovered some evidence of it. Nothing of the kind, however, occurs, and we must explain its absence as best we may. But the fact of this break in the succession—this hiatus or gap in the record—will be rendered more apparent after we have learned something as to the mode and position in which the older stone implements occur.

Implements of palæolithic age are met with in caves and in certain ancient river-gravels under circumstances which we shall presently see argue for these relics a very great antiquity.

In the limestone districts of England caverns more or less abound. These remarkable cavities are formed by the slow percolation through the rock of acidulated water in the following manner. Rain-water, after passing over and sinking through the soil, always contains a certain proportion of carbonic acid, which it has derived from decaying vegetable matter. Thus acidulated, it soaks downward through the rock by natural cracks and fissures, the walls of which it gradually dissolves, and eventually makes its exit at some lower level as a spring or springs of *hard* water. In the course of long ages this constant circulation of rain-water and springs, and consequent waste of limestone, result in the formation of caves and winding galleries, through which very frequently considerable streams have found their way as underground rivers. Thus, both by chemical action and by the tear and wear of aqueous erosion, these subterranean galleries and caves have been gradually widened and deepened. When traced from lower to higher levels they prove in not a few cases to have once communicated with the upper surface by wide apertures, formed either by the falling-in of the roof or by the water eating along some natural fissure. Many of these apertures, however, are now closed up with calcareous matter and heterogeneous débris.

After streams had flowed in such underground channels for a longer or shorter time, they were often at last compelled to abandon them, either owing to one or other of the many changes which the subterranean forces have brought about, or to some local shifting of the subaerial part of their course, such as frequently happens during heavy floods. Caves which were in this way deserted naturally became the dens of wild beasts or the abodes of savage men.

Thus in the phenomena of the English caves we have, as Sir Charles Lyell has pointed out, the following succession of changes: 1st, a period when the caves and tortuous galleries were licked out by the percolation of acidulated water; 2nd, a time when these hollows became the channels of engulphed streams; and, 3rd, a period when these streams disappeared, and the caves were occupied by wild beasts and men, whose remains are now found commingled upon the floors of the caverns.

Let us now briefly examine the mode in which these remains occur. For this purpose it will be sufficient to select one example, as this may be considered typical of all the others. Our specimen cave is the famous Kent's Cavern near Torquay, which has been systematically ransacked for some years under the direction of a committee of the British Association, the details of the investigation being from time to time communicated by Mr. Pengelly, F.R.S., who has personally superintended the operations.*

The deposits met with upon the floor of the cave are given in descending order as follows:—

1. Large blocks of limestone, sometimes cemented together by stalagmite.

2. A layer of black muddy mould, 3 in. to 12 in. in thickness.

3. Stalagmite 16 in. to 20 in. thick, reaching 5 ft. in part, almost continuous, containing large fragments of limestone, a human jaw, and remains of extinct animals.

4. Red cave-earth, varying in thickness, and containing 50 per cent. of angular fragments of limestone, with numerous bones of extinct animals, and imple-

* Full details are given in Mr. Pengelly's interesting yearly reports, *Brit. Ass. Rep.*, 1865 to 1872; the same geologist gives an admirable summary of the evidence in a Lecture to the working classes at Manchester; see also Mr. Evans's *Ancient Stone Implements;* and Sir C. Lyell's *Antiquity of Man.*

ments fashioned by the hand of man. Excavated to a depth of 4 ft.

5. Crystalline stalagmite in places 12 ft. thick, with bones of the cave-bear.

6. Breccia and red loam, with remains of the cave-bear, and some human implements.

The large blocks of limestone that cumber the floor of the cavern have of course fallen from the roof. The layer of black muddy mould which immediately underlies them has yielded portions of the human skeleton, along with fragments of pottery and articles of stone and bronze. Besides these there also occur bones of deer, oxen, sheep, pig, and other animals which are still indigenous to the country. There is nothing therefore to indicate that this upper deposit is of great antiquity; some of the remains indeed appear to belong to the Romano-British period.

The black mould rests upon a pavement of stalagmite—a deposit which is formed by the drip of water holding carbonate of lime in solution. The accumulation of this deposit must in most cases be a very slow process, nor are we in much danger of over-estimating the time required for the growth of the stalagmite in Kent's Cave. A solid cake of stalagmite, varying in thickness from 1 in. to 5 ft., and almost continuous over the whole floor of this extensive cave, implies the lapse of a very long time. We have to conceive of it forming gradually, as drop after drop of lime-water fell from the roof and evaporated upon the floor. Nor was its accumulation at all likely to have been continuous: sometimes the drip would be in one place, sometimes in another; and indeed portions of the floor might remain unvisited during lengthened periods. One may gather some notion of the time required to form a layer of stalagmite 5 ft. thick, when he reflects that some two thousand years have elapsed since the Romano-British remains were left upon the floor of the cave, and that in all that time the deposition of stalagmite has been very partial—many parts being quite free from it, while where thickest it does not exceed 6 in.; in fact the deposit occurs only in patches.*

Below the stalagmite comes a mass of red earth or loam, of irregular thickness. It has not been excavated to a greater depth than 4 ft. Both the stalagmite and the underlying cave-earths have yielded a large quantity of bones and teeth of extinct, or no longer indigenous, mammalia, commingled with which are numerous implements of flint and some of horn, "presenting," as Mr. Pengelly says, "a character so humble and so little varied as to betoken a very low type of civilisation."

Scattered throughout the whole deposit, in stalagmite and earth alike, occur many fragments of limestone, similar to those which overlaid the black mould.

Underneath the cave-earth in certain parts a lower

* Of course, the rate at which stalagmite forms depends almost entirely on the quantity of acidulated water passing through the rock. In some caves the rate will be excessive; in others again it will be very slow. Hence, even if we ascertained the rate at which the stalagmite increased in one particular cave, still that would give us no criterion by which to estimate the time required for the growth of stalagmite in any other cave, for the conditions in any two caves are never likely to be precisely the same.

bed of stalagmite appears, which reaches in places the great thickness of 12 ft. This ancient deposit rests upon a second cave-earth or breccia, in which human implements and numerous remains of the cave-bear have been found. When one reflects on the length of time required for the formation of 12 ft. of stalagmite, the great antiquity of these lower deposits cannot fail to astonish him.

There are many interesting questions suggested by the remarkable commingling of mammalian remains in these cave accumulations, but I shall reserve what I have to say upon this subject to a succeeding chapter. Meanwhile the lessons we learn from the English caves would appear to be these:—

1st. That man and certain locally or altogether extinct animals co-existed in England at some remote period.

2nd. That the long duration of this period is shown by the thickness of the stalagmitic pavements, which rest upon and are intercalated with the cave-earth; and by the evidence of drip which is more or less conspicuous all through the cave-earth itself.

3rd. That after having occupied the English caves for untold ages palæolithic man disappeared for ever, and with him vanished many animals now either locally or wholly extinct.

4th. That the deposits immediately overlying the stalagmite and cave-earth contain an almost totally different assemblage of animal remains, along with relics of the neolithic, bronze, iron, and historic periods.

5th. That there is no passage, but on the contrary a sharp and abrupt break, between these later deposits and the underlying palæolithic accumulations.

But if cavern-deposits suggest a high antiquity for our race, still more does the evidence supplied by certain river-deposits, which now fall to be described.

It has long been known to geologists that in the south and south-east of England sheets and beds of gravel frequently occur in positions which are not, and never can be, reached by the present rivers. They rest upon the gentle slopes of the valleys in those regions at a height often of many feet above the streams; nay, they even occur on some hill-tops, and are spread over more or less isolated plateaux, that terminate abruptly at the margin of the sea—in both cases occupying positions which are quite beyond the reach of any possible river-floods. Nevertheless, in the great majority of cases there can be no doubt whatever that the deposits referred to are actually the products of river-action, and were accumulated just where we see them by running water. But if this be so—and it cannot be otherwise—then great changes in the drainage-system of the south of England must have taken place since these old gravels and loams were deposited. The accompanying illustration—a diagrammatic section across a river-valley—will make this intelligible. The horizontal line 1 indicates the level at which the river flowed when the deposits *a* and *a′* were laid down, *a* being gravel and *a′* loam. The former of these deposits marks out the bed of the ancient river at its average flow, while the loam *a′* represents the finer material which the water during flood-time spread along the low-lying slopes of the valley. As the river continued to flow it gradually ate into the underlying rock, until at last it reached the level marked 2. The gravel bed *b* was then formed, and the flood-loam *b′* was deposited above the older gravel bed *a*. In the same manner, when the stream had dug down to the lowest level 3, the gravel bed *c* was accumulated, while the flood-loam *c′* over-

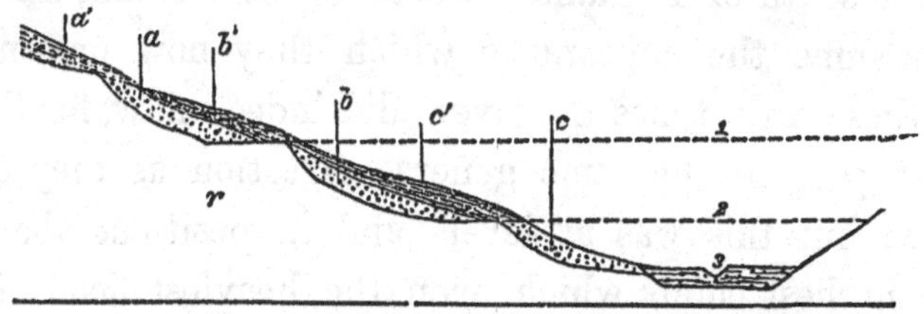

Fig. 55.—Diagram to explain formation of river-gravels and löss or flood-loam: *a b c* river-gravels, *a′ b′ c′* flood-loam; 1 2 3 successive levels of river erosion. (After Sir J. Lubbock: *Pre-hist. Times*, p. 381.)

spread the gravel of the higher level 2. Thus the deposits of gravel, sand, and loam that cover the slopes of these valleys mark out the successive levels occupied by the rivers as they slowly excavated the solid rocks to attain their present channels.

That the work of excavation required a long time for its performance no one will doubt. The erosion of solid rocks by running water, and the formation of thick accumulations of well-rounded gravel are necessarily very slow processes. Nor is it difficult to understand that the excavation of a valley thirty or forty miles in length, by some half-mile or more in breadth, to a depth of 50 or 100 ft., must have occupied a very long time indeed.

Now in almost all the old gravels which have just been described as lying quite beyond the reach of the present rivers, and as marking the successive stages of that erosion which eventually succeeded in giving to the valleys their present depth, flint implements are found in greater or less abundance. And from this it follows, as a matter of course, that man was an occupant of our country many long ages before the valleys in the south of England were hollowed out and made to assume the appearance which they now present. In those early times the rivers did indeed flow, for the most part, in the same general direction as they do now; but this was at levels, and in positions above the highest points which even the heaviest floods in our day could attain.

But this is not all. The old flood-loams and gravels have also yielded numerous remains of many extinct animals which are associated with the flint implements in such a way as to lead irresistibly to the conclusion that both were deposited contemporaneously. All these implements belong without exception to the palæolithic period. In none of the ancient river deposits do any traces of the neolithic age appear. Had polished implements of stone occurred in any of these old gravels even in small quantities, they would hardly have failed to escape detection; but when any such relics do happen to occur in alluvial deposits, these latter are invariably found to occupy positions which prove them to be of much more recent date than even the latest accumulations of what we term the ancient river-gravels. They are, in short, found at or quite close to the surface, and associated with such animals as the pig, the horse, the dog, oxen, &c.

To those who have never pondered over the subject the facts thus briefly stated may possibly not convey any vivid impression of the great antiquity of palæolithic remains. Indeed, an adequate conception of the vast lapse of time implied can only be attained by those who are willing to examine the evidence for themselves. Yet before passing to the next part of my subject, I may roughly indicate some of the geographical conditions that obtained in the south of England during the palæolithic period, in the hope that this may convey a livelier impression of the antiquity of man as an occupant of our country.

When palæolithic man lived in the south of England the Isle of Wight formed part of the mainland—a range of chalk downs, at least 600 ft. in height, running east from the Isle of Purbeck, and joining on to the Needles. The rivers that traverse Dorset, together with the Avon and the Stour, then united in one large river and flowed eastwards along a broad valley which is now occupied by the Solent and Spithead, while Southampton Water at that early period formed merely an affluent of the same great stream.* The gravel

* The Rev. W. Fox of Brixton was the first to infer the former existence of this ancient river; see *Geologist*, vol. v. p. 452. The whole subject has been worked out in great detail by Mr. T. Codrington (*Quarterly Journal of the Geological Society*, vol. xxvi. p. 528); see also Mr. Evans's *Ancient Stone Implements*, p. 605; the two last-named authors arrived independently at substantially the same conclusion.

that paved the bed of this ancient river before it had cut its way deep down into the older tertiary deposits, is now found capping the cliffs of the mainland at heights ranging from 50 ft. to 130 ft. above the level of the sea.

Again, so long a time elapsed during the palæolithic age that the Thames at London was able to excavate its valley to a breadth of four miles, and a depth of not less than 100 ft. Everywhere, indeed, throughout the south of England we have the most clear and convincing evidence that the land was worn away by streams and rivers which, at the beginning of the palæolithic age, flowed at a height of 20, 40, 60, or 100 ft., and more above their present levels; and the deepening of these valleys was completed before the neolithic age commenced. Everywhere, too, we find convincing proofs that during palæolithic times our shores stretched much farther out to sea. "When we remember," says Mr. Evans, "that the traditions of our mighty and historic capital, now extending across the valley, do not carry us back even to the close of that period of many centuries, when a bronze-using people occupied this island;—when we bear in mind that beyond that period lies another of probably far longer duration, when our barbaric predecessors sometimes polished their stone implements, but were still unacquainted with the use of metallic tools;—when to the historic, bronze, and neolithic ages we mentally add that long series of years, which must have been required for the old fauna, with the mammoth and rhinoceros, and other to us strange and unaccustomed forms, to be supplanted by a group of animals more closely resembling those of the present day;—and when, remembering all this, we realise the fact that all these vast periods of years have intervened since the completion of the excavation of the valley, and the close of the palæolithic period, the mind is almost lost in amazement at the vista of antiquity displayed." *

In fine, then, whether we have reference to the evidence furnished by the cave-deposits or to that supplied by the river-gravels, we are alike assured that after the disappearance of palæolithic man, and before the advent of his neolithic successor, there supervened a period of unknown but necessarily of great duration. There is no direct passage from the older into the newer deposits—the latter overlie the former unconformably. It now becomes our task to ascertain, if possible, what it was which interrupted the sequence—to discover what is meant by the break or hiatus that separates palæolithic from neolithic times.

* *Ancient Stone Implements*, &c., p. 622.

CHAPTER XXX.

CLIMATE OF THE PALÆOLITHIC PERIOD.

Groups of mammalian remains associated together in palæolithic deposits.—Southern, arctic, and temperate species.—Theory of annual migrations.—Condition of this country during palæolithic period.—Cause of extreme climate of Siberia.—Supposed elevation of Mediterranean area in palæolithic times, untenable.—Probable climatal effects of such an elevation.—Climate of western Europe during palæolithic period not "continental" but "insular."—Conclusions.

REFERENCE has been made in the preceding chapter to the fact that both in cave-deposits and river-gravels human implements are found associated with numerous mammalian remains belonging to species many of which are either now locally or wholly extinct. The appearance of these remains suggests many interesting inquiries, but at present I shall confine attention to one question—namely, what were the climatal conditions under which these animals occupied our country.

The species naturally arrange themselves in three groups:* the first embracing those animals which are either at present living in warm climates, or which

* In these notes I have availed myself of the lists given by Mr. Boyd Dawkins in several papers; see *Quarterly Journal of the Geological Society*, vol. xxv. p. 192; *Popular Science Review*, 1871, p. 388; *Quart. Jour. Geol. Soc.*, 1872, p. 410.

have in southern regions their nearest representatives; the second comprising animals of arctic and northern habitats; and the third containing such species as inhabit temperate latitudes.

In the first or southern group we meet with the lion, the tiger, the spotted hyæna, two extinct species of elephant, and two of rhinoceros, *Felis caffer*, an extinct species of tiger furnished with terrible sabre-like teeth, and the hippopotamus.

In the second or northern group occur the glutton, the reindeer, the musk-sheep, the pouched marmot, the alpine hare, the lemming, and the extinct mammoth and woolly rhinoceros.

The third or temperate group comprises the bison, the great urus, the grizzly bear, the cave-bear, the Irish elk, and Brown's deer (the last three being extinct species). Besides these we find also the lynx, wild cat, ermine, stoat, weasel, martin cat, otter, wild boar, horse, beaver, and other smaller animals which are still indigenous to Britain.

The panther, the fox, and the wolf also occur in British palæolithic deposits; but since these animals are capable of bearing great vicissitudes of climate, it is evident, as Mr. Dawkins remarks, that they can tell us nothing about the climatal condition of our country during palæolithic times.

Now what does this very remarkable assemblage of animals indicate? How are we to explain the occurrence in our caves and river-deposits of species belonging to such widely-separated zones? In what manner have the hippopotamus, the reindeer, and the musk-sheep become entombed in our ancient superficial deposits? I need hardly say it is quite impossible that these animals could have lived side by side. The musk-sheep, like the reindeer, feeds chiefly on lichens and grasses. In the warm months it ranges to Melville Island, and its remains were found by Dr. Kane as far north as the dreary regions of Smith's Sound. Even during the coldest season of the year it does not wander farther south than the southern borders of the Barren Grounds. The reindeer, as every one knows, is confined in Europe to Lapland and Norway. The hippopotamus, on the other hand, is restricted to the rivers of middle and southern Africa. At the same time it is highly probable that it may have ranged within comparatively recent times as far north as Nubia. Nay, we may even admit that the physical condition of Egypt is such as would be quite suited to the wants of the great river-horse, and that the present restricted range of that animal is due to other than climatal causes. But although we thus widen the limits within which the hippopotamus might live and flourish, there still intervenes a vast region up to the mountains of Norway, throughout which the conditions are equally unsuited to the reindeer and the hippopotamus. These two animals, in short, belong to distinct and widely-separated zones.

Various theories have been proposed to account for the intermingling in our superficial deposits of such discordant species. Some have held that the hippopotamus may have been covered, like the mammoth, with a thick woolly coat, to enable it to brave the severe winters of the north; but it is impossible to believe that an animal the greater part of whose time is spent in the water could have lived in a country that was occupied at the same time by musk-sheep, reindeer, and gluttons—the presence of such characteristic arctic species clearly indicating that during winter at least our rivers were frozen over.

Another view which has been largely adopted attempts to get over the difficulty by assuming that during the accumulation of our river-deposits Britain was marked by a kind of Siberian climate—by strongly contrasted summers and winters. The summers are supposed to have been so warm and genial that they wooed to the latitude of Britain the elephant and hippopotamus, which upon the approach of winter beat a rapid retreat to the south—the country they had just vacated being next invaded from the north by the arctic mammalia. A parallel to this state of things was thought to be found in Siberia, where at the present day, as is well known, great annual migrations of the fauna take place. During the intense cold of winter the reindeer roam in vast herds towards the south to gain the shelter of the forest-lands, and thus encroach upon the province of the moose-deer, or elk, where they become the prey of the tiger. As the cold of winter disappears the reindeer again pass to the north, whither they are followed by foxes, wolves, and bears. Similar migrations characterize the fauna of the northern latitudes of North America. In both cases, however, the migrations are between arctic and temperate zones, and are therefore hardly comparable to the migrations which (if this theory be true) must have characterized Europe during palæolithic times. The invasion by the reindeer of the province occupied by the elk does not astonish us, for their territories border and overlap. But it is extremely difficult, or even impossible, to believe that the musk-sheep and the hippopotamus could ever have traversed the same country in the same year; at all events, we know of no annual migration so astounding as this. Furthermore, it cannot be said that the hippopotamus, the rhinoceros, or the elephant are migratory animals. The tiger, as we know, will travel far from his usual head-quarters in pursuit of prey, but we cannot say as much for the lion and the spotted hyæna. The frequency with which remains of these animals, more especially the hyæna, occur in our caves, shows clearly that they were not mere summer visitors, but actual denizens of the country throughout the year. Looking therefore simply to the character of those species that make up the southern group, there appear to be strong grounds for rejecting the "migration theory." And we are inclined still more strongly to this opinion when we come to consider the climatal conditions under which, according to that theory, these animals inhabited Britain.

It is inferred that the southern and northern groups oscillated to and fro with the seasons at a time when the British Islands were joined to themselves and the Continent, and when the conditions were such that snow and ice still covered large areas in Wales, the north of England, Scotland, and Ireland; it is held, in short, that the palæolithic age corresponds to the latest period of local glaciers in our country.

Now it would not be difficult to show that, with the presence of perennial snow and ice in the high-grounds of Great Britain, our rivers would remain frozen over for a great part of the year. During the summer they would sometimes burst their icy bonds, and, swollen with heavy rains and the torrents derived from the melting of the snow-fields and glaciers, would overflow the low-grounds and carry devastation far and wide. Over broad areas in our valleys, therefore, there could be little vegetation.* Such areas as were not covered with snow might perhaps support scraggy birch and pine, and here and there tall grasses and dwarf willows might nestle in sheltered hollows; but mosses and lichens would most probably form the prevailing growths. If the pleistocene hippopotamus could live on such fare, he must have been a more easily satisfied animal than his modern representative. It is true that in arctic regions the short summer brings into bloom a number of pretty flowerets, and causes the grasses to shoot up with surprising rapidity; but this is due to the influence of a sun that keeps above the horizon during the greater part of summer. A glacial period in England, however, would not be tempered by the presence of a midnight sun in summer-time.

* Those who are familiar with the glacier-valleys of Switzerland and Scandinavia will understand what I mean. If even in such valleys the rivers escaping from the glaciers cover wide areas with barren heaps of gravel and shingle, how much more must this have been the case in Britain, where, according to the migration theory, each intense winter was succeeded by an excessive summer temperature.

With the poor and meagre vegetation therefore which a country of this kind would be able to support, the herbivora could not have been abundant. Indeed, only a very few mammalia could possibly have existed under the conditions supposed; and probably none of these would belong to that group which consists of such species as still inhabit the temperate zones of Europe and America. If, therefore, we eliminate the whole of this group, we have left only the true arctic mammalia—the glutton, the reindeer, the musk-sheep, the pouched marmot, the tailless hare, and the lemming—to form food for the lion and his congeners, the tiger and the hyæna. But how can we suppose it possible that these carnivora would leave the temperate zone (which, when arctic conditions supervened in Britain, must have shifted to more southern regions of Europe, where no doubt it was characterized by the presence of an abundant mammalian fauna), to prey upon the few reindeer and smaller animals that were likely to be found so far north as the frozen barrens of England? Yet we know that the hyæna at least was a regular denizen of our caves, and found matters so comfortable, that, according to Mr. Dawkins, its large size, as compared with that of the living animal, was probably due to "the abundance of food which it obtained."

Such considerations as these have led me to conclude that this theory of annual migrations is untenable. With glacial conditions in Scotland and the hilly grounds of England and Ireland, neither temperate flora nor fauna could have existed in our country. There could be nothing in such an arctic England, therefore, to tempt the herbivora away from the rich feeding-grounds of the then temperate zone, and just as little to wile the carnivora so far from their wonted haunts. And to suppose the hippopotamus, the elephant and the rhinoceros capable of migrating for enormous distances, and to such a country too, is to suppose that these animals differed entirely from their present representatives.

But it has been objected that we have no right to infer, from the fact of local glaciers having existed in Britain, that the summer temperature during that period was ungenial. Large glaciers exist in the Alps, and much greater ones fill the upper valleys of the Himalaya; yet, as every one knows, the low grounds at the base of these mountains enjoy warm and genial climates. Is not it possible, therefore, that the same conditions may have obtained in Britain during palæolithic times, and that while glaciers filled all our mountain-valleys our lowlands may have basked in summer warmth and sunshine.

Now it can hardly fail to occur to the reader that these are not cases in point. Britain is not in the latitude of northern India, neither is it in that of Lombardy or Provence, while in comparison with the Alps and the Himalaya our mountains sink into utter insignificance. Those who speculate upon the possibility of something like a Siberian climate having obtained in Britain during palæolithic times, completely ignore geographical and physical considerations. They find northern and southern mammalia in our caves and river-gravels, and jump at once to the conclusion that a period of strongly contrasted summers and winters must have characterized Britain at the time it was in the occupation of such discordant groups of animals. Yet a little consideration will serve to show that this hasty conclusion is entirely opposed to and directly contradicted by all that we know of the causes to which the various climates of the globe are owing.

It is admitted by geologists that during palæolithic times the physical features of our continent were much the same as they are now—all the great high-grounds, all the great drainage-systems, were even then in existence. Denudation, or the wearing away of the earth's surface by rains, frosts, and rivers, no doubt helped during palæolithic times to modify the features of the land, and valleys, as we have seen, were scooped out to some depth; nevertheless, when viewed on the large scale, such minor modifications of the surface may be disregarded. Then, if this be so, —if the changes referred to were so inconsiderable— we may be very sure that since the disappearance of palæolithic man, there has been even less change in the physical features of Europe. This is one important fact to remember.

Again, we know that under the present condition of things, our continent owes the equableness of its climate in great measure to the influence of the Atlantic. This is another important fact that we cannot lose sight of. Were an extensive land-surface to be substituted for that great ocean, there is good reason to believe that our summers and winters would be as excessive as those in similar latitudes of Asia and North America.

But at the present day what is the fact; why, simply this, that as we recede from the western shores of Europe, and approach the interior of that continent, the colder do the winters become, while the difference between summer and winter constantly tends to increase. In England the thermometer rarely sinks 18° below freezing-point, but in the interior of the continent, and under almost equal latitudes, we find a mean of +14°; and "it is not uncommon to see the mercury freeze at Kasan. In the interior of Siberia it often remains solid for several weeks together."* To show the contrast between the climate of western Europe and that of northern Asia, we may compare the summer and winter temperatures of Aalesund, in Norway, and Jakutsk, on the Lena. These two places are as near as may be under the same latitudes—the former being in 62°·24 N. lat., and the latter in 62°·5 N. lat. At Aalesund the summer temperature is 57° F., and in winter the thermometer registers 35° F. At Jakutsk, on the other hand, the temperature of summer reaches 62°·2 F., while that of winter sinks to 43°·8 below zero.

Now what are the causes that induce such widely divergent climates? Why is it that places situated under the same or nearly similar latitudes do not experience the same climatal conditions? If the reader will examine the charts of isotheral and isochimenal

* Kaemtz's *Complete Course of Meteorology*, p. 173.

lines, he will have no difficulty in perceiving that Europe is beholden to the presence of the Atlantic for its insular climate. Our winters are ameliorated and our summers rendered less excessive by the moist winds that are almost continually passing from west and south-west across our continent. That is a fact as well ascertained as any fact can be.

In northern Asia the conditions could hardly be more strongly contrasted. There no mild and genial ocean tempers the severity of winter. During that season every wind that blows across Siberia, no matter whence it comes, is bitingly cold. The west winds, that temper our winters with the warmth of the gulf-stream, are robbed of their moisture and cooled down before they cross the snows of the Ural Mountains to pour into Siberia. The gales from the Arctic Ocean are still colder; nor is much warmth derived from the winds that sweep northwards from the high Mongolian plateau, while at the same time the serenity of the sky favours the radiation of the ground. In summer-time the conditions are reversed. Dry and scorching winds reach Siberia from the west, and the heat of the Mongolian deserts is wafted from the south; while owing to the great clearness of the atmosphere the northern plains are soon warmed by the continuous shining of the arctic sun: and thus the temperature rises rapidly all over Siberia. In North America the seasons are also more marked than with us, and the causes for this are somewhat similar to those which induce the more strongly-contrasted summers and winters of northern Asia.

If the present climate of Europe departs so very widely from that of other regions in similar latitudes of the northern hemisphere, upon what grounds can it be supposed that a totally different state of matters existed during palæolithic times? Let it be remembered that the physical features of the land were much the same then as they are now, and that the Atlantic Ocean also was certainly in existence. No great range of mountains existed then which does not exist now, nor have any mountains been formed in Europe since palæolithic man vanished for ever. It is true that the British Islands have been separated from themselves and the continent, and that a narrow strip of land that once extended along the western borders of these islands and the mainland has been submerged; but it is idle to imagine that such changes could have had the effect which some have supposed. Nay, even if we conceived that a large part of the Mediterranean existed in the condition of dry land* during palæolithic times, still, that would not produce anything approaching to a Siberian climate in Europe.

We know that in the present economy of things the climatal influence of the Mediterranean in summer-time extends but a very short distance north into Europe. It certainly does not cross either the Alps or the Pyrenees. A line drawn from Bayonne by Viviers and Turin to the head of the Adriatic,

* This view has been brought forward by Mr. Dawkins (*Quart. Jour. Geol. Soc.*, vol. xxviii. p. 410), but, as I have endeavoured to show, the evidence he adduces is insufficient; see *Geological Magazine*, vol. x. p. 49.

corresponds to the July isothermal line of 72°·5 F. When the same line is continued towards the north-east into Hungary, it suddenly sweeps down to Constantinople; and, circling round the Black Sea, returns upon itself at Odessa, after which it continues for some distance in the normal north-easterly direction. That sudden curve to the south-east is, of course, due to the presence of the Black Sea, and seeing that the Mediterranean does not influence the same isothermal line in either France, Italy, or Austria, it is quite clear that the modifying effect of this large inland sea has nothing whatever to do with the climate of central Europe. South of the line referred to, however, we find the July isotherms violently disturbed. That of 77° F runs along the shores of the Mediterranean from Barcelona by Perpignan and Montpellier to Toulon. From this place it strikes directly south-south-east to Cape Spartivento, then curves gently on to Palermo, after which it sweeps round by Messina and the Gulf of Taranto, across Italy, and more than half-way up the Adriatic. It then doubles suddenly back upon itself, skirts the eastern shores of the Adriatic, and crosses the south of Turkey and the Ægean Sea to Mytelene and Asia Minor. This irregularity is just as surely caused by the presence of the Mediterranean as the abrupt curving of the July isothermal line of 72°·5 is due to that of the Black Sea. In short, it is perfectly apparent that the isotheral lines in the south of Europe are deflected by the inland seas from their normal direction, this deflection being confined to the immediate proximity of these basins. But in central and northern Europe, all the isotheral lines are pushed south by the overwhelming influence of the Atlantic; while in the opposite season, all the isochimenal lines are swept boldly north and north-east. (See charts at end of volume.)

Were the Mediterranean to be converted into two land-locked seas, the isotheral lines in that area would then be less disturbed, and would preserve the general south-west and north-east trend which they follow in central Europe. The summers in Italy, Turkey, and Greece would thus be somewhat hotter and drier, but the general summer temperature of Europe would remain unaffected. The winds flowing from the south would no doubt be warmer, but they would also carry less moisture. When they reached the Alps, they would part with their warmth and moisture, just as they do now; but since they would bring more of the one and less of the other than at present, the snow-fields and glaciers on the south side of the Alps would tend to shrink back. Again, the isothermal chart shows that the supposed increase of land would not produce any appreciable climatal effect beyond a short distance into Spain and France, while not a trace of its influence would reach England. The Continent, laved along its entire western borders by a wide ocean, would be cooled over an immense area by the winds blowing from west and south-west; such is the case now,* and

* See the late Professor Coffin's treatise on the "Winds of the Northern Hemisphere," *Smithsonian Contributions* for 1853.

it must have been equally so in palæolithic times. Thus, even had the warmth derived from the Mediterranean area been much greater than we can conceive it to have been, its effect would, nevertheless, be counterbalanced, and far more than counterbalanced, by the influence of the Atlantic.

The consideration of this subject brings us back to the point which, as I think, has already been sufficiently proved, namely, that during our local glacier period, there could be no great annual migrations of the mammalia in western Europe. For, since the climate required for such migrations does not exist in Europe now, and seeing that it would not obtain even if Britain were to become part of the Continent, and a large portion of the Mediterranean to be converted into land, how can we suppose it possible, that, with snow-fields and glaciers in Britain, the climate could have been other than cold and ungenial. Surely, under such geographical and physical conditions as did actually supervene in north-western Europe during our latest period of local glaciers, the summers must of necessity have been cold, and the vegetation poor and scanty. In winter-time, supposing the gulf-stream to have flowed then as it flows now, the cold would be ameliorated over all western Europe. But should it be supposed, that, during our local glacier period, the gulf-stream flowed in some other direction, then, in the absence of this great heat-bringer, our winters would indeed be excessive, and our summers dreary in the extreme.

In short, we may conclude, that, so long as Europe

exposes a vast line of coast to the Atlantic, and so long as her physical features endure, just so long will her climate continue to differ from that of either Asia or North America, no matter whether or not the British Islands become continental or the area of land in the Mediterranean basin be increased. And as it is in the present, so also it must have been in the past. No mere obliteration of our inland seas could neutralise the influence of the outlying ocean. If the summers of Europe are at present rendered less excessive by the presence of the Atlantic, the same must have been the case during the last continental condition of our islands, and that to a much greater degree, owing to the presence of more numerous and larger snow-fields. For this, if for no other reason, it seems to me that the theory of annual migrations during that period in western Europe must be abandoned.

I have discussed this theory at some length, because, as we shall presently see, the question of palæolithic climate has a direct bearing upon the antiquity of man. In order, therefore, that I may carry the reader along with me, I shall here briefly summarise the points which I have attempted to establish in this chapter:—

1. The migrations of land animals in northern Asia and equivalent latitudes in North America, take place between arctic and temperate regions. This is simply the case of adjacent provinces overlapping one another. Inasmuch, therefore, as the migration theory asks us to believe that widely-separated zones

overlapped across the whole breadth of the temperate provinces, it is unreasonable, and not supported by our knowledge of what actually occurs in nature.

2. The general character of the southern group of mammalia, as exhibited in cave-deposits and river-gravels, is non-migratory.

3. The union of Britain and Ireland to the Continent, across the up-raised beds of the English Channel and the Irish Sea, and a great increase of land within the area of the Mediterranean, could not confer upon Europe a climate in any degree approaching to that of Siberia or British America. The climate of our continent would still be *insular*, and consequently great migrations could not take place.

4. During the last continental condition of our islands, snow-fields and glaciers existed in our mountain regions—betokening a climate quite unsuited to the needs of the southern mammalia. The winters at that period must have been excessive, and the summers cold and ungenial.

5. Lastly, so long as the Atlantic continues to wash the coasts of Europe, and so long as the present configuration of the land endures, our continent must continue to enjoy an insular climate, and there is not the slightest physical evidence to show that it possessed any other kind of climate during the period that the southern mammalia inhabited Britain.

CHAPTER XXXI.

GEOLOGICAL AGE OF THE PALÆOLITHIC DEPOSITS.

Oscillations of climate.—Cold and warm periods.—Evidence of the river-deposits.—Character of the evidence cumulative in favour of former climatal changes.—No evidence of warm post-glacial climate.—Southern mammalia not of post-glacial age.—Age of cave-deposits.—No proof that they are post-glacial.—Relation of river-gravels to glacial deposits.—Age of the boulder-clay at Hoxne, &c.—Boulder-clay upon which palæolithic deposits sometimes rest belongs to the older glacial series.—Distribution of palæolithic gravels.—Comparison between palæolithic gravels of South England and river-deposits of the north.—Palæolithic deposits not met with in districts that are covered with accumulations belonging to the later glacial series, but confined to regions which we cannot prove to have been submerged during the latest period of glacial cold.—Palæolithic deposits of inter-glacial, not post-glacial age.—Bulk of them probably belong to last inter-glacial period.—Recapitulation.

THE theory of annual migrations being, as I have tried to show, untenable, we can now only explain the remarkable commingling of northern, southern, and temperate groups in our superficial deposits, by assuming that certain great oscillations of climate characterized the accumulation of our cave-earths and river-gravels. We must admit in short that the northern mammalia occupied Britain during a cold and arctic condition of things, and that on the other hand the southern forms prevailed over the same area at a time when our winters were mild and genial.

This result derives strong confirmation from a study of the deposits themselves. In the first place it has been shown by Mr. Prestwich and others that the ancient river-gravels and loams which occur at the highest levels above the present streams possess certain broad characters that serve to dis-

Fig. 56.—Generalised diagram-section across the valley of the Thames below London, greatly exaggerated vertically. (W. Whitaker): 1. Pebble-gravel (pre-glacial); 2. Boulder-clay; 3. valley-gravel (highest terrace); 4. valley-gravel and brick-earth (lower terraces); 5. the Thames and its marshes; × sea level; *r* tertiary beds and chalk.

tinguish them in a rough way from the gravels and loams that occupy the lower levels of the valleys. As a rule the former are the coarser of the two, and frequently contain large blocks of stone which could only have been transported by river-ice. Add to this that the beds often give evidence of having been subjected to a violent "push," which has tumbled them up and thrown them into confusion — phenomena which, as Mr. Prestwich has pointed out, could hardly have been produced otherwise than by the grounding of heavy ice-rafts. The general absence of these appearances, and the usually finer-grained character of the ingredients in the low-level gravels would seem to point to a milder condition of things—to a time when the rivers were less liable to flood, and ice-rafts were uncommon.

According to Mr. Prestwich * the shells met with in the high-level beds have a northerly range, and the absence of southern species tends still more to distinguish these beds from those of lower levels. In the latter there occur in great abundance two species of shells (*Cyrena fluminalis* and *Unio littoralis*), neither of which are now found living in England, but tenanting the rivers of more southerly latitudes,† a fact that seemingly corroborates the inference deduced from the appearance of the beds themselves.

Again, it cannot be denied that the northern group of mammalia is most characteristic of the high-level gravels, and the southern group of those at the lower levels. And notwithstanding that remains of both northern and southern species are not uncommonly commingled, still it is a fact, although, as Sir J. Lubbock remarks,‡ "too much importance must not be attached to the observation, that our ancient hippopotamus has been less frequently found in association with the mammoth and the hairy rhinoceros, than with *Elephas antiquus* and *Rhinoceros hemitœchus* (Falc.)," two species which had a more southerly range.

If the high and low-level gravels were deposited during different climatal conditions, this "might have produced," as Mr. Evans says,§ "some effect on the method of living, and on the implements of

* *Philosophical Transactions*, 1864, p. 278.

† *Cyrena fluminalis* is now extinct in Europe, but it still inhabits the Nile and abounds in Cashmere; *Unio littoralis* lives in the Seine and the Loire.

‡ *Pre-historic Times*, third edition, p. 299.

§ *Ancient Stone Implements of Great Britain*, p. 616.

the men of the two periods." And he thought it at one time "probable that a marked distinction might eventually be drawn between the high- and low-level implements; but so far as Britain is concerned," he is now of opinion that "this can hardly be done. Still the *facies* of a collection from two different spots is rarely quite the same, and" he thinks "there is generally a preponderance of the ruder pointed implements in the high-level gravels, and of the flat, ovate, sharp-rimmed implements in the low-level."

Now it may be admitted that none of these facts taken separately and alone is quite convincing. The character of the evidence, however, is cumulative, and when we perceive that the whole points in one and the same direction, it is impossible not to feel its weight and cogency. Under the assumption that considerable changes of climate accompanied the deposition of the palæolithic beds, the facts enumerated above have a distinct and definite meaning, but they are utterly meaningless, and must even be quite ignored if we are to accept the theory of annual migrations.

The fact that the remains of northern and southern forms occur commingled in our river-deposits, is only what one might have expected. Those who study the formation of fluviatile sediments will quickly understand how fossils, entombed at widely separated intervals, may come to occupy the same level. Rivers are constantly cutting down through their own deposits, and again filling up the excavations they make. In this way gravel and sand are banked against similar beds, which may belong to a much greater antiquity; and the line of junction it is often impossible to determine, the one deposit seeming to shade into the other. And thus beds which appear to be continuous and of contemporaneous origin may, and in point of fact do frequently deceive us in these respects. There is nothing, therefore, abnormal or extraordinary in the intermingling in our palæolithic beds of mammalian remains belonging to widely separated provinces. It would have been surprising indeed had it been otherwise.

So far, therefore, as the direct evidence of the beds themselves is concerned, there seems to be nothing against but a good deal in favour of the theory of changes of climate. That theory not only explains appearances which are left unsolved by any other hypothesis, but when we come to take a wider view of matters, and to consider the relation of the palæolithic deposits to the glacial drifts we shall find that no little light is cast upon the interesting question of the antiquity of man in Britain.

From the discussion in our last chapter it would be gathered that the palæolithic deposits are believed by certain geologists to belong to late glacial and post-glacial times: that is to say that man and his congeners, the mammals of the caves and rivers, did not enter Britain until the Great Ice Age was passing away. Now if this were so we should be forced to admit that at least one warm period had supervened after the disappearance of the last local glaciers, and before the advent of the present temperate condition of things. Such a period of mild and genial winters being necessary to account for the presence of the hippopotami, elephants, and rhinoceri of the caves and river-gravels.

Have we any evidence, then, for the former existence of a warm post-glacial period? It may be confidently answered—none whatever. There are few points we can be more sure of than this, that since the close of the glacial epoch—since the deposition of the shelly clays, and the disappearance of the latest local glaciers—there have been no great oscillations but only a gradual amelioration of climate. It is impossible that any warm period could have intervened after the last cold, and before the present temperate conditions, without leaving some notable evidence in the superficial deposits of Scotland, the north of England, Scandinavia, and North America—these being the countries in which, as we have seen, the passage from the later glacial beds to recent accumulations can be most distinctly traced. The climate of Britain is milder now than at any other period subsequent to the re-elevation of our country after the last great submergence; our winters have gradually become less intense; Britain has slowly passed from an arctic to a temperate condition of things: and Scandinavian, Swiss, and American geologists tell the same tale as regards their countries. From all this, then, it follows that the southern mammalia could not have lived in

Britain during post-glacial times. They must belong either to pre-glacial or inter-glacial ages, or to both.

When we come to ask why our cave-deposits have so frequently been relegated to post-glacial times, we get no satisfactory answer. If we put out of consideration the upper layers in certain cave-deposits, there is really nothing in the bone-earths and breccias to limit the age of these accumulations to such a recent period. Accordingly, many geologists have been of opinion that the mammalian remains occurring in the caverns were introduced at various epochs, and may belong to pre-glacial equally with post-glacial times. Buckland thought that the great mammals existed in Britain before the period of the diluvium or drift; and his belief is shared by Mr. Godwin-Austen and some of our leading geologists. Professor Ramsay is decidedly of opinion that "caves such as those in which mammalian remains occur must have existed in pre-glacial times; and therefore it would be strange," he adds, "if none of those explored contained pre-glacial remains." But between pre-glacial and post-glacial times there intervened several mild inter-glacial periods, during which, as I have shown, the climate in Scotland was suited to the wants of the mammoth, the reindeer, the Irish deer, the urus, and the horse; and I have also pointed out that, so far as geological evidence is concerned, there is nothing to show that some inter-glacial periods may not have been warm enough to cause all the snow and ice to vanish from Britain. If such was the case as regards Scotland, England must have been characterized by similar climatal conditions. It is, therefore, not necessary to suppose that any large portion of the bone-deposits was pre-glacial; for during inter-glacial periods the caves would form dens for wild beasts, just as they must have done in pre-glacial times. To some such mild and genial inter-glacial period or periods I would refer the hippopotamus and the other southern forms met with in English caves. The conditions under which these animals lived need not have been comparable to those that characterize the tropics. All that we are entitled to infer is that the winter temperature of Britain during certain inter-glacial periods must have been mild and genial. It is by no means necessary to suppose, however, that the summers were much warmer than they are now. An equable and genial climate, with no great difference between the seasons, would nourish an abundant vegetation, and render the country habitable by a prolific mammalian fauna.

Now let us turn our attention to the river-gravels, for the purpose of ascertaining what relation they bear to the glacial deposits. It has been shown that, in some valleys at least, the river-gravels are of more recent date than certain accumulations of glacial age. Thus, for example, at Hoxne, in Suffolk, fresh-water beds containing flint implements were seen by Mr. Prestwich to overlie boulder-clay. From this it has been inferred that these beds, and indeed all palæolithic deposits, are of post-glacial age—a conclusion which was reasonable and indeed inevitable, so long as the existence of only one deposit of boulder-clay was known. But in order to prove the post-glacial age of the Hoxne river-beds, it now becomes necessary to show that the boulder-clay at that place can only belong to the close of the glacial epoch: for as we have seen boulder-clay may be of various ages. There are two boulder-clays with intervening sands and gravels in East Anglia; in Lancashire and the north-west a similar succession of probably more recent date occurs, and the same is the case in Ireland. In Scotland, again, we find fresh-water beds intercalated with the till, and, further, ancient river gravels are seen lying above that stony clay, and covered by yet more recent glacial accumulations. Finally, distinct breaks representing mild periods occur in the glacial formations of Scandinavia, of Switzerland, and America.

To what age, then, does the boulder-clay at Hoxne belong? There can be no doubt whatever that it is older than the upper boulder-clay of Lancashire and the north-west of England, and certainly dates back to a much earlier period than the shelly clays and local moraines of Scotland. Indeed, judging from its position, so far removed from the great centres of glaciation, I had no hesitation in correlating it with the chalky boulder-clay of Mr. Searles V. Wood—a correlation with which my friend and colleague, Mr. Whitaker agrees. There can be no doubt, he says, that the boulder-clay at Hoxne forms part of the widespread sheet that covers Essex, Herts, Beds, Cambridge, Suffolk, Norfolk, &c. In short, it occupies the same geological horizon as the boulder-clay that one may trace along the north-east coast of England into Scotland at Berwick—a deposit formed at the time when the great ice-sheet was shrinking back. It is evident, then, that the ancient river-gravels at Hoxne belong to a later date than either the till (No. 1)* or boulder-clay (No. 2). This fact, however, does not enable us to say that the fresh-water gravels in question are of post-glacial age. We know that after the deposition of the boulder-clay (No. 2) a mild inter-glacial period supervened, and that this period was succeeded by a new advance of the glaciers and an age of floating ice. This is a case in which mere superposition fails to tell us all that is necessary to be known, before we can conclude that the overlying beds are of post-glacial age.

Before we can make up our minds as to the geological age of the river-gravels in question, it is evident that we must consider the distribution of similar accumulations throughout the country, and their relation to the drift-deposits belonging to the last cold period. When we have done so, the first noteworthy fact that we shall remark is that palæolithic river-gravels are exclusively confined to the south and south-east of England. They occur in the valleys of the Ouse, the Waveney, the Thames, the Avon, and their numerous tributaries, and at various places along the southern coast of England.

* The figures refer to the numbers in the Tables given at p. 426.

North of the Ouse and west of the valley of the Axe, no river-gravels have yielded any palæolithic implements.* Nor has a single trace of gravels belonging to this age been anywhere recognised in Wales, the northern counties of England, Scotland, or Ireland. The tool-bearing drifts are thus singularly confined to a somewhat narrow and circumscribed area.

North of this area we find river-gravels of undoubted post-glacial age lying in the bottoms of the valleys, and occupying positions that seldom rise more than a few feet or yards above the present levels of the streams. They are also, as far as bulk is concerned, of inconsiderable importance. The palæolithic gravels of the south, however, not only attain a great thickness, but from their position we can see that since the time of their formation very considerable derangement of the drainage-systems has taken place. Mr. Prestwich remarks: "One feature of these deposits is, that although closely related to the present configuration of the surface, yet they are always more or less independent of it. They are often near present lines of drainage, yet could not as a whole possibly have been formed under their operation." In short it holds generally true that while the palæolithic gravels of the south began to be laid down at a time when the streams were commencing to hollow out their valleys, the gravels of the north were for the most part not deposited until after the valleys in which they occur had come to assume much of their present

* *Ancient Stone Implements of Great Britain*, p. 477, *et seqq.*

appearance. This is very significant, but I shall come to that presently; meanwhile let us glance at the distribution of the old mammalia with which palæolithic man was contemporaneous.

Not for the moment taking account of cave-deposits, we find that beds containing remains of the fauna characteristic of palæolithic times do not occur at the actual surface in any district that lies beyond the limits reached by the implement-bearing beds. To this rule there are very few exceptions, nevertheless it is true that remains of the hippopotamus were obtained from an ancient river accumulation near Leeds; and some of the northern forms have been occasionally met with here and there in different parts of the country. Underneath the glacial deposits of Norfolk the old mammalia occur in great abundance; and, as we have seen, similar fossils appear in the inter-glacial beds of Scotland. But in the superficial river-gravels north of the palæolithic area, they are conspicuous by their absence, and this is specially the case with the southern forms.

How then are we to explain this anomalous distribution? It cannot be said that the mammalia may never have occupied the midland and northern districts. The fact that their bones occur abundantly in caves that lie beyond the limits of the palæolithic beds, shows that the animals were by no means confined to a narrow area in the south-east of England, and the occurrence of the hippopotamus near Leeds is further proof in the same direction. If the climate was suited to the mammalia that swarmed in the south — to elephants, rhinoceri, and hippopotami, to lions, hyænas, and tigers—it surely could not have been other than genial in the north of England and in Scotland. Yet in neither region do any of these animals occur in the superficial river-gravels.

Now, take in connection with all this the remarkable fact that the south-east of England, where palæolithic and mammaliferous river-gravels and loams are so well developed, is precisely that part of the country in which we find no traces of the later glaciation, and over which the sea did not prevail during the last great submergence of the British area. We have seen that some of the palæolithic gravels rest upon a boulder-clay which we can recognise as a deposit partly of land-ice, partly of marine formation, and belonging to the times that preceded the last inter-glacial period. After that boulder-clay had been laid down, dry land appeared in Scotland and the north of England, and at a later date, a movement of subsidence ensued which resulted in drowning Wales to a depth of probably not less than 2,000 ft. How far south the deposits belonging to this great submergence go, has not been definitely ascertained—the southern shores of the old sea that washed the slopes of the Welsh mountains yet remain to be traced out. All we know is that Scotland, Wales, and the north of England were largely submerged; but, when the deposits that belong to this period are followed south, they become less clearly distinguishable, and are apt to be confounded with the marine deposits (sands and gravels) of the older glacial ages. We must, therefore wait the results of more detailed investigations before we can attempt to indicate the southern boundaries reached by the sea in the midland counties of England, during the deposition of the middle sands of Lancashire, &c. (See No. 4 in Tables, p. 426.) But, although there is this uncertainty as to the southern limits of these deposits, there is none as to the fact that true palæolithic gravels do not extend northwards into those districts over which the sea is known to have prevailed at the time of the last great submergence. So, that, were we to colour on a map that portion of the British Islands which is understood to have been submerged at the period referred to, the colour would also serve to indicate the regions in which no palæolithic implements have occurred, and where the old fauna characteristic of palæolithic times is either entirely absent from the river-gravels, or represented by only a few scattered examples, all with one exception (that of the Leeds hippopotamus) consisting of northern forms. On the other hand, that part of the map left uncoloured would be coextensive with the districts in which palæolithic implements have been discovered, and where, in the ancient river-gravels, remains of the old mammalia often occur in great abundance.

If palæolithic deposits have a very limited range, such is not the case with those of neolithic age. Implements belonging to this later age, occur everywhere throughout the British Islands. From Caithness to Cornwall, and from the east coast of England to the western borders of Ireland, they are being continually picked up. Even in the bleak Orkney and Shetland Islands, and all over the Inner and Outer Hebrides relics of neolithic times have been met with. So that the wide distribution of these implements is in striking contrast to the limited range of palæolithic remains.

We know that neolithic man was accompanied by a mammalian fauna that differed very much from that with which palæolithic man was associated. Dogs, horses, pigs, several breeds of oxen, the bison, the red deer, the Irish elk, and such like were the characteristic forms of neolithic times. Remains of these animals have occurred again and again in recent river- and lake-deposits in almost every part of England, Scotland and Ireland. It is doubtful, however, whether the mammoth and the woolly rhinoceros lived in Britain down to neolithic times. Further evidence on this head is most desirable. It seems, however, to be beyond all reasonable doubt, that the southern group of animals had, as far as Britain is concerned, utterly vanished before the advent of the neolithic period.

How, then, are all these facts to be accounted for. Why are palæolithic river-gravels restricted to the south-east of England, while neolithic remains occur broadcast throughout these islands? What is the reason for the limitation of the southern mammalia to one small area in the south-east, and why should the mammoth and woolly rhinoceros occur so abundantly in the valley-gravels of that district, while they appear so seldom, and that only at wide intervals, in the valley-gravels of the north? Beyond the palæolithic area, the great storehouses of mammalian remains, both of southern and northern forms, are caves, and certain beds which, from their position, we can recognise as being of pre-glacial and inter-glacial age.

The answer which I give to all these queries is simply this—the palæolithic deposits are of pre-glacial and inter-glacial age, and do not, in any part, belong to post-glacial times. They are either entirely wanting or very sparingly represented in the midland and northern counties, in Wales, Scotland, and Ireland, because all those regions have again and again been subjected to the grinding action of land-ice, and the destructive influence of the sea. But in those districts which were never overwhelmed by the confluent ice-masses, and in such regions as were not submerged during the last great depression of the land in late glacial times, the valley-gravels form a continuous series of records from pre-glacial times to the present day. In short, the palæolithic beds dovetail into the glacial drifts, and are overlapped by the marine deposits thrown down during the final cold period. To the last inter-glacial period, then, we must refer the great bulk of the palæolithic river-gravels of the south-east of England. They are contemporaneous with those ancient valley-gravels of Scotland which overlie the till and boulder-clay, and which are themselves partially re-arranged and covered with marine deposits belonging to the time of great submergence. No doubt, however, portions of the ancient tool-bearing gravels, especially in the districts south of the Thames, may date back to the earlier warm periods of the glacial epoch, and thus be contemporaneous with the fresh-water beds in the Scottish till; while some may go back to even pre-glacial ages. All that we can say for certain, is, that no palæolithic bed can be shown to belong to a more recent date than the mild era that preceded the last great submergence.

When this view of the succession is accepted, many apparent anomalies receive a simple, and, as it seems to me, a satisfactory explanation. After the great ice-sheet shrank back, and the till and boulder-clay had been deposited, a land-surface existed—rivers flowed down the valleys, and plants and animals clothed and peopled the country. In Scotland the fluviatile deposits belonging to that period have been subjected to great denudation, but in one place at least they have yielded animal remains—frogs and water-rats. But if the country had never been submerged after the withdrawal of the ice from the low-grounds, there is good reason to believe that the presence of relics of palæolithic man, and remains of the mammalia with which he was associated, would have occurred in the valley-gravels of Scotland, Ireland, and the northern and midland counties of England, just as in those of the south-east.

Again, the high-level or older valley-gravels of the palæolithic area, contain a predominance of northern forms, and afford other indications of a cold climate having prevailed at the time of their formation, while the gravels at the lower levels are characterized by the prevalence of southern types, and otherwise yield proofs of having been deposited under mild conditions of climate. This is precisely what we might have anticipated. As the land-ice retired into the deep mountain-glens, the country would be invaded first by the arctic mammalia; the rivers would be liable to be frozen over, and, upon the breaking-up of the frost in summer-time, ice-rafts would run aground and thrust the fluviatile deposits into confused heaps. As the climate ameliorated, and the rivers excavated their valleys to a greater depth, the arctic mammalia would gradually retire, and be succeeded by animals belonging to more temperate zones. The southern mammalia would thus come later in time than the northern forms. By-and-by a process of submergence ensued, and (as the evidence derived from the glacial and inter-glacial deposits in both this and foreign countries clearly shows), this depression of the land took place at a time when the mild climate still endured. Step by step the whole land, as far south, probably, as the middle of England, was submerged to a very considerable extent, and then a glacial condition of things supervened for the last time. While an icy sea flowed over so large a portion of the British Islands, it is obviously impossible that any of the temperate or southern mammalia could have lingered on in the south-east of England. They must have disappeared before the cold had reached its climax, and although, for aught we can tell, they may have been succeeded for a time by the arctic mammalia, yet even these would eventually be compelled to migrate southwards as the severity of the climate increased. When, at last, the process of submergence ceased, and was converted into a movement of elevation, we know that the British Islands became connected to themselves and the Continent, as had doubtless been the case during the preceding mild inter-glacial period. Local glaciers yet filled our mountain-valleys, and the land was again clothed and peopled with plants and animals. Reindeer at first roamed over the country, but eventually an abundant mammalian fauna stocked our plains and valleys—consisting of oxen, red deer, Irish elks, horses, and other animals characteristic of temperate zones. It is extremely uncertain, however, whether the mammoth and the woolly rhinoceros returned to their old haunts. Certainly none of the southern mammalia ever did—these had vanished for ever, and with them palæolithic man had also disappeared—his neolithic successor being now the lord of the soil.

It is thus, then, that I would explain the anomalous distribution of the palæolithic deposits, and that remarkable gap or hiatus which separates the neolithic from the palæolithic age.

CHAPTER XXXII.

GEOLOGICAL POSITION OF NEOLITHIC, PALÆOLITHIC, AND MAMMALIFEROUS DEPOSITS OF FOREIGN COUNTRIES.

Palæolithic deposits wanting in Switzerland.—Mammalia of inter-glacial beds. —Swiss post-glacial deposits belong to neolithic, bronze, and more recent periods.—Post-glacial deposits of northern Italy of neolithic or more recent age.—Mammalian remains of Piedmont.—Palæolithic tools and remains of southern mammalia nowhere found in superficial deposits overlying the great northern drift.—Wide distribution of neolithic implements over northern Europe.—Inferences.—Palæolithic man and the southern mammalia not post-glacial.—Distribution of the old mammalia over Siberia and North America.—Proofs of mild climates within Arctic Circle.—Trees in Greenland.—Mammalia absent from districts covered with the later glacial deposits, but abound in the districts beyond.—General conclusions.

IF the conclusions now arrived at in regard to the pre-glacial and inter-glacial position of palæolithic remains and the old mammalia be reasonable, they ought not to contradict, but, on the contrary, should receive confirmation from the evidence supplied by other regions. I propose, therefore, to discuss very briefly the relation borne by the great mammal-bearing beds of Switzerland, Italy, and North America to the glacial deposits of those countries.

In Chapter XXVII. it will be remembered that some account is given of certain lignite beds which rest upon the ancient ground-moraine of the Swiss lowlands, and are covered by accumulations belonging to the last great advance of the glaciers in that country.* With those lignites are associated the remains of elephant and rhinoceros—both southern forms—and other mammalia, none of which are northern. There is no doubt that the underlying ground-moraine corresponds to our till, and just as little that the overlying glacial débris is the Swiss equivalent of our kames, ice-floated erratics, and shelly clays; in other words, the Swiss lignites occupy exactly the horizon of our palæolithic gravels.

With this singular notable exception, nowhere in the Swiss low-grounds do any remains of the southern mammalia occur. Relics of the neolithic and bronze ages are plentiful—bones of dogs, pigs, deer, sheep, horses, &c., turn up in every recent alluvium—but not a single trace of any of the southern mammalia or of palæolithic man has ever been discovered. The mammoth and the woolly rhinoceros have been disinterred from the old loams of the Rhine and other rivers, but lions, tigers, hippopotami, elephants, and hyænas are nowhere visible. Now, on the assumption that these animals and palæolithic man did not live in Europe after the last great increase of the Swiss glaciers, their absence from the surface alluvia and gravels is precisely what we should have anticipated. The deposits which may once have contained them have been well-nigh obliterated—the only fragments left being the lignite beds of Dürnten and Wetzikon. It may be, however, that underneath the vast deposits of löss belonging to the last cold period, palæolithic deposits lie concealed.

* See also Appendix, Note F.

The same facts meet us on the Italian side of the Alps. Marl-beds and peat-bogs occur in many places, particularly in Piedmont, where the latter occupy depressions in the surface of the ancient glacial moraines. Traces of lake-dwellings (*Palafitte*) are found beneath these old peat-mosses, and answer in every respect to the similar relics met with in Switzerland (*Pfahlbauten*). The animal remains associated with the *palafitte* are the dog, horse, ox, goat, sheep, stag, roebuck, boar, bear (*Ursus arctos*), &c.* Nowhere have the mosses, alluvia or marl-beds (which are all clearly of later date than the moraines belonging to the last extension of the alpine glaciers) yielded any trace of the old southern mammalia or of palæolithic man. The human relics all belong to neolithic, bronze, or still more recent times.

It is well-known, however, that beds of lignite and river accumulations containing remains of the mastodon, the rhinoceros, the hippopotamus, &c., occur in Piedmont. These deposits are older than the moraines of Ivrea and the great heaps of similar débris lying at the mouths of the lake-valleys of Northern Italy. According to the Italian geologists they must be correlated with the lignite beds of Dürnten and Wetzikon in Switzerland, and hence are not of pre-glacial or post-glacial but of inter-glacial age.

Palæolithic implements are said to occur in the gravels of the Tiber,† but they have not been discovered in fluviatile deposits in any other part of Italy.

* Gastaldi, *Lake Habitations and Pre-historic Remains in Italy.*

† *Ancient Stone Implements*, &c., p. 571.

So also in beds of sand near Megalopolis, in Greece, palæolithic implements are associated with bones of the great pachyderms;* and palæolithic valley-gravels containing remains of the African elephant occur near Madrid.† It is in France, however, in the valleys of the Somme and the Seine, where such deposits are most typically represented.

In Scandinavia, Denmark, Holland, northern Germany and Russia no palæolithic gravels have been detected, neither have the remains of the more characteristic southern genera—*Hippopotamus*, *Elephas antiquus*, *Rhinoceros megarhinus*, and *R. leptorhinus* (Owen) —ever occurred in any superficial river deposit throughout that vast area. The lion and hyæna, however, have left their traces in the caves of Germany.

Now when one thinks of it, this distribution of palæolithic and mammalian remains is a very remarkable one. For when we put aside the caves we find that no palæolithic implements and none of the southern mammalia occur in those regions which are more or less thickly covered with the sand, gravel and erratics belonging to the northern drift. Holland, Denmark, northern Russia, and the whole of northern Germany, from the borders of the Baltic down to near the base of the Hartz Mountains and the Riesengebirge, are abundantly covered with plateaux and heaps of drift in the hollows of which occur numerous lakelets and pools. These deposits form what has long

* *Op. et loc. cit.*

† *Op. et loc. cit.*; see also Casiano de Prado's *Descripcion fisica y geológica de la Provincia de Madrid*, p. 186.

been known as the Northern Drift, and they undoubtedly belong to the last cold period of the glacial epoch. Hence, as we have seen, they are the equivalents of the kames, shelly clays and ice-floated erratics of Britain. Now over all this area the superficial river-deposits have yielded only remains of the present temperate fauna, with human implements belonging to neolithic or still more recent times. Of the earlier palæolithic age they contain no trace whatever. Similarly, in the great valleys that drain down from the Central Alps—the Rhine, the Rhone, the Danube, and the Po—we find immense accumulations of flood-loam or löss, the upper portions of which at least must belong to that last cold period during which the glaciers advanced upon the lowlands of Switzerland and overwhelmed the forest-lands of Zurich. Now it is a fact that hitherto the only human relics found resting upon these deposits appertain to neolithic and more recent times. Neither palæolithic implements nor bones of the southern mammalia have there been detected. It is only when we get fairly beyond the limits of the "alpine diluvium" belonging to the latest era of great glaciers, that palæolithic deposits come on in force, as in the north-east of France.

No one will be inclined to believe that, at a time when the hippopotamus and the southern forms of elephant and rhinoceros were joint tenants of England and the north-east of France, these animals never strayed into similar latitudes of Germany. During the palæolithic period Britain was united to the Continent, and it is in the highest degree unlikely that man and his southern congeners were then restricted to the few river-basins in Western Europe, where their remains are now met with. Besides, as we know from the evidence of the caves, palæolithic man was a denizen of Germany, and the lion and the hyæna were his congeners there.

No sooner, however, do we admit the inter-glacial age of the palæolithic river-deposits than all our difficulties vanish, and the manner in which the gravels are distributed becomes full of meaning.

After the great mer de glace that extended from Scandinavia to the plains of Germany had retired, and the massive Rhone glacier had retreated from the low-grounds of France and once more shrunk into its mountain valley, and when the mighty ice-streams that invaded the plains of Piedmont had melted away, vegetation followed the retreating steps of the ice, and palæolithic man, accompanied by the arctic mammalia, wandered over Europe. As the climate grew milder these latter gradually migrated northwards, and were succeeded by the temperate and southern groups; and, seeing that the hippopotamus ranged as far at least as Leeds in England, there is no reason why it should not have followed some of the great European rivers down to the shores of the Northern Ocean. The Elbe, the Weser, the Rhine and the Meuse were surely as likely to tempt the old river-horse to a bath as the smaller rivers of England. Nor were such waters as the Rhone and the Po likely to be despised.

This period of mild and genial winters eventually

passed away, but before it did so certain great changes in the geography of Europe took place. A large part of the British Islands and Scandinavia sank down below the sea, and Denmark, Holland, the plains of Germany, and northern Russia also disappeared below the waves. But as this wide-spread depression reached its climax, the last cold period of the glacial epoch began. Ice-rafts and icebergs floated south from Scandinavia and the British mountains, and dropped their blocks and rubbish over the drowned countries. At the same time huge glaciers descended all the valleys of the Alps, and, advancing upon the low-grounds of Switzerland, overwhelmed the forests and here and there buried beneath their débris deposits containing remains of the southern and temperate mammalia. The rivers, greatly swollen, swept down vast quantities of fine glacial silt, underneath which were concealed such of the palæolithic beds as the rivers themselves had not demolished and rearranged. The southern mammalia had left Europe, and in their place came the northern forms—mammoths and woolly rhinoceri, the mammoth ranging into southern Italy—and at a still later date herds of reindeer, which lived south as far as the slopes of the Pyrenees. Glaciers then existed in the Vosges, and the Black Forest, and those of the Pyrenees exceeded in size their present puny descendants.

After this glacial condition of things had lasted for a long time, the climate began to ameliorate. In the north of Europe the land again slowly rose out of the sea, but the old fresh-water deposits which had been laid down in palæolithic times, had, in the interval of depression, suffered extensive denudation. Over wide areas they had entirely disappeared—waves and currents had ploughed them up, rearranged their materials, or buried them under accumulations of gravel and sand. Again, in the great valleys of the Alps, the glaciers had entirely demolished them—the only traces of them preserved being the lignites of Dürnten, and the bone-beds and lignites of northern Italy. In the lower reaches of the valleys that drain from the Alps, vast sheets of löss had been distributed—obscuring and concealing any palæolithic beds which may have existed at the surface before the last cold period commenced. But beyond the glaciated regions, and south of the area which had been submerged, the old valley-gravels, with their interesting memorials of man and his associates the southern mammalia, were left undisturbed.

When the re-elevation of northern Europe had been so far completed as again to join the British Islands to themselves and the Continent, the climate still continued cold and ungenial. The mammoth and its associate the woolly rhinoceros, seem to have lived down to this time in central Europe, although, as I have said, it is doubtful whether they ever revisited Britain. But palæolithic man and the southern mammalia not only did not re-enter Britain, but would even appear to have become extinct in Europe.

It was neolithic man, who, after the last ice period had passed away, became the sole occupant of Europe

—his relics being found everywhere throughout the continent, accompanied by the remains of that group of mammalia which is still characteristic of these latitudes.

We must now glance at the distribution of the old mammalia in the higher latitudes of Asia and North America. The great plains of Siberia, extending from the base of the Ural Mountains to Behring's Straits, are traversed by several large rivers, which in many places have exposed fine sections of those extensive alluvial beds that almost everywhere throughout this vast region form the subsoils. These alluvial deposits are often literally packed with the remains of mammoth, woolly rhinoceros, bison, and horse. So abundant, indeed, are remains of the mammoth, that for many years they have actually been quarried for the sake of the ivory—in 1821, no less a quantity than 20,000 lbs. of this product having been obtained from New Siberia alone. The Liakhov Islands and New Siberia have evidently formed at one time a portion of the Asiatic continent—indeed, they appear to be chiefly composed of ancient river alluvia, which are scarped into a series of low cliffs along the coast—the resort of the ivory hunters, where they may be seen every summer digging the mammoth-tusks out of the frozen soil which has partially thawed. The fossils are usually well-preserved. Indeed, on one occasion, the actual carcass of a mammoth was exposed in one of the cliffs in so fresh a state, that the dogs ate the flesh.

The presence of these numerous animal remains indicates the former prevalence of a milder climate in Siberia than now. For we can hardly doubt that the animals actually occupied the low-lying tracts through which the rivers of northern Asia flow. At the same time it is evident, that, during winter, carcasses would frequently be frozen into the ice in the upper reaches of the rivers, and, when summer returned, would often be floated down for long distances before they became finally entombed. It is impossible to believe, however, that all the remains which occur so abundantly along the whole borders of the Arctic Sea, have been floated down in this way from lower latitudes. By far the larger proportion must belong to creatures that lived and died in the latitudes where their bones are now found.

It is remarkable, that, nowhere in the great plains of Siberia do any traces of glacial action appear to have been observed. If cones and mounds of gravel and great erratics like those that sprinkle so wide an area in northern America and northern Europe had occurred, they would hardly have failed to arrest the attention of explorers. Middendorff does indeed mention the occurrence of trains of large erratics which he observed along the banks of some of the rivers, but these, he has no doubt, were carried down by river-ice.* The general character of the tundras is that of wide flat plains, covered for the most part with a grassy and mossy vegetation, but here and there bare and sandy. Frequently nothing inter-

* *Reisen in den äussersten norden und osten Sibiriens*, Band iv. Theil i. p. 269.

venes to break the monotony of the landscape. The eye wanders over a sealike expanse that stretches far away until it seems to blend with the blue distance. Here and there, perhaps, a slight roll of the ground makes a faint low arch against the pale horizon, and serves as a landmark to Samojede and Ostjak, but otherwise the ocean of the tundra extends without interruption—an interminable plain.*

It would appear, then, that in northern Asia representatives of the glacial deposits which are met with in similar latitudes in Europe and America do not occur. The northern drift of Russia and Germany; the åsar of Sweden; the kames, eskers, and erratics of Britain; and the iceberg drift of northern America have apparently no equivalents in Siberia. Consequently we find the great river-deposits with their mammalian remains, which tell of a milder climate than now obtains in those high latitudes, still lying undisturbed at the surface.†

We must now take a rapid glance at the distribution of mammalian remains in North America.

In the regions lying to the west of the Rocky Mountains (Alaska), we have a continuation of the same physical conditions that characterize the more northerly latitudes of Asia, namely, great plains intersected by large rivers. Along the banks of these rivers, north of Mount St. Elias, numerous mam-

* Schrenck's *Reise durch die Tundren der Samojeden,* Theil i. p. 271. Some good descriptions of Tundra landscapes will be found in G. Kennan's *Tent Life in Siberia.*

† It is by no means improbable that the mammoth and woolly rhinoceros may have survived in northern Asia down to a comparatively recent date.

malian remains (especially the mammoth) have been detected. In Kotzebue Sound, Captain Beechey found that the wasting of the frozen cliffs was continually exposing the bones and tusks of mammoths and other quadrupeds—among which were urus, reindeer, musk-ox, a large deer (perhaps the moose), and others. But in the northern latitudes east of the Rocky Mountains, no such mammalian remains have been detected. According to Sir J. Richardson, "none have hitherto been found in Rupert's Land, though the annual waste of the banks of the large rivers and the frequent land-slips would have revealed them to the natives or fur-traders had they existed even in small numbers. They are rare also, or altogether wanting, in Canada."*

Nevertheless, proofs are not wanting of a former mild condition of things having prevailed within comparatively recent times in the far north of British America. Sir Edward Belcher brought away from the dreary shores of Wellington Channel (lat. 75° 32′ N.) portions of a tree which there can be no doubt whatever had actually grown where he found it. The spot where it was found lay about a mile and a half inland. I give the account in Sir Edward's own words. He says: "I at once perceived that it was no spar, and not placed there by human agency: it was the trunk and root of a tree, which had apparently grown there and flourished, but at what date who will venture to say? It is, indeed, one of the questions involved in the change of this

* *Journal of a Boat Voyage through Rupert's Land*, vol. ii. p. 210.

climate. As the men proceeded with the removal of the frozen clay surrounding the roots, which were completely cemented, as it were, into the frozen mass, breaking off short like earthenware, they gradually developed the roots, as well as what appeared to be portions of leaves and other parts of the tree, which had become embedded where they fell, and now were barely distinguishable—at least, not so much as some impressions on coal—to the casual observer. . . . When a warmer climate prevailed here, this tree possibly put forth its leaves and afforded shade from the sun: most fervently did I just now wish for its return. . . . Two neighbouring mounds were also dug into, but they proved to be peat—doubtless other stumps and vegetable matter, the only remaining traces of what might at some distant period have been a forest. All the surrounding earth and tufts of grass indicated this spot to have been the bottom of some lake or marsh."* Dr. Hooker, who carefully examined the piece of wood brought by Sir Edward, pronounced it to belong to a species of pine, probably to *Pinus* (*Abies*) *alba*, the most northern conifer, which advances as far north as the sixty-eighth parallel. The structure of the wood was found to differ remarkably in its anatomical characters from that of any other conifer with which Dr. Hooker was acquainted, and the peculiar conditions of an arctic climate seemed to him to afford an adequate explanation of the appearances presented.†

* *The Last of the Arctic Voyages*, vol. i. p. 380.

† *Op. cit.* p. 381; *British Association Report* for 1855, p. 101. Captain

With the exception of this discovery of Sir Edward Belcher's, and possibly also of those mentioned in the note below, we have no direct evidence whatever that any milder climate than the present has prevailed in British America since the disappearance of the great ice-sheet.* The whole surface of the country, from the shores of the Arctic Ocean down to the latitudes of the great lakes, and even considerably farther south, is more or less abundantly sprinkled with drift deposits—with till, and heaps and hummocks of sand and gravel, and numerous erratics. Yet nowhere over this wide area, down to the borders of the United States, do the extinct mammalia appear in any post-glacial deposits. In the neighbourhood of the great lakes they occur in fresh-

McClure discovered in Banks's Land, in lat. 74° 48′, many fossil-trees, as well as fragments not fossilised, lying over a wide extent of ground (*Discovery of a North-West Passage*, p. 208). Again, trunks of trees, which had evidently grown *in situ*, were detected in Prince Patrick's Island, in lat. 76° 12′ N. long. 122° W., by Lieutenant Mecham. Two of these measured respectively 4 ft. in circumference, and a third 2 ft. 10 in.; one of the former reached 30 ft. in length. Unfortunately the species was not determined, but according to the ship carpenter it resembled larch. "When comparatively dry, it was tried as fuel, but its virtue had gone; it threw out little or no flame, but smouldered rather than burnt, like so much tinder" (*Voyage of H.M.S. Resolute*, McDougall, pp. 292, 293). It is highly probable that all these trees belong to the same period as the pine described by Sir Edward Belcher, but without a geological examination of the ground it would be hazardous at once to conclude that they do. The remains of trees, &c., of miocene age occur so plentifully within the Arctic Circle, that it is just possible that the trees discovered by McClure and Lieutenant Mecham may be referable to miocene times.

* I am aware that several arctic explorers are of opinion that the climate of Greenland has altered for the worse within quite recent times. The huts of Esquimaux have been met with in places which are now not visited by the natives. And this, taken in connection with other evidence, points as some think to a somewhat milder climate having prevailed in those regions within even historical times. But the succession of a few unusually mild years would possibly explain all the appearances referred to.

water clays, along with abundant vegetable remains, and these clays are clearly overlaid by glacial deposits. It is only when the southern limits of the northern drift are approached, that the extinct mammalia begin to be found in any numbers at the very surface; and their remains occur in greatest profusion in the regions which have not been reached by the drift.

Professor J. D. Whitney has given a remarkable illustration of these phenomena in his description * of the "driftless region" of Wisconsin, Iowa, and Minnesota. He tells us that a considerable area, chiefly in Wisconsin, and near the Mississippi River, is quite destitute of drift, and this is the more remarkable seeing that the regions to the north, west, east, and south are all more or less deeply covered with such deposits.

Whitney describes the surface of the rock within the driftless region as being uneven and irregular, and bearing the marks of chemical rather than of mechanical erosion. This is especially the case where limestone or dolomite forms the rock in place. No glacial furrows or striæ, no drift scratches, and no evidence of the rock having been planed down to a level are anywhere visible. A variable thickness of surface-wash overlies the rocks, which is evidently the result of weathering and chemical action. From these and other facts Whitney concludes that the region under review "must have formed an island at the time when the great currents from the north were bringing down the detrital

* *Report of the Geological Survey of Wisconsin*, vol. i. p. 114, *et sqq.*

materials, which are spread over so vast an area in the northern hemisphere."

Now throughout this remarkable region the remains of numerous extinct mammalia have again and again been detected. They occur promiscuously embedded in the surface-wash, or in cracks and crevices of the limestone. The animals mentioned are mastodon, elephant, buffalo, extinct species of peccary, racoon, and several rodents, &c. Many of these were got in clayey loam at considerable depths from the surface, as much as 40 ft. in some cases, indicating the lapse of a long period since the time of their entombment.

Beyond the driftless region, however, in those tracts that are thickly covered with the gravel, sand, and boulders laid down during the "iceberg" period (see table No. 5, p. 428), no mammalian remains occur in superficial or post-glacial deposits. In such districts they are only met with occupying an inter-glacial position.

Thus, in the western as in the eastern hemisphere, we are confronted with precisely the same phenomena. In regions which can be proved never to have been overridden by the great continental glaciers, and in districts which give no evidence of submergence during the latest period of glacial cold, the extinct mammalia occur in less or greater abundance at the very surface. In Britain, and central Europe, the old ossiferous alluvia, when traced from the low-grounds to the mountains, disappear as soon as the iceberg-drift, moraines, and alpine diluvium are reached. Nowhere in morainic turbaries or alluvium which can be demonstrated to be of post-glacial age, do any of the extinct southern mammalia or palæolithic implements appear. But when the hippopotamus, the elephant and their congeners do occur, in regions that are covered by the latest glacial deposits, they invariably occupy inter-glacial or infra-glacial positions. The great plains of Siberia never could have nourished glaciers. We cannot conceive that even during the most intense cold of the glacial epoch, conditions similar to those which characterized Britain, Scandinavia, Switzerland, and North America, could have existed in northern Siberia: the absence of high-grounds, and the comparative dryness of the climate, must have prevented any accumulation of glacier-ice. Nor can I learn that marine deposits similar to our shelly marine clays and iceberg-drift cover any portion of northern Asia. If morainic débris, mounds and cones of sand, and large erratics, like those of North America, occurred in Siberia, travellers would hardly have failed to notice them. Siberia would thus appear to have escaped the glacial and marine erosion which overtook a large part of Europe and North America, and consequently river-deposits belonging to mild inter-glacial, and pre-glacial ages have been preserved. Hence the great bone accumulations of northern Asia appear only where we might have anticipated they would. It is otherwise, as we have seen, with the corresponding latitudes in British America. Over all that vast region the evidences of glacial action are most conspicuous, and nowhere do the extinct mammalia occur. The trees discovered by Sir E. Belcher and others, while belonging to a later date than the great glaciation, are in all probability older than the period of submergence and floating ice.

The anomalous distribution of the extinct mammalia appears inexplicable on the assumption that the ossiferous beds are all of post-glacial age; but if they belong for the most part to inter-glacial times, the mode of their occurrence is precisely what might have been expected. It seems indeed impossible to resist the conclusion that—at the time palæolithic man and the southern mammalia frequented the lower latitudes of Europe (where their remains occur so abundantly in river-gravels and cave-deposits), and while mammoths, horses, buffaloes, and oxen roamed over northern Siberia—Scotland, Ireland, Denmark, Scandinavia, and other regions of northern Europe also supported an abundant mammalian fauna, and that the mastodon and its congeners likewise occupied what are now the wooded regions and barrens of North America. And the remains of these creatures seldom or never occur in the regions referred to, because the deposits which once contained them have either been obliterated by the action of ice and the sea, or they are covered up and concealed by drift accumulations.

CHAPTER XXXIII.

CONCLUSION.

SO many diverse threads of evidence have now been followed, that it may be well rapidly to catch these up, and so weave them into one connected whole. Hitherto we have followed the analytical method, we must now in conclusion pursue the synthetical, and endeavour to build up the story of that chequered past, whose records we have just been perusing.

Upwards of 200,000 years ago the earth, as we know from the calculations of astronomers, was so placed in regard to the sun that a series of physical changes was induced, which eventually resulted in conferring upon our hemisphere a most intensely severe climate. All northern Europe and northern America disappeared beneath a thick crust of ice and snow, and the glaciers of such regions as Switzerland assumed gigantic proportions. This great sheet of land-ice levelled up the valleys of Britain, and stretched across our mountains and hills down to low latitudes in England. Being only one connected or confluent series of mighty glaciers, the ice crept ever downwards and outwards from the mountains, following the direction of the principal valleys; and pushing far out to sea, where it terminated at last in deep water, many miles away from what now forms the coast-line of our country. This sea of ice was of such extent that the glaciers of Scandinavia coalesced with those of Scotland, upon what is now the floor of the shallow North Sea, while a mighty stream of ice flowing outwards from the western seaboard obliterated the Hebrides, and sent its icebergs adrift in the deep waters of the Atlantic. In like manner massive glaciers, born in the Welsh and Cumbrian mountains, swept over the low-grounds of England, and united with the Scotch and Irish ice upon the bottom of the Irish Sea. At the same period the Scandinavian mountains shed vast icebergs into the northern ocean, and sent southward a sheet of ice that not only filled up the basin of the Baltic but overflowed Finland, and advanced upon the plains of northern Germany: while from every mountain-region in Europe great glaciers descended, sometimes for almost inconceivable distances, into the low countries beyond.

Ere long this wonderful scene of arctic sterility passed away. Gradually the snow and ice melted and drew back to the mountains, and plants and animals appeared as the climate ameliorated. The mammoth and the woolly-coated rhinoceros roamed in our valleys, the great bear haunted our caves, and pine-trees grew in the south of England; but the seasons were still well marked. In winter-time frost often covered the rivers with a thick coat of ice, which the summer again tore away, when the rivers, swollen with the tribute of such receding glaciers as still lingered in our deeper glens, rushed along the valleys and spread devastation far and wide. By slow degrees, however, the cold of winter abated, while the heat of summer increased. As the warmth of summer waxed the arctic mammalia gradually disappeared from our valleys, and sought out northern and more congenial homes. Step by step the climate continued to grow milder, and the difference between the seasons to be less distinctly marked, until eventually something like perpetual summer reigned in Britain. Then it was that the hippopotamus wallowed in our rivers, and the elephant crashed through our forests; then, too, the lion, the tiger, and the hyæna became denizens of the English caves.

Such scenes as these continued for a long time; but again the climate began to change. The summers grew less genial, the winters more severe. Gradually the southern mammalia disappeared, and were succeeded by arctic animals. Even these, however, as the temperature became too severe, migrated southward, until all life deserted Britain, and snow and ice were left in undisputed possession. Once more the confluent glaciers overflowed the land, and desolation and sterility were everywhere.

During these great oscillations of climate there were not infrequent shiftings in the distribution of land and sea; but such vicissitudes, although doubtless producing local effects, certainly do not seem to have been the causes of the chief climatal changes. It is much more likely that the mild inter-glacial periods were induced by eccentricity of the earth's orbit, combined with precession of the equinoxes.

We cannot yet say how often such alternations of cold and warm periods were repeated; nor can we be sure that palæolithic man lived in Britain during the earlier warm intervals of the glacial epoch. But since his implements are met with at the bottom of the very oldest palæolithic deposits, and since we know that the animals with which he was certainly contemporaneous did occupy Britain in early inter-glacial ages, and even in times anterior to the glacial epoch itself, it is in the highest degree likely that man arrived here as early at least as the mammoth and the hippopotamus.

Be this, however, as it may, the evidence appears to be decisive as to the presence of man in Britain during the last mild inter-glacial period. And this being so, it is startling to recall in imagination those grand geological revolutions of which he must have been a witness.

During the last inter-glacial period he entered Britain at a time when our country was joined to Europe across the bed of the German Ocean; at a time when the winters were still severe enough to freeze over the rivers in the south of England; at a time when glaciers nestled in our upland and mountain valleys, and the arctic mammalia occupied the land. He lived here long enough to witness a complete change of climate—to see the arctic mammalia vanish from England, and the hippopotamus and its congeners take their place. At a later date, and while a mild and genial climate still continued, he beheld the sea slowly gain upon the land, until little by little, step by step, a large portion of our country was submerged—a submergence which, as we know, reached in Wales to the extent of some 2,000 ft. or thereabout. We know, further, that simultaneously with the partial drowning of the British Islands a vast area in northern Europe also sank down below the waves.

When this great submergence commenced, the climate, as I have said, was genial; and it continued so up to a time when the subsidence had reached, or nearly reached a climax. Then it was that the last cold period began. Intense arctic cold converted the rocky islands which then represented Britain into a frozen archipelago. From the ice-foot that clogged the shores fleets of rafts set sail, and as they journeyed on dropped angular stones and rubbish over the bottom of the sea. At the same time icebergs floated away from the Scandinavian mountains, and strewed their burdens over the submerged districts of northern Europe, while the alpine glaciers crept out upon the low-grounds of Switzerland and overwhelmed the forest-lands of Zurich and Constance.

A similar succession of changes transpired in North America. After the continental ice-sheet had retired for the last time great lakes appeared, and a luxuriant forest-growth overspread the land, which became the resort of a prolific mammalian fauna—mastodons, elephants, buffaloes, peccaries, and other animals. By-and-by, however, depression ensued, and icebergs, issuing from the frozen north, scattered over the site of the old forest-lands, erratics, and heaps of rubbish.

During this latest cold period of the glacial epoch, palæolithic man, for aught that we can say, may have occupied the south of Europe; but it is in the highest degree unlikely that he lived so far north as the unsubmerged portions of southern England.

Another great change now ensued. Those mysterious forces by which the solid crust of the globe is elevated and depressed now again began to act—the sea gradually retreated, and our hills and valleys eventually reappeared. Step by step the British Islands rose out of the waters, until for the last time they became united to the Continent. Snow, however, still covered our loftier mountains, and glaciers yet lingered in a few of our mountain-valleys. The treeless land was now invaded by the reindeer, the moose-deer, the arctic fox, the lemming, and the marmot, and neolithic man likewise entered upon the scene: his palæolithic predecessor had, as far as Britain and northern Europe are concerned, vanished for ever.

Thus the palæolithic and neolithic ages are separated by a vast lapse of time—by a time sufficient for the submergence and re-elevation of a large part of Europe and a very considerable change of climate.

In early neolithic times the climate was somewhat excessive, but as ages passed away it gradually became ameliorated. A strong forest-growth by-and-by covered the country, and herds of oxen wandered in its grassy glades; but the southern mammalia never returned to their old haunts, and it is even doubtful whether the mammoth and the woolly rhinoceros again appeared in Britain. They seem, however, to have still lingered on for a time in central Europe.

As years rolled on the sea again stole in between our islands and the Continent, until a final severance was effected. It is beyond my purpose, however, to trace the later changes. From early neolithic times a gradual improvement and progress attended the efforts of our barbaric predecessors, until at length a period arrived when men began to abandon the use of stone implements and weapons, and for these to substitute bronze. And so, passing on through the age of bronze and the days of the builders of Stonehenge, we are at last brought face to face with the age of iron and the dawn of history.

[Postscript.—A remarkable discovery has just been announced. Mr. Tiddeman writes to *Nature*, Nov. 6, 1873, that amongst a number of bones obtained during the exploration of the Victoria Cave, near Settle, Yorkshire, there is one which Mr. Busk has identified as *human*. Mr. Busk says: "The bone is, I have no doubt, human; a portion of an unusually clumsy fibula, and in that respect not unlike the same bone in the Mentone skeleton." The interest of this discovery consists in the fact that the deposit from which the bone was obtained is overlaid, as Mr. Tiddeman has shown, "by a bed of stiff glacial clay containing ice-scratched boulders." Here, then, is direct proof that man lived in England *prior to the last inter-glacial period*. I have said above (p. 507) that it is highly likely that man may have occupied Britain in early inter-glacial or pre-glacial times; but I hardly looked for so early and complete a confirmation of views which I first published in the beginning of 1872.]

APPENDIX.

NOTE A.

TABLE OF SEDIMENTARY STRATA.

POST-TERTIARY . .	QUATERNARY OR PLEISTOCENE.*
CAINOZOIC or TERTIARY	PLIOCENE.* MIOCENE.* EOCENE.*
MESOZOIC or SECONDARY	CRETACEOUS.* JURASSIC.* TRIASSIC.*
PALÆOZOIC or PRIMARY	PERMIAN.* CARBONIFEROUS.* DEVONIAN OR OLD RED SANDSTONE.* SILURIAN.* CAMBRIAN. LAURENTIAN.

THE Formations marked with an asterisk have all been considered to yield evidence, more or less satisfactory, of the former action of ice. But in some cases the proofs are hardly convincing. It must be remembered that the records of mild and genial climates are more likely to be preserved than are traces of cold and glacial conditions. The former will usually be represented by abundantly fossiliferous marine and fresh-water deposits. In the case of the latter, a few ice-floated stones and boulders are all the relics that are likely to be handed down. The ice-markings on the rocks, and the morainic accumulations of mountain-valley and lowland, are almost certain sooner or later to be obliterated. Take, for example, the Glacial or Drift Formation of Quaternary times. Even now, the action of the weather, of frost, and rain, and rivers, is slowly but surely effacing the marks left by the old glaciers. And should our islands eventually become submerged, it might well be that, as the land sank down, what the atmospheric forces had failed to obliterate would succumb to the action of the sea. Should the land be afterwards re-elevated, it is very doubtful indeed if a single recognisable trace of former glacial work would remain. The farther back we go in time, therefore, the more difficult must it become to detect evidences of ice-action. The older formations consist for the most part of deposits which gathered on the floors of ancient oceans. Very few land-surfaces have been preserved. Consequently, if we are to find in the older formations any traces of former glacial cold, it will consist for the most part of scattered stones and boulders embedded in the heart of oceanic accumulations. Neverth less, there are not wanting, even in some of the palæozoic formations, deposits which bear the strongest resemblance to morainic débris. Of course, when we are dealing with formations so far removed from us in time, and in which the animal- and plant-remains depart so widely from existing forms of life, we can hardly expect to derive much aid from the fossils in our attempts to detect traces of cold climatal conditions. The arctic shells in our post-tertiary clays are convincing proofs of the former existence in our latitude of a severe climate; but when we go so far back as palæozoic ages, we have no such clear evidence to guide us. All that palæontologists can say regarding the fossils belonging to these old times is simply this, that they seem to indicate, generally speaking, mild, temperate, or genial, and even sometimes tropical, conditions of climate. Many of the fossils, indeed, if we are to reason from analogy at all, could not possibly have lived in cold seas. But, for aught that we know, there may have been alternations of climate during the deposition of each particular formation; and these changes may be marked by the presence or absence, or by the greater or less abundant development of certain organisms at various horizons in the strata. Notwithstanding all that has been done, our knowledge of the natural history of these ancient seas is still very imperfect; and therefore, in the present state of our information, we are not entitled to argue, from the general aspect of the fossils in our older formations, that the temperature of the ancient seas was never other than mild and genial.

It is beyond my purpose in this Note to do more than jot down a few instances of what have been considered as indications of former ice-action.

The oldest, or Laurentian, formation consists of rocks which have been so highly altered from their original condition, that, even if they had ever contained any evidence of old ice-action, it must have been obliterated.

I am not aware that any trace of ice-action has ever been recorded in connection with the Cambrian, but I have sometimes thought that the conglomerates belonging to this formation, in the north-west of Scotland, might possibly have had a glacial origin.

In the Lower Silurian of the south of Scotland (Glen App and Dalmellington) we find large blocks and boulders (from one foot to five feet in diameter) of gneiss, syenite, granite, &c., none of which belong to the rocks of that neighbourhood. Indeed, no such rocks, of older age than the Silurian, occur nearer than the Laurentian rocks of the north-western Highlands and islands. Possibly the boulders may have come from some ancient Atlantis; and, considering the great size of the blocks, and the considerable distance they may have travelled, it is not unreasonable to conjecture that ice may have had something to do with their transport. No ice-markings, however, have been observed upon any of the stones; nor, when we reflect upon the vast age of this deposit, could these be expected to have resisted the long-continued action of percolating water. Professor Dawson records a somewhat similar instance in the Lower Silurian of Maimanse, Lake Superior, where a conglomerate occurs with boulders two feet in diameter. Again, in the Upper Silurian of Nova Scotia, the same author has detected beds of angular stones and chips, "the materials of which seem precisely similar to that which is at present produced by the disintegrating action of frost on hard, and especially schistose and jointed rocks."

The conglomerates belonging to the Old Red Sandstone formation in the north of England and in Scotland have appeared to several competent observers closely to resemble a consolidated boulder-drift. In the south of Scotland, however, and probably in north of England also, the conglomerates hitherto assigned to the Old Red Sandstone formation really appertain to the base of the Carboniferous. My colleague, Mr. R. Etheridge, jun., informs me that in certain localities near Victoria, Australia, he and his colleagues observed a conglomerate which in some places reached at least one hundred feet in thickness, and which Mr. Selwyn was inclined to regard as probably of glacial origin. The latter author says: "The character of the conglomerate beds is such as almost to preclude the supposition of their being due to purely aqueous transport and deposition. It is, however, very suggestive of the results likely to be produced by marine glacial transport; and the mixture of coarse and fine, angular and water-worn material, much of which has clearly been derived from distant sources, would also favour this supposition." The deposit occurs, Mr. Etheridge informs me, at many localities throughout the colony, with a slightly varying matrix. Grooved and scratched stones have not been observed. Mr. Selwyn classed the deposit provisionally as Devonian. Great beds of conglomerate occur at the bottom of the Carboniferous, in various parts of Scotland, which it is difficult to believe are other than ancient morainic débris. They are frequently quite unstratified, and the stones often show that peculiar blunted form which is so characteristic of glacial work. These are confusedly huddled together in a dull, tough arenaceous matrix; and the whole deposit will sometimes continue to exhibit these appearances over a wide area. Ever and anon, however, we detect traces of water-action; the stones become more rounded, and are spread out in more or less regular layers or beds. The coarse, unstratified portions so closely resemble till, that only a practised eye can distinguish the difference at a glance; and many geologists who had not previously tried their hammers on the deposit, might well mistake it for a post-tertiary glacial accumulation. It is typically developed in the Lammermuir Hills district, and appears also in Ayrshire, Arran, and other regions in the west of Scotland. None of the stones, as far as I am aware, have ever shown glacial striæ. Erratic blocks have occurred in certain French Carboniferous strata, and Mr. Godwin-Austen has attributed the transport of these to floating ice. In North America the Carboniferous deposits appear also to have yielded what seems to be evidence in the same direction. Thus, in Ohio for example, a boulder of quartzite, 17 in. by 12 in. in diameter, was found embedded in a seam of coal and overlying shale; and Professor Newberry has suggested that it may have been transported by ice down some ancient Carboniferous river, and so dropt into its present position, where eventually the sediment covered it up. Similar boulders in coal have been met with elsewhere in America. Professor Dawson also mentions the occurrence, in the Lower Carboniferous of Nova Scotia, of angular fragments and chips of various hard rocks cemented together, which he thinks may fairly be regarded as evidence of somewhat intense winter cold, in the same way as the angular débris detected by him in the Upper Silurian of Nova Scotia. But he cites a still more remarkable example from the coal formation of that colony. This he describes as a gigantic esker of Carboniferous age, "on the outside of which large travelled boulders were deposited, probably by drift-ice; while in the swamps within, the coal-flora flourished, and fine mud and coaly matter were accumulated." At the recent meeting of the British Association at Bradford (September, 1873), Mr. Blandford described what he took to be evidence of ice-action in certain Carboniferous deposits in India.

Professor Ramsay has given a detailed account of the occurrence, in Permian conglomerate, of blunted and well-scratched stones, which seems conclusively to prove the existence of glaciers and icebergs in England at some time during that far-distant age. The same author has further suggested that similar accumulations in Germany, belonging to the Permian, betray like traces of old ice-action.

The New Red Sandstone (Triassic) of Devonshire has yielded scattered blocks or erratics, which Mr. Godwin-Austen was the first to attribute to the action of floating ice. Mr. Pengelly, however, thinks that the blocks need not have travelled far, but might have been moved along an ancient shore-line by breakers.

In the north of Scotland, a coarse boulder conglomerate is associated with the Jurassic strata in the east of Sutherland, the possibly glacial origin of which long ago suggested itself to Professor Ramsay and other observers. Recently, Mr. Judd has come to the conclusion that the boulders were floated down by ice from the highland mountains at the time the Jurassic strata were accumulating.

The Cretaceous formation has likewise yielded what Mr. Godwin-Austen has suggested are ice-borne stones and boulders. These erratics have been detected not only in the chalk of England, but also in the Cretaceous strata of the Alps. It is possible, however, that some of these may have been carried seawards, attached to the roots of drifted trees; and others, again, may have been floated away by some of the larger sea-weeds. But according to Mr. Godwin-Austen, with whom Sir C. Lyell agrees, only coast-ice could have transported the block and stones got in the white chalk near Croydon.

In the Eocene strata of Switzerland erratics have been met with, some of them angular and others rounded. They often attain a large size; blocks ten feet long being not uncommon, and one even measured 105 ft. in length, 90 ft. in breadth, and 45 ft. in height. Some of the blocks consist of a kind of granite, which is not known to occur anywhere in the Alps.

The Miocene of the Moncalieri-Valenza hills, in the north of Italy, has been shown by Professor Gastaldi to contain enormous blocks of alpine rocks, which require us to call in the agency of floating ice to account for their presence.

The Quaternary embraces, of course, the glacial and associated deposits which form the subject of this volume.

Now, if we were to judge only from the general aspect presented by their organic contents, we should be forced to admit that none of these formations, from the Silurian down to the Miocene, afforded any trace whatever of cold or glacial conditions. Yet that very cold conditions did supervene during the continuance of some of these formations, seems indisputable. Geologists are staggered by the appearance of glacial deposits in the Permian—a formation whose fossils indicate mild and genial rather than cold climatal conditions. The occurrence in the Eocene, also, of huge, ice-carried blocks seems incomprehensible, when the general character of the Eocene fossils is taken into account, for these have a somewhat tropical aspect. So likewise the appearance of ice-transported blocks in the Miocene is a sore puzzle. The fossils embedded in this formation speak to us of tropical and sub-tropical climates having prevailed in Central Europe; nay, more, Miocene deposits have been detected in high arctic latitudes. Species of sequoia, coniferæ, poplar, willow, beech, oak, plane-tree, walnut, plum or prunus, buckthorn, Andromeda, Daphnogene, and several other evergreens, grew during Miocene times in North Greenland! Even in Spitzbergen, abundant traces of the same kind of vegetation have been preserved. Yet it was precisely during the continuance of this period or age that the great erratics were carried down from the Alps, and dropped on what was then the sea-bottom in the North of Italy. These apparently contradictory appearances, may, however, be satisfactorily accounted for by inferring a former alternation of cold and warm climates, like that which I have tried to show prevailed during Quaternary times. It is singular, to say the least of it, that the beds which contain those great erratics (Eocene and Miocene) are wonderfully barren of fossils; and there is nothing, therefore, in the palæontological evidence that need cause us to hesitate in attributing the presence of the large erratics in the Eocene and Miocene to former severe conditions of climate. Indeed, the very fact that evergreens found during Miocene times a congenial habitat in North Greenland and Spitzbergen shows, on the principles explained in Chapters viii., ix. and x., that the eccentricity of the earth's orbit had then attained a high value, and that a series of genial and glacial climates must consequently have alternated in our hemisphere at that time. For the same reason, I see nothing contradictory in concluding, with Professor Ramsay, that glaciers existed in England during the Permian period, even although the plants and shells, &c., usually met with in Permian strata seem to indicate mild climatal conditions. The more characteristic Permian conglomerates are unfossiliferous; so also are those coarse boulder-beds belonging to the Carboniferous, the Old Red Sandstone, and the Lower Silurian. It might quite well be that during the continuance of each and all of those periods, cold and mild climates alternated. Silurian and Carboniferous fossils are obtained in high arctic latitudes, and we should not therefore feel surprised if those formations in our own or analogous latitudes should occasionally exhibit traces of severe climatal conditions.

A few references to papers describing traces of ice-action in the older formations may not be without use:—

Explanation of Sheet 14, *Ayrshire*, p. 7, and *Explanation of Sheet* 7, *Ayrshire, Geol. Surv. of Scot.* p. 8; *Canadian Naturalist*, vol. ii. p. 6, and vol. vi.; *Physical Geology and Geography of Victoria*, p. 16; *History of the Isle of Man* (Cumming); "Geology of Eastern Berwickshire," *Mem. Geol. Surv. Gt. Brit.* p. 41; *Report Geol. Survey of Ohio for* 1870; *Acadian Geology*, p. 324; *American Naturalist*, vol. vi. p. 439; *Quart. Jour. Geol. Soc.* 1855, p. 185: *Philosophical Magazine*, April, 1865 (Ramsay: Reply to Sir C. Lyell); *Quart. Jour. Geol. Soc.* 1873, p. 402; *Ibid.* 1850, p. 96; *Ibid.* vol. xiv.; *Ibid.* vol. xxix. p. 187; *Petrifakten und erratische Jurablöcke im Flysch*, &c. (Bachmann); *Sugli Elementi che compongono i Conglomerati Mioceni del Piemonte* (B. Gastaldi); *Principles of Geology* (Lyell), vol. i. chap. x. and xi.

NOTE B.

Quaternary Deposits of the British Islands, with some of their Equivalents in other Countries.

	Deposits.	Fossils.	Physical Conditions.	Foreign Equivalents.
Recent period,	13. Alluvium; peat; raised beaches.	13. Sub-fossils.	13. Partial re-elevation; the present.	13. Alluvium; loam; marl; peat; bog iron ore; calcareous tufa, etc.
Post-glacial period,	12. Peat; buried forests; river and cave deposits in part.	12. Relics of man; recent and extinct or no longer indigenous mammalia; *Reindeer*, *Megaceros Hibernicus*, *Bos primigenius*, *B. longifrons*.	12. Continental condition of British islands followed by partial submergence; seasons towards close of this period more marked than at present; in the earlier stages climate more severe than now. Neolithic man; and passage to bronze and iron periods.	12. Denmark—Peat and buried trees in part; Kjökken-möddings. Switzerland—Pfahlbauten. France, Belgium, etc.—River and cave deposits in part, with reindeer, aurochs, etc. Italy—Palafitte, etc. Neolithic man; passage from stone to bronze and iron periods.
	11. Raised beaches, etc.	11. Littoral shells, etc.	11. Elevation of the land; climate cold-temperate.	11. Scandinavia—Raised beaches, and deposits with Baltic shells (Sweden).
Last Glacial period,	10. Valley-moraines and river-gravels (diluvium); river-gravels in south of England.	10. (?) Arctic mammalia, mammoth, Siberian rhinoceros, etc.	10. Local glaciers in mountain-valleys; land probably rising.	10. Germany—Terminal moraines in BlackForest and the Vosges; löss of the Rhine and Neckar in part, with *Elephas primigenius*, *Rhinoceros tichorhinus*, etc. France and Belgium—River and cave deposits in part. North America—Valley moraines of White Mountains, etc.
	9. Sand and brick-clay.	9. Boreal and arctic shells.	9. Arctic climate; land of less extent than now, but gradually rising.	9. Scandinavia—Sand and clay, with arctic shells, etc. Switzerland—Moraines overlaying older glacial and inter-glacial deposits; alpine diluvium, etc. France, Belgium, etc.—River and cave deposits, with arctic mammalia; mammoth, Siberian rhinoceros, etc. North America—Clay, etc., with arctic shells (Leda clays).
	8. Erratic blocks and earthy *débris*.	8. No fossils.	8. Period of floating ice; land deeply submerged, but rising.	8. Scandinavia, Denmark, and northern Europe generally—Erratics. Switzerland—Great extension of glaciers; terminal moraines. Carinthia—Moraines near the Raiblersee, etc. Italy—Moraines of Rivoli, Ivrea, etc. France and Southern Europe—River and cave deposits, with arctic mammalia and palæolithic implements. North America—Iceberg-drift of Ohio, etc.
	7. High-level beaches and marine drift.	7. Shells indicating a somewhat cold sea.	7. Submergence of the land to a depth in Wales of about 2,000 feet; land south of the Thames not submerged; climate passing from temperate to arctic; land rising.	7. Scandinavia—High-level marine deposits. Switzerland, Italy, and Carinthia—Advance of glaciers; older alpine diluvium; gravel overlying lignites of Dürnten, etc.; river-gravel with rhinoceros, etc., near Klagenfurt.

					France and Southern Europe—River and cave deposits. North America—Sand and gravel of interior in part.
The Glacial Epoch.	Last Inter-glacial period, ...	6. Sand, gravel, etc.; remodelled drift; kames and eskers.	6. Marine shells.	6. Land sinking; climate temperate.	6. Scandinavia and Northern Europe—Åsar and plateaux of marine gravel and sand. France—Diluvium of plateaux. Germany—Gravel of Vosges and Black Forest. Switzerland—Dürnten beds in part. North America—Cones and ridges of gravel and sand in the interior of continent, in part.
		5. Cave-deposits; river-gravels, etc., in part.	5. Palæolithic implements; extinct and no longer indigenous mammalia; hippopotamus, rhinoceros, elephant, etc.; mammoth, Siberian rhinoceros, etc.	5. Britain at first insular, with cold climate; next continental, with climate changing from cold to temperate and genial, and again to temperate; in early stages of continental conditions the arctic mammalia invade Britain; subsequently these disappear, and are succeeded by the hippopotamus, etc.; afterwards submergence ensues, and insulation perhaps effected before the climate again becomes suited to arctic mammalia.	5. Scandinavia—Peat, with palæolithic implements and cave-bear below Jära-wall (?). Switzerland—Dürnten beds, with *Elephas antiquus*, *Rhinoceros Merkii*, etc. Italy—Bone-beds and lignites of Piedmont, etc. Europe generally—Cave and river deposits, with palæolithic implements, and extinct or no longer indigenous mammalia. North America—Lacoustrine deposits and buried forests of Ohio.
	Great Cycle of Glacial and Inter-glacial periods, ...	4. Morainic rubbish, perched blocks and "diluvium."	4. No fossils.	4. Arctic conditions passing away; local glaciers.	4. Switzerland—Moraine rubbish. Scandinavia—Perched blocks, etc. North America—Perched blocks, etc. Formation of löss in river-valleys generally = fine glacial silt.
		3. Boulder-earth and clay of maritime districts [in inland parts, till with no fossils]; perched blocks at high levels.	3. Arctic shells in part.	3. Arctic climate; mountainous parts of Britain covered with snow and ice; glaciers cease to be confluent.	3. Switzerland and Carinthia—Grundmoränen in part and erratics at high levels. Scandinavia—Till in part; perched blocks at high levels. North America—Till in part; boulder-clay with shells in maritime districts (Canada); perched blocks at highest levels.
		2. Till and boulder-clay, with intercalated and subjacent beds of silt, sand, clay, gravel, etc.; cave-deposits; river-gravels, etc.	2. Arctic shells in part; oak, birch, pine, hazel, alder, willow, etc.; extinct and no longer indigenous mammalia, both of arctic and southern forms; palæolithic implements.	2. Intense glacial conditions, with great confluent glaciers; intermediate mild, and warm periods; arctic and southern mammalia visit Britain alternately, according as climatal conditions become suited to their needs.	2. Scandinavia—Upper and lower till, with intercalated fresh-water deposits, and subjacent gravel and sand. Switzerland and Carinthia—Grundmoränen. Europe generally—River and cave deposits, with palæolithic implements, and arctic and southern mammalia. North America—Till or unmodified drift with subjacent and intercalated beds, containing remains of trees and mammalia.
	Pre-glacial period, .	1. Norwich crag.	1. *Elephas*, *Mastodon*, etc., some northern species of shells.	1. Indications of approaching cold.	1. North America—Peaty bed, River of Inhabitants in Cape Breton.

NOTE C.

Traces of a Glacial Period in the Southern Hemisphere.

It is well known that the evidence of extensive ice-action in post-tertiary times is not confined to the lands within our hemisphere. Mr. Darwin, many years ago, showed that large glaciers came down within geologically recent times, to low levels in the Cordillera of South America, and his observations have been supplemented by Mr. D. Forbes, who informed Mr. Darwin that he had seen ice-worn rocks and scratched stones at about the height of 12,000 feet in various parts of the Cordillera, between lat. 13° and 30° S. Yet no true glaciers, Mr. Darwin says, "now exist even at much more considerable heights along this whole space of the Cordillera. Further south on both sides of the continent, from lat. 41° to the southernmost extremity, we have the clearest evidence of former glacial action, in numerous immense boulders, transported far from their parent source." Recently these observations have been extended by Professor Agassiz, the publication of whose investigations glacialists await with impatience. Drs. Haast and Hector have shown also that in New Zealand colossal glaciers at one time descended to low levels in that island; and some traces of glacial action have been observed in the mountains of the south-eastern corner of Australia. According to the theory supported in this volume, glacial and mild climates would alternate in the two hemispheres, and consequently the glacial deposits and ice-markings detected south of the equator will be either a little older or younger than the similar memorials met with in our hemisphere; while at the same time they must of course belong to the same great epoch. At present, however, the glacial deposits of the south are not sufficiently known, and we cannot tell whether they contain any records of inter-glacial mild climates. But a study of the distribution of animals and plants throws much curious light upon the subject; and I would commend to the reader's attention that most interesting chapter in Mr. Darwin's work, in which he discusses the geographical distribution of plants and animals. He will there find some striking evidence brought together in favour of alternations of mild and glacial conditions having occurred in the north and the south. (See *Origin of Species*, Sixth edit., p. 316, *et sqq.*)

NOTE D.

Map and Sections of Loch Lomond.

This map and the accompanying sections are the work of my friend, Mr. R. L. Jack. The sections are drawn on a true scale (same as the map), and are designed to give a clear idea of what is meant by a rock-basin. It will be observed that the lake is deepest in its narrow upper reaches, where, half-way between Inversnaid and Tarbet, it attains a depth of 100 to 105 fathoms. In its lower and wider reaches it shallows to 20, 12, 5 and 1 fathom. But so gradual is this shallowing, that were the lake to be drained of all its water, we should hardly be able to discover, without levelling, which was the deepest part of the hollow. The horizontal section brings out this feature in a striking manner. When, therefore, mention is made of a rock-basin,

PLATE XIII

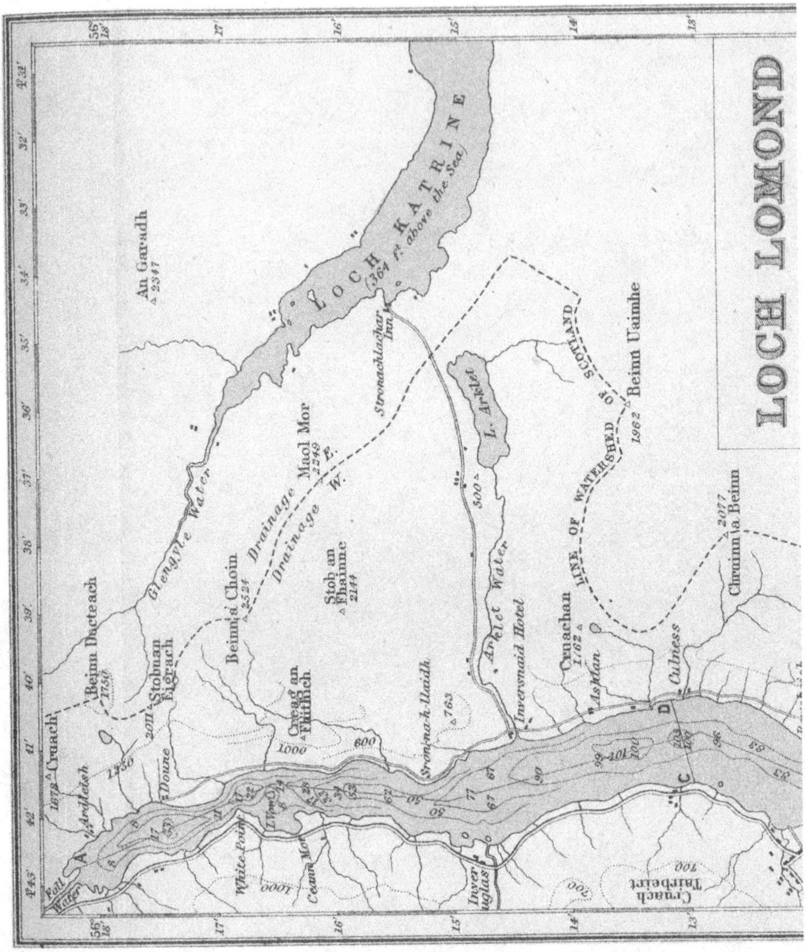

LOCH LOMOND

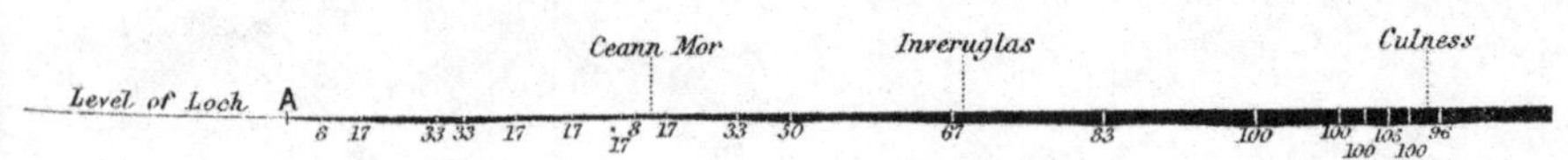

SECTION OF LOCH

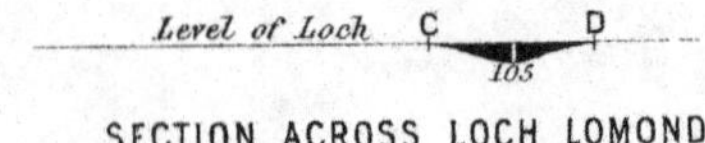

SECTION ACROSS LOCH LOMOND

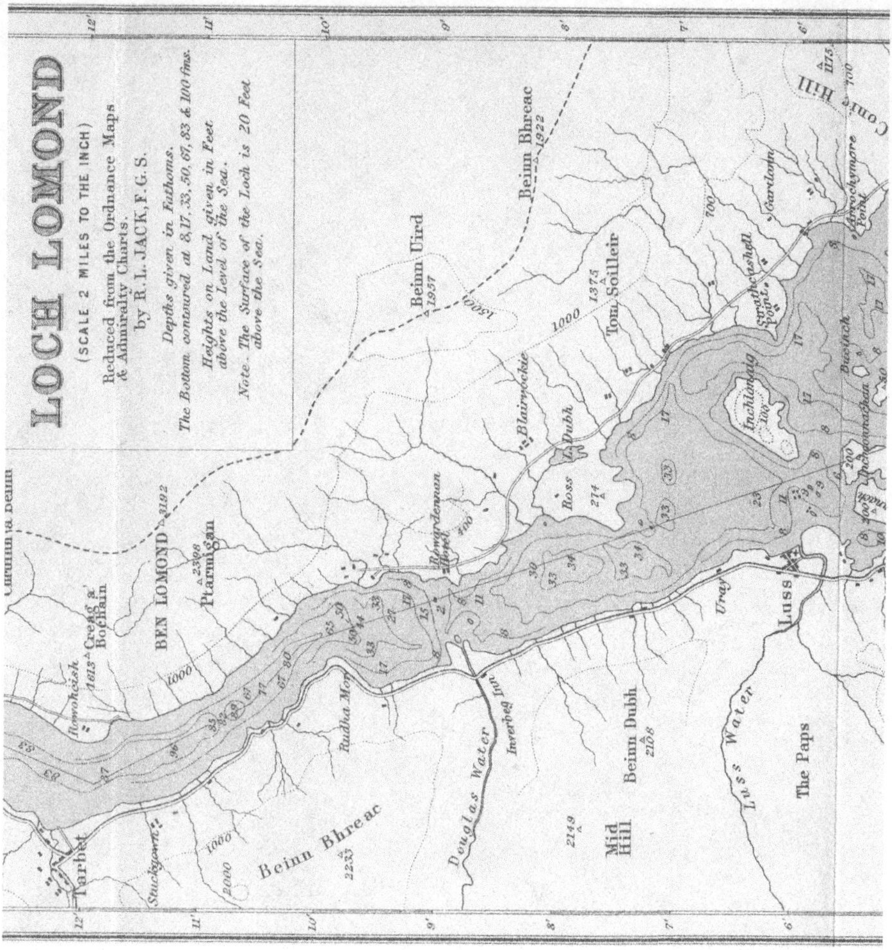

LOCH LOMOND

(SCALE 2 MILES TO THE INCH)

Reduced from the Ordnance Maps & Admiralty Charts.
by R. L. JACK, F. G. S.

Depths given in Fathoms.
The Bottom contoured at 8, 17, 33, 50, 67, 83 & 100 fms.

Heights on Land given in Feet above the Level of the Sea.

Note. The Surface of the Loch is 20 Feet above the Sea.

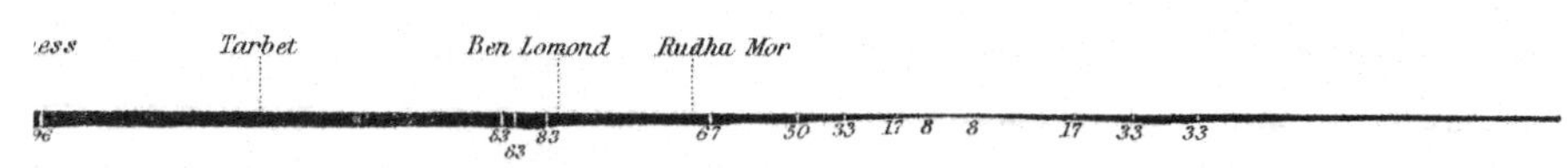

LOMOND ON THE SCALE OF HALF AN INCH TO THE MILE THROUGH THE

Depths in Fathoms - Height of Islands in Feet

ON SAME SCALE

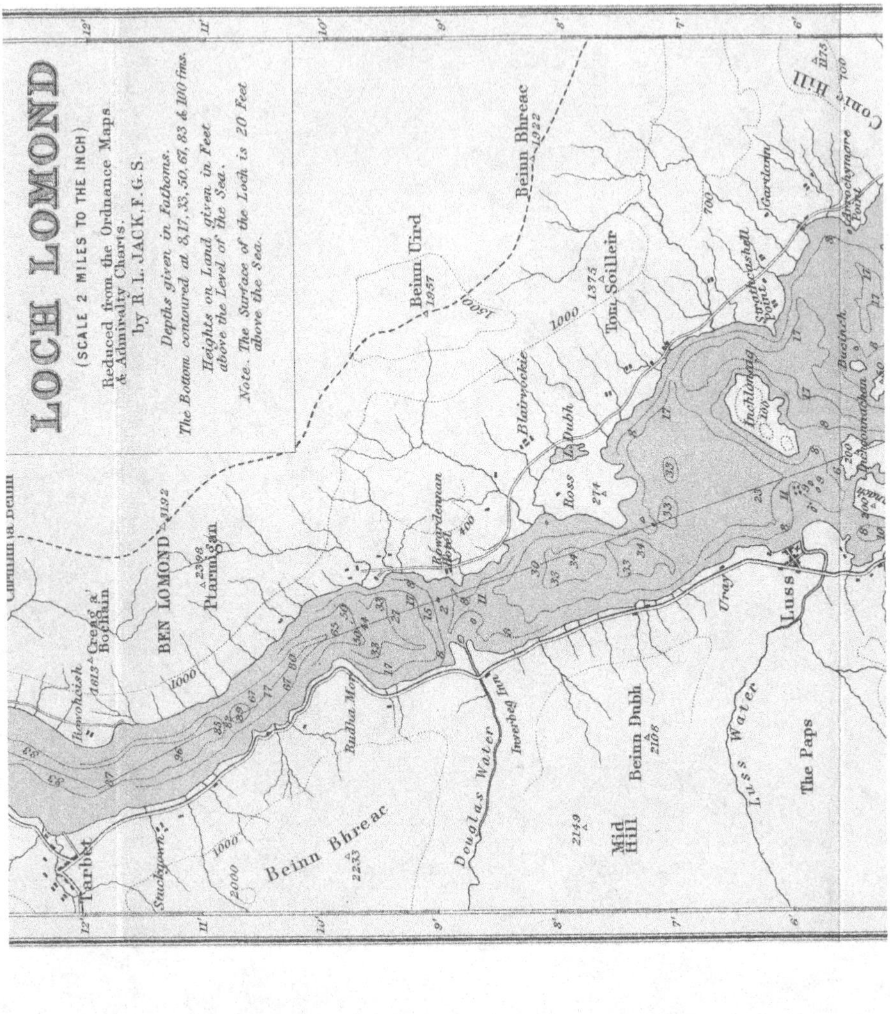

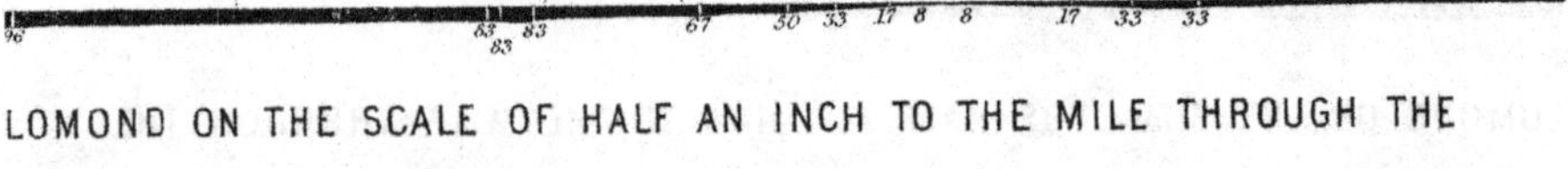

LOCH LOMOND
(SCALE 2 MILES TO THE INCH)
Reduced from the Ordnance Maps & Admiralty Charts.
by R. L. JACK, F.G.S.
Depths given in Fathoms.
The Bottom contoured at 8, 17, 33, 50, 67, 83 & 100 fms.
Heights on Land given in Feet above the Level of the Sea.
Note. The Surface of the Loch is 20 Feet above the Sea.
BEN LOMOND
Ptarmigan
Beinn Uird
Beinn Bhreac
Tom Soilleir
Inchlonaig
Luss
The Paps
Luss Water
Beinn Dubh
Mid Hill
Douglas Water
Inverbeg Inn
Rudha Mor
Tarbet
Stuckgown
Rowardennan Hotel
Ross
Conic Hill

ess
Tarbet
Ben Lomond
Rudha Mor
96
83
83
83
67
50
33
17
8
8
17
33
33

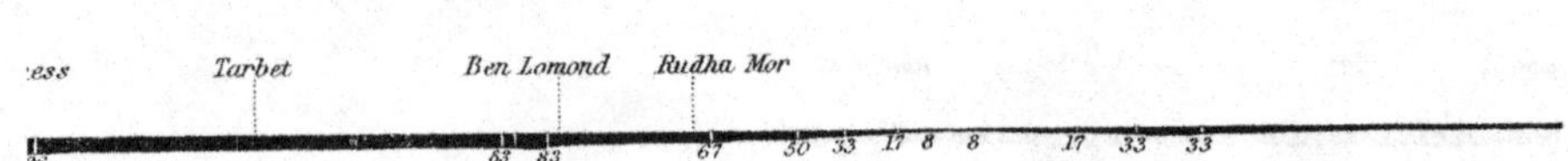

LOMOND ON THE SCALE OF HALF AN INCH TO THE MILE THROUGH THE

Depths in Fathoms - Height of Islands in Feet

D

ON SAME SCALE

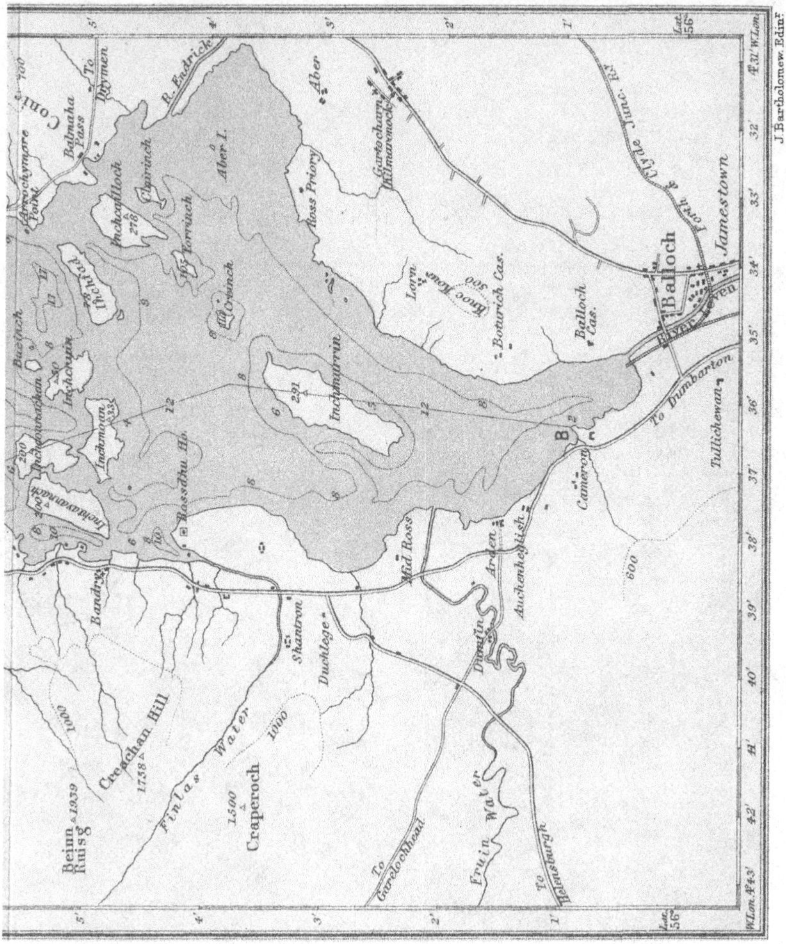

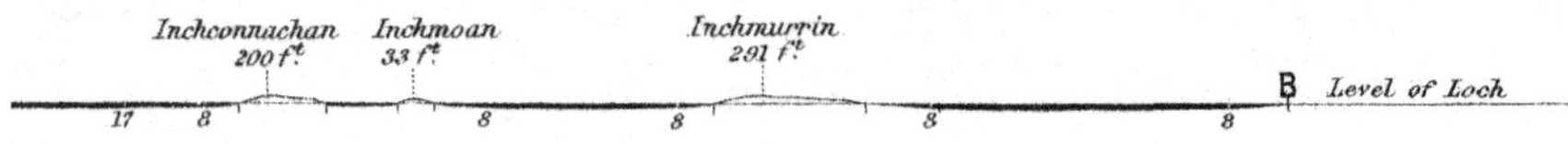

LINE A B ON MAP

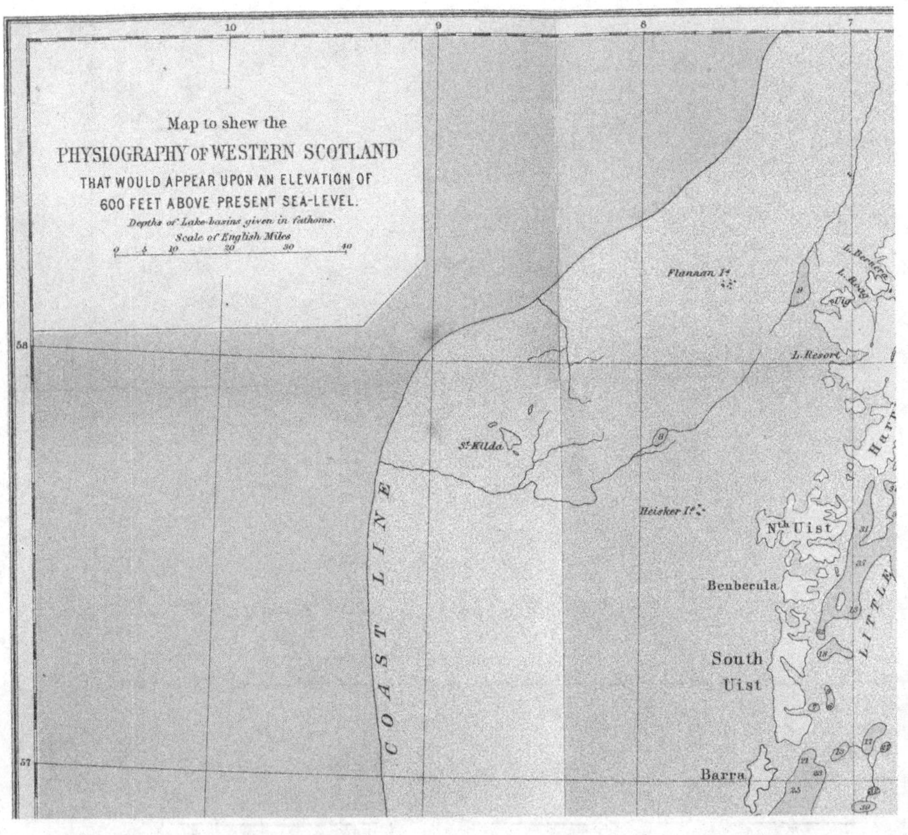

Map to shew the
PHYSIOGRAPHY OF WESTERN SCOTLAND
THAT WOULD APPEAR UPON AN ELEVATION OF
600 FEET ABOVE PRESENT SEA-LEVEL.
Depths of Lake-basins given in fathoms.
Scale of English Miles
0 5 10 20 30 40
10
9
8
7
58
57
Flannan Is
L. Bernera
L. Roag
Uig
L. Resort
St Kilda
Harris
Heisker Is
Nth Uist
Benbecula
South
Uist
Barra
COAST LINE
LITTLE

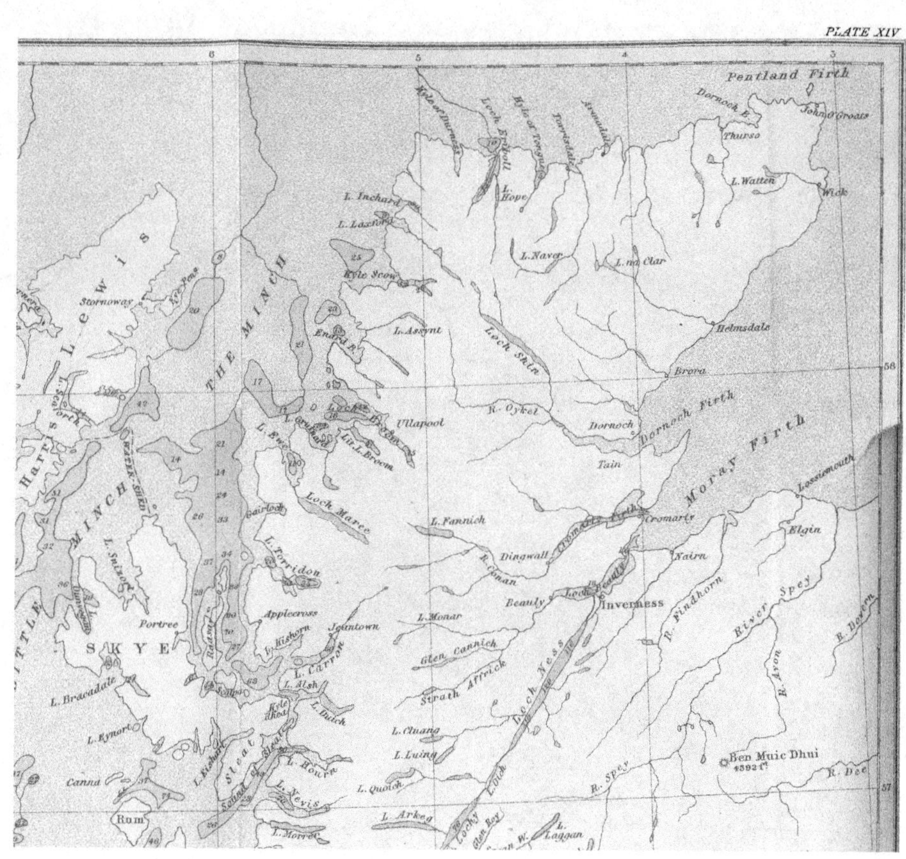
PLATE XIV
Pentland Firth
John O'Groats
Thurso
L. Watten
Wick
L. Inchard
L. Laxford
Kyle Scow
L. Hope
L. Naver
L. na Clar
L. Assynt
Loch Shin
Helmsdale
Brora
R. Oykel
Dornoch
Dornoch Firth
Tain
Moray Firth
Lossiemouth
Ullapool
Lit. L. Broom
L. Ewe
Loch Maree
L. Fannich
Cromarty
Elgin
Nairn
Dingwall
Beauly
Inverness
R. Findhorn
River Spey
Stornoway
Lewis
THE MINCH
Harris
MINCH
Gairloch
L. Torridon
Applecross
L. Kishorn
Jeantown
L. Carron
L. Alsh
L. Duich
L. Monar
Glen Cannich
Strath Affrick
Loch Ness
R. Avon
R. Deveron
Portree
SKYE
L. Bracadale
L. Eynort
Canna
Rum
L. Hourn
L. Nevis
L. Morrec
L. Cluang
L. Luing
L. Quoich
L. Arkeg
Glen Roy
L. Laggan
R. Spey
Ben Muic Dhui
R. Dee

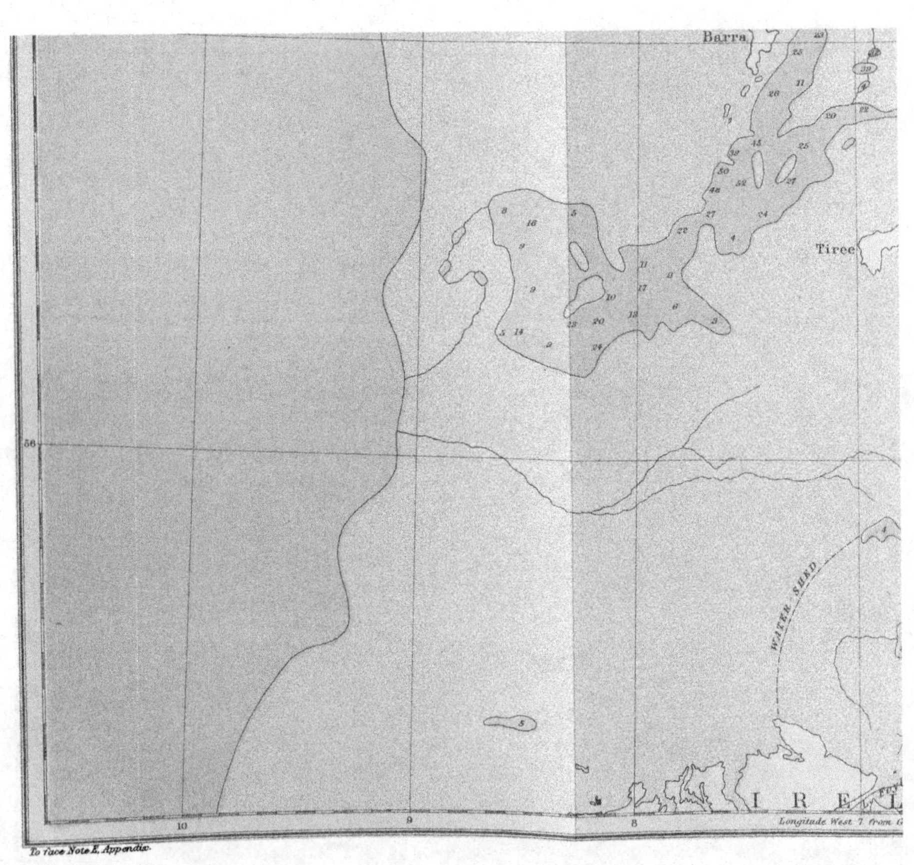

To face Note E, Appendix.

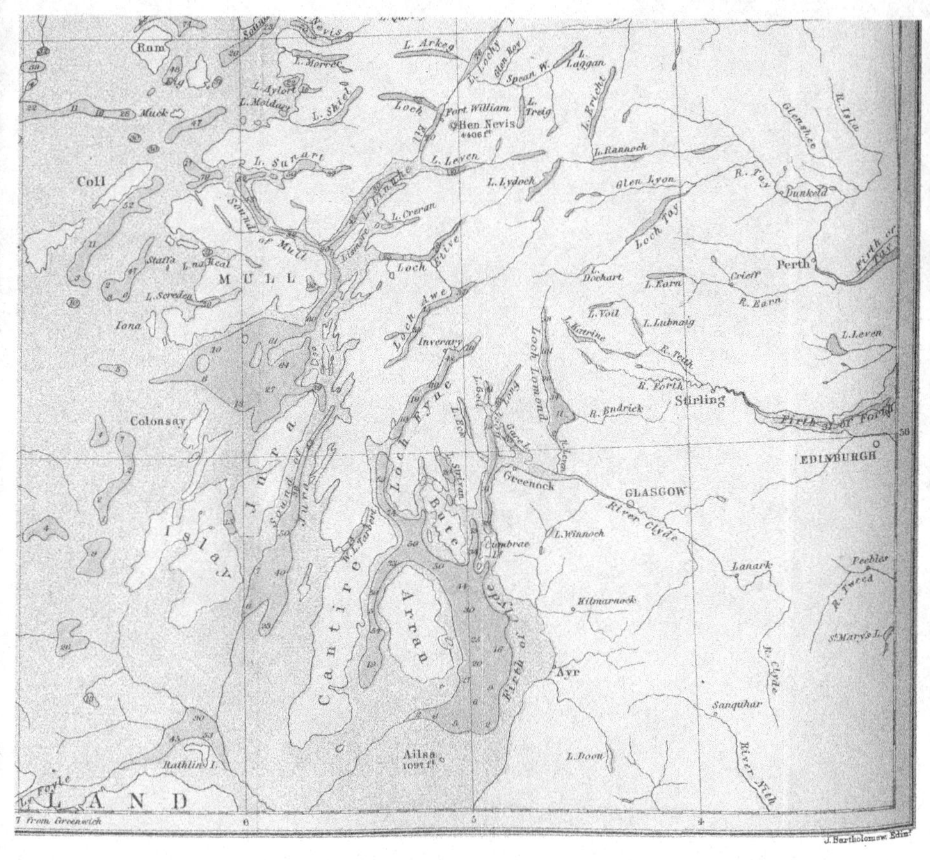

Rum
Muck
Coll
MULL
Staffa
Iona
Colonsay
Islay
Jura
Sound of Mull
Sound of Jura
L. Sunart
L. Arkeg
L. Lochy
Glen Roy
Spean W.
L. Laggan
L. Ericht
L. Treig
Fort William
Ben Nevis
4406 ft
L. Leven
L. Linnhe
L. Creran
Lismore
Loch Etive
Loch Awe
Inverary
Loch Fyne
L. Rannoch
L. Lydoch
Glen Lyon
Loch Tay
R. Tay
Dunkeld
Glenshee
R. Isla
Perth
Crieff
R. Earn
L. Earn
L. Dochart
L. Voil
L. Lubnaig
L. Katrine
R. Teith
R. Forth
Stirling
Firth of Forth
L. Leven
EDINBURGH
Loch Lomond
R. Endrick
L. Long
L. Eck
Gare L.
Greenock
GLASGOW
River Clyde
Bute
Cumbrae
L. Winnoch
Kilmarnock
Lanark
Peebles
R. Tweed
St. Mary's L.
R. Clyde
Sanquhar
River Nith
Ayr
L. Doon
Arran
Cantire
W. L. Tarbert
Firth of Clyde
Ailsa
1097 ft
Rathlin I.
L. Foyle
LAND
from Greenwich
6
5
4
J. Bartholomew, Edin.

100 fathoms deep, we are not to think of a profound hole like a huge pit, but of an elongated cavity, overhung it may be on both sides with more or less steep mountains or hills, and sloping in from both ends at a degree of inclination so slight as to be imperceptible to the eye. Were the Lake of Geneva to be drained, its bed would have merely the appearance of a great plain (yet that lake reaches a depth of 980 feet); and the cavity of Loch Lomond would not be more conspicuous.

Loch Lomond is a very interesting and satisfactory example of a rock-basin. We are quite sure of its depth, because it has been sounded all over by the officers of our navy, and we know that it does not lie in a line of dislocation or gaping fissure, neither is it crossed by any such fractures or displacements. It is as excellent a specimen of an excavated basin as the heart of a glacialist could desire.

NOTE E.

Map showing the Physiography of West of Scotland that would appear upon an Elevation of 600 feet.

This map has been prepared with much care and skill by my friend and colleague, Mr. R. L. Jack, who was good enough to relieve me of the labour of reducing the work from the Admiralty Chart of the West Coast of Scotland (No. 2,635).

Some years ago I was struck with the fact that the deepest parts of the Scottish seas appeared precisely in those places where a glacialist who held Professor Ramsay's views might readily have expected to find them. Not only did deep rock-basins occur in all the sea-lochs or fiords, but they also made their appearance again and again off the coasts of many islands in such positions as could not but be highly suggestive to a glacialist. In connection with these facts, it was also singular to observe that, while deep submarine hollows were so abundantly developed along the wild western shores of Scotland, they were almost entirely wanting in the corresponding latitudes on the other side of the island. And, then, one could not fail to notice that, with the exceptions of the Friths of Forth, Tay, Inverness, Cromarty, and Dornoch, no fiord-valleys open out upon the German Ocean, and no such islands as the Inner and Outer Hebrides appear off the east coast. Fiord-valleys and islands abound in the west, and there rock-basins are numerous; hardly any fiord-valleys or islands exist in the east, and there submarine hollows are rarely to be found. As far as I can make out from the Admiralty charts, only one deep submarine basin occurs along the whole stretch of coast-line between Duncansby Head and Berwick, and that is in the upper reaches of the Frith of Forth, between St. Margaret's Hope and a point east of Kinghorn and north-east of Inch Keith. The hollow is a long narrow trench, gradually opening out as it shallows to the north-east of Inch Keith. It is deepest near Inch Garvie, where its bottom is 246 ft. below the surface of the sea, or 186 ft. lower than the lip of the trench. It shallows passing east, but deepens again to 168 ft. between Inch Colm and the Oxcars Rocks, shallowing once more, and again deepening to 138 ft. before it finally shelves away. It is certain, however, that this basin must at one time have been more extensive. Immense quantities of silt and sand are borne down into the estuary of the Forth, and great banks of sand and mud have accumulated, especially in the upper reaches of the estuary. There cannot be much doubt, therefore, that the submarine hollow has been greatly silted up.

No hollow so deep appears in any of the other friths that open into the German Ocean, but each is characterised by the presence of great sand- and mud-banks, which in many cases impede navigation. In the Frith of Tay the mud-banks are specially noticeable; above Dundee the Frith at low-water shows little more than a series of slimy banks, with winding water-lanes; and below Dundee, at the mouth of the estuary, the mud and sand are pushing out seaward, so as to form a well-defined submarine delta. In short, it is evident that all the friths on the east side of the island have been and are still being gradually silted up. Yet we may still trace elongated hollows in these friths. There is one 48 ft. deep opposite Broughty Ferry; two occur in Beauly Loch, 108 ft. and 72 ft. deep respectively; Cromarty Frith is 120 ft. deeper than the sea outside, and although Dornoch Frith is very shallow, it is still 36 ft. deep above the Bar, which is only 12 ft. below the sea.

To return to the west coast: I would first direct the reader's attention to the general slope of the sea-bottom. It will be observed that (putting rock-basins for the moment out of account) the bottom of the sea in the North Minch falls away towards the north, a river being inserted to show the direction the drainage would take were an elevation of the whole west coast to supervene. Between the north end of Skye and the Shiant Isles the soundings indicate the existence of a ridge which would form a low water-shed between the country of the North Minch and that lying to the south. The configuration of this latter, however, is exceedingly irregular, and it is difficult to ascertain from the charts in what direction the lakes in the Little Minch would drain; most likely, however, it would be south-west into the large lake which is represented as sweeping from South Uist round Barra Head, and sending a river out to the sea. West from the Island of Islay another stream is inserted to show the slope of the land in that direction. South of the same island it will be noticed that the drainage would be south-east by the North Channel. It must not be supposed that these rivers are put down at random. The charts have been closely followed, and it is believed that the lines indicated are as near as possible those that would be taken by the streams and rivers upon an elevation of 600 feet. A greater number might have been inserted, but it was thought better to give only such as would suffice to indicate the general slope of the sea-bottom.

A glance at the map will show that the chief submarine basins occupy certain well-defined positions, and form two distinct groups. The first group embraces what may be termed *fiord-basins*. Enough has been said in the text regarding the rock hollows which are known to occur in our sea-lochs. These of course agree in direction with the sea-lochs, with which they are sometimes almost co-extensive, as in the case of Loch Fyne; or entirely so, as in the case of Loch Etive. But an examination of the Admiralty charts proves the existence of numerous submarine basins which lie beyond the sea-lochs, and run parallel to the course of sounds, channels, and straits. As examples we may take the basins of Raasay Sound, the Inner Sound, Sleat Sound, the Passage of Tiree, the Frith of Lorn, and Jura Sound. Now, these basins occur in what are simply the continuations of fiord-valleys. If the land were elevated for 600 ft. it would be seen that all these "Sounds" and "Passages" only formed the lower-reaches of mountain-valleys. The Sound of Jura, for example, would appear as a wild mountain-valley continuous with that of Loch Craignish. In the same manner, Sleat Sound would be continuous with the valleys that now hold Lochs Alsh and Hourn. And each of these valleys, as the map shows, would contain deep fresh-water lakes. We may therefore define the *fiord-basins* as those hollows which occupy the beds of, and extend in the same direction as submerged mountain-valleys. They therefore follow the general slope of the sea-bottom, as the map itself sufficiently indicates.

In the text reasons are given for concluding that the rock-basins in our sea-lochs were excavated by glaciers which once filled all those now submerged land valleys. We may now examine one or two of the rock-basins that appear in the Sounds and Straits, for the purpose of ascertaining whether appearances are such as to indicate a similar origin for them.

The map represents a large lake as occupying the sites of Raasay Sound and the Inner Sound, and stretching northwards to a point opposite Loch Broom. This is one of the deepest areas on the west coast of Scotland, the lip of the submerged basin being 50 fathoms, and its deepest part no less than 138 fathoms below the surface of the sea. Were the land to be sufficiently elevated, we should have here a fresh-water lake 88 fathoms, or 528 ft. in depth; so that, even were the land to be upheaved for 600 ft., the bottom of the Raasay Lake would still be 38 fathoms below the level of the sea. Its deepest part trends along the east coasts of Raasay and Rona, and it shallows gradually away towards the north; that is to say, it is deepest where the channel is narrow—while, on the contrary, it begins to shallow as it expands into the North Minch. Now, if we examine the map of glacial striæ, we shall find that this large submerged basin was at one time occupied by a massive glacier that flowed in precisely the same direction as the trend of the basin itself, that is towards the north. Note further, that the striæ on the shores of Lochs Carron and Kishorn show that the glacier-ice which once filled those lochs swept over the low-grounds of Skye between Broadford and Loch Eishart, where also it has left marks of its passage. This was doubtless at the same time that Raasay Sound and the Inner Sound were choked with glacier masses streaming outwards from Skye, Gairloch, Loch Torridon, and Loch Carron itself. For the reasons given in the text, the erosion produced by this ice would be most excessive where the latter was strangled, or compressed and heaped up. Consequently we find that it is between the mainland and the islands of Raasay and Rona that the basin attains its greatest depth. As the glacier crept out into the Minch it had room to expand, and therefore its erosive action became weaker in that direction.

Take another example. Between Knapdale and the Island of Jura it will be observed that an elongated basin, somewhat resembling Loch Lomond in outline, extends from Loch Craignish down the Sound of Jura. The deepest part of this basin lies between Loch Crinan and a point nearly opposite the extreme south end of Jura Island. In this long narrow section the depths are very irregular; in fact, the hollow here consists of a string of small rock-basins, ranging from 102 to 110 fathoms in depth, the lip of the basin itself being 60 fathoms from the surface; so that in actual depth the upper reaches of the basin attain at the most 42 or 50 fathoms. From the south end of Jura the basin widens out, and as it does so it gradually shallows, attaining, however, an exceptional depth (40 fathoms) immediately opposite the north point of Gigha Island. Now, these appearances are precisely such as might have been expected; the narrow and deep portions of the basin occur just in those places where the erosive power of the ice would be greatest; and, on the other hand, the basin shallows as the fiord-valley opens out, for the simple reason that here the glacier had room to extend itself and shelve off.

I have selected for illustration two of the simplest cases. When we come to examine other fiord-basins, we not infrequently find that they are mixed up with a set of basins which cannot be said to coincide with mountain-valleys. These form our second group, one or two simple examples of which I shall describe first, and thereafter point out how the two groups sometimes coalesce.

The basins of the second group not infrequently extend at right angles to the trend of the fiord-basins, and are most typically developed along the inner shores of islands, especially when these are placed opposite the mouths of sounds and sea-lochs As typical examples, I may mention the basin lying north of Rathlin Island, which faces the Sound of Jura, and the great series of basins that stretches along the inner shores of the Outer Hebrides. For reasons which will be given presently, these may be conveniently termed *deflection-basins.*

The Rathlin Island basin is one of the simplest, and at the same time one of the most striking examples of a deflection-basin. It attains the great depth of 133 fathoms, and, taking its lip at 80 fathoms from the surface, we have 53 fathoms, or 318 feet, as the actual depth of the excavation. It will be noticed that this basin does not rest in a fiord-valley; there is no sea-loch or deep land-valley opening out upon it from the Irish coast. If excavated by ice, that ice could not have flowed from Ireland. How, then, is the basin to be accounted for? It will be remembered that during the period of the till the Scottish mer de glace advanced upon the coasts of Antrim and Donegal, and became confluent with the Irish ice-sheet. This is well shown by the manner in which the glacial markings in the extreme north of Ireland turn away towards the west and south-east, instead of pointing right out to sea. The Scotch ice split upon Ireland, and flowed westward into the Atlantic, and south-east by the North Channel into the Irish Sea. [I shall not be surprised, if the glacial striæ in the extreme north of Ireland prove to have been the work rather of Scotch than of Irish ice.] Bearing in view the facts thus briefly referred to, there can be no difficulty in understanding why a deep hollow came to be ground out at Rathlin Island. Here it was that the immense glacier mass discharging by the Sound of Jura met with resistance to its progress. Rathlin Island, in fact, behaved like a large boulder in the bed of a stream; it stemmed the current, which was thus forced to flow east and west, and the usual result followed—a hollow was dug out in front. If the linear trend of the *fiord-basin* in the Sound of Jura indicates the former path of the glacier that formed it, not less does the crescent-shaped *deflection-basin* at Rathlin Island point out where the ice-current divided to flow in opposite directions.

Another good example of a deflection-basin will be observed circling round the north of Rum. It reaches its greatest depth opposite Loch Eishart, where the excavation on the sea-bottom is as much as 74 fathoms, the bottom of the basin being 139 fathoms from the surface. Now it is certain that this is precisely where, during the climax of the glacial epoch, there would be immense erosion caused by the stemming of the ice that streamed out from the Coolin Mountains and Loch Eishart. Note how the basin is continued into Canna Sound, where it attains a depth of more than 50 fathoms, its bottom being 130 fathoms below the surface of the sea. A similar deep excavation makes its appearance between Eigg and Rum, which has an actual depth of not less than 48 fathoms, although the bed of the sea is only 86 to 88 fathoms deep at that place. Although this latter basin is separated from the one lying north of Rum, they were doubtless formed by the same glacier-mass, which, splitting upon Rum, would pour round that island, and exert excessive erosive power in the channels that separate Rum from Eigg and Canna.

Let us take yet another example. Mention has already been made of the deep basin that extends north from the Inner Sound into the North Minch, where it ends against the Shiant Isles and a bank known to fishermen as the Shiant East Bank. It will be observed that facing the end of the Raasay basin (which is a fiord-basin), another deep submarine hollow extends itself along the shore of the Long Island, opposite Loch Shell. This, there can be no doubt, belongs to our group of deflection-basins. When the ice which ploughed out the Raasay basin flowed out so far as to reach the Shiant East Bank, it would have a tendency to creep along the general slope of what now forms the bed of the sea; that is, it would tend towards the north. But as the whole of the North Minch became at the same time choked with glaciers descending from the wilds of Sutherland, it is evident that its passage in that direction would necessarily be blocked up. It would therefore be compelled to abut upon Lewis. Now we know that the ice which filled the North Minch attained so great a thickness that its upper strata were enabled to overflow the whole of Lewis from south-east to north-west, to a height of not less than 1,250 feet and probably even higher than that. This is shown by the abundant traces of glacial erosion all over the island. But while the upper strata of ice were grinding across Lewis, there would necessarily be an "undertow" tending along the coast both to the north-east and the south-west. The greatest pressure would be exerted close in shore, where the high ground opposed the direct passage of the ice; and hence deflection-basins would be scooped out in such places. The process, indeed, would be precisely the same as in the cases of Rathlin and Rum. The map represents a whole series of similar basins, extending along the inner margin of the Outer Hebrides. None of these are fiord-basins, but off the mouth of Loch Dunvegan, in Skye, there appears to be a union of basins belonging to both groups. South of Benbecula, however, the hollows which trend along the coast of the Hebrides seem certainly to be deflection-basins. This will become apparent when we reflect that, during the climax of the glacial epoch, the comparatively open space lying between Benbecula, South Uist, and Barra, on the one hand, and Skye, Rum, Coll, and Tiree on the other, must have been filled with glacier-ice. From Loch Bhracadail, Loch Eynort, Loch Bhreatal, and Loch Scavaig, thick masses descended and became confluent with the ice that carved out the deep rock-basin lying north of Rum. At the same time glaciers streaming out from the Kyles of Skye, Loch Hourn, and Loch Nevis united in Sleat Sound, and swept past Eigg in the same general direction, namely towards south-west by west, until the mer de glace abutted upon the Outer Hebrides. Here, then, there would be intense grinding power exerted; and while the upper strata of the ice would overflow in a westerly or north-westerly direction such portions as were not too lofty, the lower strata of the glacier-mass would sweep south-west by south, until, as the ice rounded the opposing high ground, it found freedom to extend itself more to the west, and so to shelve off into deep water. Thus the trend of many of the submarine basins, as shown upon the map, indicates the direction followed by the undertow of the great mer de glace, and will not always be found to run parallel with the marks of glacial erosion upon the contiguous land. For example, the deflection-basins lying off the east coast of Lewis trend from south-west to north-east, whereas the glacial markings on the land go across the island from south-east to north-west.

The two groups of basins which I have thus briefly described frequently become confluent, as one would naturally have expected. The upper reaches of Loch Carron, for example, occupy a fiord-basin, but where the hollow expands from the Kyles of Skye to north-west it forms a deflection-basin; it is along its lower margin, indeed, where this hollow attains its greatest depth. If the land were elevated for 600 ft. we should find the sea-bottom deeply scooped and hollowed in front of all the islets that stood right in the way of the ice-flow. But the map only shows such hollows as would form rock-basins and become fresh-water lakes. Yet if we examine the Admiralty charts, we shall observe that a deep horse-shoe-shaped excavation would circle round the north end of Eigg, being evidently the work of the ice that came down Sleat Sound, and so with other islets; but when these are not very high, the erosion in front of them has been less excessive. In short, if the sea-floor were exposed to view, we should find that wherever abrupt ground rose opposite the mouth of a mountain-valley, a hollow of greater or less depth would circle round it like a collar. The Island of Arran would afford a splendid example; the Island of Mull, opposite Loch Linnhe, would be another hardly less striking; so would Rum, Eigg, Coll opposite the Sound of Mull, and many others. Thus the two groups of basins ever and anon coalesce, and in fact graduate into each other. Nevertheless, they must be distinguished, for while the fiord-basins invariably indicate the direct route taken by the mer de glace, the deflection-basins frequently indicate only the trend of the undertow, the upper strata having often overflowed the opposing land, and so swept on in the original direction.

Besides these two groups of basins, a number of small ones are indicated as scattered about at a distance both from fiord-valleys and islands. These are comparatively shallow, scarcely exceeding seven or eight fathoms in depth, and not often attaining even that. As a rule, their longer axis coincides in direction with what appears to have been the path of the mer de glace. Similar small hollows often occur in low-lying tracts on the land, as, for example, in the low-grounds of Lewis, where they are seen to coincide in direction with the lines of glacial erosion. Some of these are rock-basins, and others are mere depressions in the glacial deposits.

In fine, it seems to me that the distribution of submarine basins round the coasts of Scotland strikingly confirms the conclusions we had arrived at from an examination of the glaciated aspect of the land itself; namely, that the whole country—with the exception, perhaps, of the higher hill-tops—was at one time deeply smothered in ice, which flowed out by all our sea-lochs, overflowing the islands off our coasts, and only stopping at last in the deep waters of the Atlantic. And, to my mind at least, it is no less evident that the remarkable distribution of our deep submarine hollows can only be accounted for by Professor Ramsay's theory of the glacial origin of rock-basins.

NOTE F.

Glacial and Inter-glacial Deposits of Northern Italy.

The following note I had at one time intended to insert in the text immediately after the chapter that treats of the Superficial Deposits of Switzerland; but as much of it is of the nature of a discussion, it has been thought better to place it here. The reasoning employed, and the results sought to be arrived at will, however, be better appreciated if the reader, before perusing this note, will first glance his eye over Chapter XXVII.

Just as in Switzerland we have central low-grounds, with the Alps rising in the south and the Jura in the north, so also in Italy we find the great plains of Piedmont and Lombardy flanked by the Alps on the one hand, and bounded by the hills of Turin and the northern spurs of the Apennines on the other. With two exceptions, every great valley that opens out from the Alps upon the plains of Northern Italy contains a lake, as is the case with similar valleys in Switzerland. The two exceptions referred to are the valleys of the Dora Riparia and the Dora Baltea.

Thus, at the first glance there appears to be a broad general resemblance between the regions on both sides of the Alps. The geologist might therefore expect to meet with a like similarity in the glacial phenomena of the two regions; and up to a certain point his expectations would no doubt be realised. But when he came to correlate the Swiss with the Italian deposits, he would find the task by no means so easy, and the resemblance between the two not nearly so striking as he anticipated. So long as he confined his attention to the mountain valleys, he would observe precisely the same appearances as present themselves in the mountain valleys of Switzerland. Rounded and polished rocks, morainic débris, and perched boulders he would see everywhere; and at the lower ends of the great lakes he would encounter huge terminal moraines. But out upon the broad plains he would find only wide-spread deposits of gravel and loam, the stony glacial clay so often met with upon the low-grounds of Switzerland nowhere appearing upon the plains of Northern Italy.

Of all the glacial deposits of Italy, perhaps the most striking are the moraines of the Dora Baltea. They form a huge semicircular embankment opposite the mouth of the large valley of Aosta, and some idea of their vast extent may be gathered from the simple statement that they rise out of the plains of Piedmont as steep hills, to a height of 1,500 ft., and even in one place to very nearly 2,000 ft. Measured along its outer circumference,* this great morainic mass is found to have a frontage of at least fifty miles, while the plain which it encloses extends for some fifteen miles from Andrate

* That is from Andrate by Mongrando, Saluzzola, Cavaglia, and Caluso to the bridge over the Chiusella.

southwards, with a breadth of about eight miles. Two lakelets (the largest of which is little more than two miles in length by one in breadth) occur within the moraine.

MM. Martins and Gastaldi have shown * that the moraine matter rests upon beds of coarse gravel, and that these again repose upon deposits of sand, and the succession given is as follows:—

3. Moraine.
2. Alpine diluvium.
1. Marine sands.

The upper deposit (No. 3) forms the great bulk of the semicircular range of hills above referred to. It exactly resembles the moraines of the Swiss Alps, being composed of a pell-mell heap of angular blocks and débris, with some admixture of earth and sand.

The bed (2) also answers precisely to those wide accumulations of gravel which cover so large an area in the low-grounds of Switzerland. It shows no trace of fossils, and the rounded stones of which it consists have evidently been derived from the Alps. None of these stones are striated, and no angular blocks occur among them.

The underlying marine sands contain a number of fossils, many of which belong to species still living in the Mediterranean. Out of ten shells which are said by MM. Martins and Gastaldi to be characteristic of these sands, eight are even now denizens of the neighbouring sea, one is doubtful, and only one is said to be extinct.

Resting upon the sands occurs here and there an ancient alluvium, which is considered by Martins and Gastaldi to be of older date than the alpine diluvium. This deposit has yielded remains of the mastodon, the rhinoceros, the hippopotamus, the elephant, along with recent land and fresh-water shells. The bottom of the marine sands is not always seen; in some places, however, these beds may be observed resting upon the solid rocks of the Alpine districts, while in other places they repose upon certain loose accumulations of older tertiary age. According to the Italian geologists, both the marine sands and the alluvium with bones belong to the Pliocene period, and are considered therefore to date back to pre-glacial times.

In a recent publication,† however, Signor Gastaldi explains that he has termed these bone-bearing beds "*pliocenic alluvia*," not so much because he wished to make them a constituent part of the pliocene formation, but rather to discriminate them from the alpine diluvium, which of course is a later accumulation. Besides the "pliocenic alluvia," there occur at various places in Northern Italy, as in the environs of Carignano, at Lanzo near Stura; at Gifflenga in the valley of the Cervo; at Boca, Maggiora, &c., certain beds of lignite which the same eminent observer is inclined to consider as being the precise equivalents of the ossiferous alluvia; and he quotes the opinions of MM. Comalla and Stoppani, who have no hesitation in saying that the lignites of Leffe (Gandino), in which occur remains of the elephant (*E. meridionalis*), the beaver, the emys (not distinguishable from the recent *Cistudo europea*), deer, and goats, really belong to post-pliocene, and not to pliocene

* *Bulletin de la Société géologique de France*, tom. vii. 2me serie, p. 554; Professor Favre's *Recherches géologiques*, tom. i. p. 169.

† "Appunti sulla Memoria del Sig. J. Geikie, *On Changes of Climate, &c.*" *Atti delle Reale Accademia delle Scienze di Torino*, vol. viii. Aprile 1873.

times. In short, as Professor Gastaldi remarks, the lignites rest upon pliocene deposits, and are covered by diluvium, and thus occupy the same relative position as the Swiss lignites (Utznach, Wetzikon, &c.), which, it will be remembered, lie at the base of the Alpine diluvium.

But while Professor Gastaldi is clearly of opinion that the Italian and Swiss lignites belong to one and the same age, he does not agree with Professor Heer that the Swiss lignites are inter-glacial, and objects to the suggestion which I had ventured to make,* namely, that the ossiferous beds of Northern Italy mark an inter-glacial period. I feel somewhat sure, however, that if Professor Gastaldi were to study the Swiss deposits, he could come to no other conclusion than that arrived at by the eminent Swiss botanist. It is beyond question that the Swiss beds rest in some places upon true glacial deposits, upon unmistakable ground-moraine—*i.e.* clay holding scratched stones and boulders. If, therefore, the Italian lignites be of the same age as those of Switzerland, they can only be referred to inter-glacial times. It is true that, so far as is known, no glacial deposits underlie the Italian lignites; but the same is the case with not a few of the Swiss lignite beds, as for example those at Utznach, where the beds rest directly on highly disturbed deposits of Miocene age. The mere absence of underlying moraine matter is no proof, therefore, that the Italian lignites are of pre-glacial age; the Swiss lignites were supposed to be so, until at Wetzikon and Dürnten they were found to repose upon a true erratic deposit belonging to post-pliocene, or glacial times.

In the publication referred to I not only suggested the inter-glacial age of the ossiferous alluvia of Piedmont, but I even went so far as to state that the underlying marine sands might probably prove eventually to belong to inter-glacial times also. I was quite aware that this suggestion would appear bold to Italian geologists, and it was not without some trepidation that I ventured to express my views upon the subject. Nevertheless, holding as I did and still do, decided convictions concerning the great interval of time represented by the Swiss lignites, and by their equivalents, in Northern and Western Europe, I was prepared to risk the charge of boldness, in the hope that the whole subject would be thoroughly ventilated. In this hope I have not been disappointed. Signor Gastaldi, in a most interesting communication to the Academy of Sciences, Turin, has taken up the question. After giving his arguments the careful study which they deserve, I am compelled still to disagree, which I the more regret as it may appear presumptuous in me—whose personal acquaintance with Northern Italy has only been obtained during a few short holiday excursions—to differ from the opinion of one who has made that region a lifelong study. The question, however, is not one of the geological succession of strata. There is no dispute as to what relative position the "marine sands" of Piedmont occupy. They are clearly of older date than any recognizable morainic or diluvial deposits in Northern Italy; and if it were simply a question of local geology, one could have no good reason for doubting their pre-glacial age. But then the question is not one of local geology alone; the Italian deposits must be considered in the light of the evidence derived from contiguous regions. If it be true that certain oscillations of climate accompanied the deposition of the glacial deposits and

* *On Changes of Climate during the Glacial Epoch.* London, Trübner & Co.

their equivalents in every region of the northern hemisphere in which these accumulations have been studied, it is not unreasonable to hope that in Italy also we shall find some indications of the same great world-changes. What then, let us ask, is the evidence furnished by the "marine sands" of Piedmont? Do they afford us any definite proof that they are of pre-glacial age?

Of the shells which occur in these sands a certain percentage are not known as living species. The great majority, however, still occupy European seas. My friend Mr. Etheridge, who has been kind enough to examine for me some lists of the fossils obtained from the marine sands which are exposed here and there at the base of the Alps between Lake Maggiore and the Ticino, tells me that in his opinion these deposits do not date back to so old a time as the pliocene beds of England. It is difficult, however, to ascertain the proportion of living to extinct species in these Italian pliocenes, and the results obtained during the recent dredging cruise of the *Porcupine*,* make it doubtful whether many of the shells which are now only known in a fossil state in the Italian tertiary deposits may not eventually prove to be still living species.

Signor G. Michelotti, well known for his works on the pliocene and miocene faunas of Piedmont, examined for Professor Gastaldi a series of fossils from the marine sands at the base of the Alps, and informed him that not a single characteristic miocene shell appeared amongst the number.† But even if it should eventually prove that the extinct or apparently extinct species in the so-called pliocene sands at the base of the Alps are in the proportion of 15 or 20 per cent., still that will not prove these deposits to be of pre-glacial age. Nay, it would not even follow from this that the yellow sands of Piedmont were accumulated at the same period of time as those English deposits that contain a similar percentage of apparently extinct species. The mode adopted by M. Deshayes and Sir Charles Lyell for ascertaining the relative antiquity of Tertiary deposits is no doubt most excellent, so long as the deposits we examine happen to form a more or less continuous series, and are confined to some definite geographical area. The Norwich Crag, for example, contains about 18 per cent. of extinct or apparently extinct species of sea-shells. Now, if certain other English Tertiary deposits are found to contain a greater percentage of extinct forms than this, it is legitimate to infer that these must be older than the Norwich Crag, just as on the contrary we should consider those beds to belong to later times which happen to exhibit a smaller number of extinct species; for, as Sir C. Lyell remarks, "the greater number of recent species always implies the more modern origin of the strata." But when we pass into a different geographical area, it is evident that although we there detected superficial accumulations in which the proportion of extinct to recent species was the same as in the Norwich Crag, still it would not follow that these accumulations had been deposited at the same time as the English beds referred to. We have to take into consideration the fact that marine faunas must in the course of time be subjected to very different conditions, and that, owing to geological and geographical changes, species characteristic of certain areas may die out and

* *Depths of the Sea*, p. 183 *et sqq.*

† "Studii Geologici sulle Alpi Occidentali," *Mem. del R. Comitato geologico d'Italia*, vol. i., 1871.

become extinct at a more rapid rate than the contemporaneous life-forms of other regions. Hence, deposits laid down at one and the same time in different latitudes and in separated districts, may come to envelope and contain assemblages of shells amongst which the proportion of extinct to recent species may vary indefinitely. And the difficulty of identifying contemporaneous deposits becomes the greater the nearer these approach in age to recent accumulations. In short, it may well be that the newer pliocene of one country may be either older or younger than the newer pliocene of another. We may conclude, therefore, that the evidence supplied by the organic remains in the marine sands of Piedmont is not sufficient to prove that these sands are of pre-glacial age.

But it will be said that, since the sands in question are overlaid by the great moraines, they must necessarily date back to pre-glacial times. Now this would certainly follow, if it could be shown that these moraines mark the furthest limits reached by the glaciers during the climax of the glacial epoch. There are several considerations, however, which lead to the inference that the moraines referred to do not mark the southern limits of the ancient glaciation, but belong indeed to a more recent date.

On the north side of the Alps there is distinct evidence to show that Switzerland experienced at least two glacial periods, separated by an intervening period of milder conditions. During the first cold period the glaciers increased to such an extent that all the ice-streams issuing from the mountain-valleys coalesced upon the low-grounds to form one gigantic mer de glace that rose some 2,000 feet high upon the flanks of the Jura. Towards the north-east, the ice would appear to have overflowed these hills, and thereafter to have descended the Frickthal to the Rhine at a point some twelve miles below the confluence of that river with the Aar. How much further west it may have gone we cannot say, but there is good reason to believe that the Swiss mer de glace united with the glaciers of the Black Forest. Again, it is certain that from that part of the mer de glace which flowed to the south-west a great glacier crept outwards upon the plains of France, over the dreary Dombes, and descended the valley of the Rhone, as far at least as Valence, in the department of Drome.

Now it will readily be admitted that during the greatest extension of the ice on the north side of the Alps, gigantic glaciers must at the same time have filled all the mountain valleys of Northern Italy. In proof of this, we are referred to the great moraines of the Dora Baltea and those of the Dora Riparia, and the similar heaps of débris which occur at the lower ends of all the great Italian lakes—Orta, Maggiore, Lugano, Como, Lecco, Isea, and Garda. These moraines indicate, no doubt, the former presence of very large ice-streams, yet it is hardly conceivable that they can be the equivalents of the old grundmoränen of the Swiss low-grounds. When we picture to ourselves the condition of Switzerland and the adjoining tracts of France and Germany during the climax of glacial cold—when we think of the Rhone glacier after its egress from the low-grounds of Switzerland, flowing for 130 miles out upon the plains of France—when, further, we conceive of its northern branch uniting with the glaciers of the Rhine and its tributaries, and thereafter pouring over the end of the Jura to coalesce with the ice-fields of the Black Forest, it is impossible to believe that on the southern side of the Alps the glaciers could have been, comparatively speaking, so insignificant, that they never succeeded in getting well out of their mountain-valleys. The more southerly latitude of Italy will not enable us to explain this anomaly.

It is perfectly true that the present glaciers on the south side of the Alps are quite insignificant when compared to those occupying similar positions in Switzerland, and during the climax of the glacial epoch it is more than likely that the Swiss glaciers would much surpass those of Italy in importance. Still, those geologists who consider that Sahara existed at that time as a vast inland sea, will perhaps admit that the difference between the climates of the opposing slopes of the Alps would not then be so marked as it is now. We may, indeed, believe that the Italian glaciers would be arrested in their downward course sooner than those of Switzerland, yet the vast extent of the latter indicates a former intensity of cold which must needs have given rise to glaciers in Italy of even greater magnitude than those that occupied the lake-basins, and dropped their superficial moraines on the low-grounds beyond. In short, we are led to infer that when the Rhone glacier was depositing its moraines in the plains of France, the glaciers of the Dora Riparia and Dora Baltea must have advanced far beyond the mouths of their mountain-valleys, and may even have traversed the plains of Piedmont, and abutted upon the hills of Turin.

I am well aware that there are no deposits on the plains of Piedmont which can be referred to this great extension of the glaciers, and so far there is no direct and positive evidence in favour of such an extension. But the great valley of the Po, like that of the Danube, and that of the Rhine between the Vosges and the Black Forest, is everywhere covered by river-deposits of comparatively recent origin. It is quite possible, therefore, that a deposit of moraine-profonde or till may lie concealed at a greater distance from the Alps than the conspicuous moraines of the Baltea, &c. No inference can be drawn either one way or the other from the fact that no terminal moraines are known to occur further south than the mouths of the Alpine valleys in Italy. Terminal moraines do not exist on the plains of Germany or the low-grounds of France to mark the limits reached by the ice during the coldest periods of the glacial epoch; yet is it not unlikely that the Scandinavian ice-sheet reached into Northern Germany, and the Rhone glacier certainly flowed south as far as the low-grounds of Dauphiné.

The slopes of the Moncalieri-Valenza Hills are sprinkled with boulders and large erratics of alpine rocks, which were at one time supposed to have been carried across Piedmont by the ice of the Glacial Epoch. But subsequent and more detailed observations* have led Gastaldi to the opinion that the blocks in question are merely the denuded wreck of certain great beds of conglomerate belonging to the Miocene formation. No one who has visited the ground is likely to dispute this conclusion. One sees embedded in the miocenic conglomerate large erratics of precisely the same character as those that are lying loose on the hill slopes; and the conclusion seems irresistible that these latter are but the relics of those portions of the conglomerate which the denuding forces have carried away. At the same time, it must be remarked that if the glaciers, during the Glacial Epoch, ever did reach the

* "Sugli Elementi che compongono i Conglomerati Mioceni del Piemonte;" *Memoria della Reale Accademia delle Scienze di Torino*, serie ii. vol. xx.

Hills of Turin, the erratic blocks which they would then have left behind would now be indistinguishable from the denuded remains of the older erratic formation which Gastaldi has clearly shown to be of miocene age.

While, therefore, it must be admitted that there is no positive evidence * to show that the Italian glaciers ever crept further south than the limits reached by the terminal moraines which now circle round the mouths of the Alpine valleys, it is on the other hand equally true that no proof is forthcoming to show that they did not. We are not, however, without some indirect evidence in favour of the great extension which I have inferred.

Let me ask the reader to go back with me in imagination to the miocene period—that period which preceded in time the pliocene, and during which the great conglomerates of the Moncalieri Hills were deposited. At that time the Adriatic Sea extended up the great valley of the Po, and in all probability communicated with the Mediterranean across that low range of hills which now serves to connect the Maritime Alps with the Apennines. The Alpine valleys then formed long fiords, and the waves rose high on the northern slopes of the Apennines. Such conditions were maintained during many long ages, so as to allow vast heaps of sediment to gather upon the bed of the old sea. In some places these accumulations now form considerable hills. We find them fringing the northern flanks of the Apennines,† and extending in unbroken succession from Moncalieri to Valenza, forming those great deposits of conglomerate (and associated beds of gravel, sand, and marl), of which I have already spoken. An examination of these deposits shows that they have been derived from the degradation of the Alpine mountains. The sea in which they gathered washed the base of the Alps, and extended into the great valleys. But although this must have been the case, yet it is remarkable that no trace of miocene deposits can be anywhere detected along that ancient coast-line between Lake Maggiore and the River Ticino. We cannot doubt that at the time the thick beds of conglomerate, gravel, sand, &c. (which are so conspicuous in the Moncalieri-Valenza Hills, and which occur in those hills at a higher level than the base of the Alps) were being deposited, similar materials were also gathering on the sea-bottom along the shores of the northern mountain-land. Yet no trace of these now exists. Again, it is to be noted that while the pliocene sands that fringe the Moncalieri-Valenza Hills rest upon deposits of miocene and eocene age, yet the pliocene of the Dora Baltea and the Sessia recline upon the solid rocks of the mountains. It is clear, therefore, that before the pliocene beds were laid down, the pre-existing miocene and eocene deposits had been removed. In short, it is evident that after the close of the miocene period, and before the yellow marine sands of Piedmont were accumulated, there must have been enormous denudation along the base of the Alpine mountains. Whether this great erosion is to be referred in chief part to the action of gigantic glaciers I do not say, but it is difficult to find a simpler and more satisfactory explanation.

Reference has already been made to the fact that, with two remarkable

* It seemed to me, however, that the mountains behind Nomaglio (Val d'Aosta) were glaciated at least half way up from the surface of the great lateral moraine to their summits. If this be so, then the glacier that flowed at that level must have attained greater dimensions than the glacier which brought down the lateral moraine that extends from Andrate to Cavaglia. The glaciation at the higher level was much less distinctly marked than that at and below the summit level of the moraines, indicating that the former was probably effected at a much earlier age than the latter.

† They occur also upon the southern slopes of that range, but I confine attention to the valley of the Po.

exceptions, all the great mountain valleys of Northern Italy contain lakes at their lower ends. But no lake occurs at Rivoli, and no large sheet of water, but only two inconsiderable lakelets, appear opposite the mouth of the Val d'Aosta. Now, it does appear singular that just where we might have expected large rock-basins to appear, we should find nothing of the kind. If glaciers dug out the basins of Maggiore, Garda, and the other lakes in Italy, and those of Constance, Lucerne, Zurich, &c., in Switzerland, why should not the colossal glaciers of the Val d'Aosta and the Dora Riparia have excavated similar hollows near Ivrea and Rivoli? The answer is that they did do so, but the basins so scooped out were subsequently filled up again with aqueous deposits.

Since I ventured this suggestion to explain the absence of a rock-basin at Ivrea, Signor Gastaldi has reinvestigated the matter, and now gives it as his opinion that such a buried rock-basin does really exist, and that portions of it are still visible in the little lakes of Candia and Viverone. He thinks that the basin is of no great depth, as the rocks in which it has been excavated are of a more durable character, and must therefore have yielded less easily to erosive action than those which contain the great lakes of the other valleys of Northern Italy. As the great glacier slowly retired up the valley of the Dora, the river swept down vast heaps of gravel, sand, and silt (diluvium), with which it gradually filled up the ancient lake.

To draw these scattered remarks together, I shall now, in a few paragraphs, endeavour to correlate the Swiss and Italian deposits:—

1. At the period of most intense cold all the Alpine valleys were filled with glaciers, which in some cases reached a depth of 2,000 ft., and even more. These coalesced upon the low-grounds of Switzerland to form a great mer de glace, from which two principal ice-flows extended—the one into the valley of the Rhine, where it became in all probability confluent with the glaciers of the Black Forest; the other into the low-grounds of France, over which it advanced as far south as Valence, in the department of Drome, a distance of 130 miles, at least, from Geneva. At the same time, vast glaciers descended the southern valleys of the Alps into the plains of Northern Italy. There is no positive evidence forthcoming to show how far they invaded these low-grounds, but taking into consideration the colossal proportions attained by the glacier of the Rhone, it is highly probable that the Italian glaciers may have crossed the valley of the Po so as to abut upon the Hills of Turin, a distance from the base of the Alps of only 25 miles.

2. Owing to a change of climate, the ice gradually melted back until it had retired from the low-grounds, and shrunk into the deep mountain-valleys. During this retreat great perched blocks were stranded along the mountain-slopes, and masses of sand and gravel were strewn over all the low-grounds to which the water from the melting ice had access.

3. The climate becoming still more ameliorated, Switzerland assumed a vegetation similar to that which now characterises it; while at the same time the elephant, rhinoceros, urus, and other animals, became denizens of the country. On the south side of the Alps, similarly, a strong forest growth sprung up, and numerous mammalia inhabited the land. [The deposits containing these remains rest upon, and are therefore older than certain beds of sand of marine origin, which Italian geologists recognise as being of pliocene and pre-glacial age. The fossil contents of these sands, however, do not

demonstrate their pre-glacial age. So far as direct evidence goes, there is nothing to show that the sands may not be of inter-glacial age.]

4. The climate again becoming cold, the glaciers began another advance. Large rivers flowed down the valleys, and distributed vast heaps of gravel and sand, just as they had done during the retreat of the earlier glaciers. These deposits gathered over the site of the ancient forests in Switzerland, and overlaid the marine sands and ossiferous alluvia of Piedmont, spreading far and wide over the whole valley of the Po. Eventually, the glaciers themselves crept out upon the low-grounds of Switzerland, overriding the river gravels and ancient forests, and dropping their moraines often miles below the lower ends of the lakes, as at Spreitenbach, in the valley of the Limmat, and below Mellingen, in the valley of the Reuss. At the same time, the glaciers that occupied the Italian valleys deployed upon the low-grounds at the base of the Alps, and deposited their moraines above the great gravel beds that rest upon the marine sands.

5. Finally, the glaciers again retired, until at last they assumed their present proportions.

NOTE G.

List of Organic Remains found Fossil in the Glacial Deposits of Scotland.

(By Robert Etheridge, Jun., F.G.S.)

The succeeding list of organisms found fossil in the glacial deposits of Scotland has been prepared with a view of demonstrating how far such deposits have proved fossiliferous.

The list does not pretend to be other than it is: a compilation from the published results of those workers who have devoted much time and energy to the elucidation of the subject, a list of whose papers and publications is appended.

The present compilation does not, perhaps, contain *every* species recorded from the glacial deposits of Scotland; it is hardly possible indeed that it could, for notwithstanding the great strides made of late towards the reconciliation of the various names used by some of the earlier writers, yet so much doubt exists regarding the correctness of some of their determinations that it is almost impossible to avoid here and there, either on the one hand an omission, or on the other a recapitulation, under a different designation, of a previously expressed species.

It was at first contemplated to arrange the species indicated under the various divisions of Till, Boulder-clay, Brick-clays, &c., but for reasons similar to those just expressed in regard to species, this idea was abandoned, and the present general arrangement substituted.

The first column of the list is devoted to the name of the genus and species, the second to the principal localities at which each species has been found, together with a few of the better known synonyms.

The terminology of the Mollusca followed is that of Jeffreys' "British Conchology."

Succeeding the list of fossils will be found a few short notes on each of the principal localities, with the conditions under which the organic remains were found.

The chief papers and publications consulted for this compilation are as follows, foremost amongst them being those of Messrs. Crosskey and Robertson:—

1841. Rev. D. Landsborough. Description of Newer Pliocene Deposits at Stevenston. *Pro. Geol. Soc.*, vol. iii. p. 444.

1846. E. Forbes. The Fauna and Flora of the British Islands, &c. *Mems. Geol. Survey*, vol. i. pp. 406–432.

1850. J. Cleghorn and J. Smith. On the Till, near Wick, in Caithness. *Quart. Journ. Geol. Soc.*, vol. vi. p. 385.

1853. Forbes and Hanley. A History of British Mollusca. London, 1853, 4 vols. 8vo.

1857. J. A. Smith. Horns of *Cervus tarandus* in Dumbartonshire. *Pro. Roy. Phys. Soc. Ed.*, vol. i. p. 247.

1858. T. F. Jamieson. On the Pleistocene Deposits of Aberdeenshire. *Quart. Journ. Geol. Soc.*, vol. xiv. p. 509.

1862. J. Smith. Researches in Newer Pliocene and Post Tertiary Geology. Glasgow, 1862, 8vo.

1866. J. Haswell. Glacial Clay at Cornton, near Bridge of Allan. *Geol. Mag.*, vol. ii. p. 182.

Rev. H. C. Crosskey. On the *Tellina calcarea* bed at Chapelhall, near Airdrie. *Quart. Journ. Geol. Soc.*, vol. xxi. p. 219.

T. F. Jamieson. History of the last Geological Changes in Scotland. *Quart. Journ. Geol. Soc.*, vol. xxi. pp. 161–203.

J. Bryce. Order of Succession of the Drift Beds in Arran. *Quart. Journ. Geol. Soc.*, vol. xxi. pp. 204–213.

Rev. T. Brown. Glacial Beds at Elie, Fifeshire. *Trans. Roy. Soc. Ed.*, vol. xxiv. p. 617.

1867. Rev. H. C. Crosskey. Fossils collected at Windmillcroft. *Trans. Geol. Soc. Glasgow*, vol. ii. p. 115.

C. W. Peach. Fossils of the Boulder-clay of Caithness. *Pro. Roy. Phys. Soc. Ed.*, vol. iii. p. 38 and p. 396.

J. Bennie. *Bos longifrons* and *B. primigenius* in the ancient drift of the Clyde. *Trans. Geol. Soc. Glasgow*, vol. ii. p. 152.

1868. A. Geikie. Glacial Drift of Scotland. *Trans. Geol. Soc. Glasgow, App.*, vol. i.

J. C. Howden. Superficial Deposits of the South Esk. *Trans. Geol. Soc. Ed.*, vol. i. p. 141.

J. Geikie. Discovery of *Bos primigenius* in the Lower Boulder-clay of Scotland. *Geol. Mag.*, vol. v. p. 393.

S. P. Woodward. Manual of Mollusca, 2nd edition. London, 1868.

1869. J. A. Mahoney. Organic Remains found in Clay near Crofthead, Renfrewshire. *Geol. Mag.*, vol. vi. pp. 390–393.

W B. Dawkins. Distribution of the British Post-Glacial Mammalia. *Quart. Journ. Geol. Soc.*, vol. xxv. p. 192.

Prof. Turner. On the Bones of a Seal found in Red Clay near Grangemouth. *Pro. Roy. Soc. Ed.*, 1869–70, p. 105.

1871. J. Young and R. Craig. On the Occurrence of Seeds of Freshwater Plants and Arctic Shells, &c., in beds under the Boulder-clay at Kilmaurs. *Trans. Geol. Soc. Glasgow*, vol. iii. p. 310.

Rev. H. C. Crosskey. On Boulder-clay. *Trans. Geol. Soc. Glasgow* vol. iii. p. 149.

R. Craig. Section in Cowden Glen, &c. *Trans. Geol. Soc. Glasgow,* vol. iv. p. 17.

1867–1873. Crosskey and Robertson. A series of papers on the Post Tertiary Fossiliferous Beds of Scotland, contained in the *Transactions of the Geological Society of Glasgow,* vols. ii. to iv.

The compiler has also to acknowledge very material assistance kindly rendered by the following gentlemen, Messrs. David Robertson, F.G.S., H. B. Brady, F.G.S., James Bennie, C. W. Peach, A.L.S., and R. Etheridge, F.R.S.

PLANTÆ.

CRYPTOGAMIA.

Thallogens.

Diatomaceæ and Desmidaceæ.

In the Cowden Glen Inter-glacial Beds, Renfrewshire, eighteen genera, comprising thirty-one species, of *Diatomaceæ,* and three genera and species of *Desmidaceæ,* have been recorded by Mr. Mahoney.*

Algales.

Nullipora (Melobesia) polymorpha. *Linn.*	Caithness, in boulder clay; Dalmuir; West Tarbert; Paisley; Garvel Park New Dock, Greenock.
Corallina officinalis. *Linn.*	Paisley.
Jania rubens. *Lamx.*	Paisley.

Acrogens.

Musci.

Accompanying the Diatoms of the Cowden Glen deposit are six genera and eleven species of Mosses. Further information regarding these will be found in Mr. Mahoney's paper previously cited, and also in another by the same author "On the Botany of the Windmillcroft Beds." †

PHANEROGAMIA.

Exogens.

Amongst various remains of Phanerogamous plants recorded from Scotch Glacial Deposits the following appear to be the more worthy of notice :—

Corylus Avellana. *Linn.*	Roots and stems in the inter-glacial beds of Cowden Glen, Renfrewshire.
" " sp. ind.	In brick-clay at Portobello.
Cratægus Oxycanthus. *Linn.*	Do. do.
Betula alba. *Linn.*	Inter-glacial beds of Cowden Glen.
" sp. ind.	In brick-clay at Portobello.
Galium palustre. *Linn.*	Inter-glacial beds of Cowden Glen.
Hippuris, sp.	Kilmaurs, in a peaty deposit with the remains of *Elephas primigenius.*

* *Geol. Mag.,* vol. vi. p. 391.

† *Proc. Nat. Hist. Soc. Glasgow,* vol. i. p. 159.

Myriophyllum spicatum. *Linn.*	Inter-glacial beds of Cowden Glen.
Pedicularis palustris. *Linn.*	Do. do. do.
Pinus sylvestris. *Linn.*	Bark in the inter-glacial beds of Cowden Glen; brick-clay at Portobello.
Potamogeton lucens. *Linn.*	Inter-glacial beds of Cowden Glen.
,, sp. ind.	Seeds at Kilmaurs, in a peaty deposit with remains of *Elephas primigenius.*
Quercus, sp. ind.	In brick-clay at Portobello.
Ranunculus aquatilis. *Linn.*	Inter-glacial beds of Cowden Glen.
Salix alba. *Linn.*	Leaves in the beds of Cowden Glen.
? Scutellaria galericulata. *Linn.*	Inter-glacial beds of Cowden Glen.
Taxus baccata. *Linn.*	In brick-clay at Portobello.
Vaccinum Myrtillus. *Linn.*	Twigs in the Cowden Glen inter-glacial beds.

Endogens.

Hordeum distichum.	A cereal, apparently closely allied to this species, was found by Dr. Howden at Montrose, in a bed of peat resting on glacial marine clay beneath estuary beds.*
Scirpus lacustris. *Linn.*	Inter-glacial beds of Cowden Glen.

ANIMALIA.

INVERTEBRATA.

Sub-kingdom: PROTOZOA.

Class: Rhizopoda.

Foraminifera.

Genus Biloculina. *D'Orbigny.*	
B. elongata. *D'Orb.*	Dalmuir; Tangy Glen, near Campbeltown, in laminated clay overlaid by boulder clay; Kilmaurs.
B. depressa. *D'Orb.*	Lochgilp; Duntroon; Kilchattan Tile-work, Bute.
B. ringens. *Linn.*	Caithness, in boulder clay; Lochgilp; Duntroon; Paisley; Garvel Park New Dock, Greenock; Kilchattan Tile-work, Bute; Tangy Glen, near Campbeltown.
Genus Bolivina. *D'Orb.*	
B. punctata. *D'Orb.*	Duntroon; West Tarbert.
Genus Bulimina. *D'Orb.*	
B. marginata. *D Orb.*	Caithness, in boulder clay; Lochgilp; Cumbrae College; Duntroon; East and West Tarbert; Garvel Park New Dock, Greenock; Crinan.
B. pupoides. *D'Orb.*	Duntroon; Garvel Park New Dock, Greenock; Caithness, in boulder clay; Kilchattan Tile-work, Bute.
Genus Cassidulina. *D'Orb.*	
C. crassa. *D'Orb.*	Lochgilp; Cumbrae College.
C. lævigata. *D'Orb.*	Caithness, in boulder clay; Lochgilp; Kilchattan Tile-work, Bute; Tangy Glen.

* *Trans. Geol. Soc. Edinburgh,* vol. i. p. 144.

Genus CORNUSPIRA. *Schultze.*	
C. foliacea. *Philippi.*	Lochgilp; East Tarbert; Duntroon; Garvel Park New Dock, Greenock; Tangy Glen, near Campbeltown; Annochie, in brick clay.
Genus CRISTELLARIA. *Lamarck.*	
C. rotulata. *Lamk.*	Caithness, in boulder clay; Crinan; Garvel Park New Dock, Greenock; Kilchattan Tile-work, Bute.
Genus DENTALINA. *D'Orb.*	
D. communis. *D'Orb.*	Caithness, in boulder clay; Cumbrae College; Crinan; Duntroon; Old Mains, Renfrew; Garvel Park New Dock, Greenock; Kilchattan Tile-work, Bute.
D. guttifera. *D'Orb.*	Duntroon.
Genus DISCORBINA. *Parker & Jones.*	
D. globularis. *D'Orb.*	Duntroon; Tangy Glen, near Campbeltown, in laminated clay overlaid by boulder clay.
D. rosacea. *D'Orb.*	Caithness, in boulder clay; Lochgilp; Paisley.
Genus GLOBIGERINA. *D'Orb.*	
G. bulloides. *D'Orb.*	Caithness, in boulder clay; Lochgilp; Duntroon; Tangy Glen, near Campbeltown, in laminated clay overlaid by boulder clay.
Genus LAGENA. *Walker & Jacob.*	
L. apiculata. *Reuss.*	Garvel Park New Dock, Greenock.
L. biconica. *Brady, M.S.*	Lochgilp; Duntroon.
L. caudata. *D'Orb.*	West Tarbert.
L. distoma. *P. & J.*	West Tarbert; Crinan; Duntroon; Paisley; Garvel Park New Dock, Greenock; Kilchattan Tile-work, Bute.
L. globosa. *Montg.*	Caithness, in boulder clay; Lochgilp; Cumbrae College; East Tarbert; Duntroon; Old Mains, Renfrew; Paisley; Garvel Park New Dock, Greenock; Kilchattan Tile-work, Bute; Tangy Glen, near Campbeltown.
L. gracillima. *Seg.* var.	Garvel Park New Dock, Greenock; Tangy Glen, near Campbeltown.
L. Jeffreysii. *Brady.*	East Tarbert; Tangy Glen, near Campbeltown: Caithness, in boulder clay.
L. lævis. *Montg.*	Cumbrae College; West Tarbert; Crinan; Duntroon; Old Mains, Renfrew; Garvel Park New Dock, Greenock; Kilchattan Tile-work, Bute; Tangy Glen, near Campbeltown; Paisley.
L. Lyelli. *Sequenza.*	Duntroon.
L. marginata. *Montg.*	Caithness, in boulder clay; Cumbrae College; Duntroon; Old Mains, Renfrew; Paisley; Tangy Glen, near Campbeltown; Kilchattan Tile-work, Bute.
L. semistriata. *Will.*	Caithness, in boulder clay; Lochgilp; West Tarbert; Old Mains, Renfrew.
L. squamosa. *Montg.*	Caithness, in boulder clay; Lochgilp; Duntroon; Tangy Glen, near Campbeltown.
L. striata. *Montg.*	Cumbrae College; Lochgilp; West Tarbert; Duntroon; Paisley; Garvel Park New Dock, Greenock; Kilchattan Tile-work, Bute.
L. sulcata. *W. & J.*	Caithness, in boulder clay; Cumbrae College; Lochgilp; E. Tarbert; Duntroon; Garvel Park New Dock, Greenock; Tangy Glen, near Campbeltown. Synonym: *L. costata*, Williamson.

Genus LITUOLA. *Lamarck.*	
L. ornata. *Brady.*	Lochgilp.
L. scorpiurus. *Montf.*	Paisley; Kilchattan Tile-work, Bute.
Genus NODOSARIA. *Lamarck.*	
N. humilis. *Roemer.*	Monreith Tile-works, Wigtownshire (*Geo. Surv.*)
N. raphanus. *Linn.*	Caithness, in boulder clay.
N. scalaris. *Batsch.*	Duntroon.
Genus NONIONINA. *D'Orb.*	
N. asterizans. *F. & M.*	Caithness, in boulder clay; Cumbrae College; Tangy Glen, near Campbeltown; Towncroft Farm, near Grangemouth, in red clay with seal bones.
N. depressula. *W. & J.*	Caithness, in boulder clay; Dalmuir; Crinan; Paisley; Garvel Park New Dock; Kilchattan Tile-work, Bute; Tangy Glen, near Campbeltown; Monreith Tile-work, Wigtownshire (*Geol. Survey*).
N. turgida. *Will.*	Lochgilp; Kilchattan Tile-work, Bute; Duntroon.
Genus PATELLINA. *Williamson.*	
P. corrugata. *Will.*	Paisley; Garvel Park New Dock, Greenock.
Genus PLANORBULINA. *D'Orbigny.*	
P. Mediterranensis. *D'Orb.*	Caithness, in boulder clay.
Genus POLYMORPHINA. *D'Orbigny.*	
P. communis. *W. & J.*	Cumbrae College.
P. gibba. *D'Orb.*	East Tarbert.
P. lactea. *W. & J.*	Caithness, in boulder clay; Dalmuir; Lochgilp; Kilchattan Tile Work, Bute; Tangy Glen.
P. oblonga. *Brown.*	East Tarbert.
P. compressa. *D'Orb.*	Caithness, in boulder clay; Dalmuir; Cumbrae College; East and West Tarbert; Duntroon; Old Mains, Renfrew; Paisley; Garvel Park New Dock, Greenock; Kilchattan Tile-work, Bute; Tangy Glen, near Campbeltown; Towncroft Farm, near Grangemouth, in red clay with seal bones.
Genus POLYSTOMELLA. *Lamarck.*	
P. arctica. *P. & J.*	Caithness, in boulder clay; East and West Tarbert; Kilmaurs.
P. crispa. *Linn.*	Caithness, in boulder clay; Lochgilp; West Tarbert; Duntroon; Tangy Glen, near Campbeltown, in laminated clay below boulder clay.
P. striato-punctata. *F. & M.*	Caithness, in boulder clay; Dalmuir; Lochgilp; Bucklivie; Annochie, in brick clay; East and West Tarbert; Crinan; Duntroon; Old Mains, Renfrew; Paisley, in laminated clay under a true glacial clay; Kilchattan Tile-works, Bute; Tangy Glen, near Campbeltown; Windmillcroft; Terally Brickwork, near Drummore, and Monreith Tile-work, Wigtownshire (*Geol. Survey*).
Genus PULVINULINA. *Parker & Jones.*	
? P. Caracalla. *Roemer ?*	Caithness, in boulder clay.

Genus QUINQUELOCULINA. *D'Orbigny.*	
Q. agglutinans. *D'Orb.*	Paisley; Garvel Park New Dock, Greenock.
Q. bicornis. *W. & J.*	Duntroon.
Q. seminulum. *Linn.*	Caithness, in boulder clay; Dalmuir; Cumbrae College; Lochgilp; East and West Tarbert; Duntroon; Paisley; Garvel Park New Dock; Kilchattan Tile-work, Bute; Tangy Glen, near Campbeltown, in laminated clay below boulder clay; Windmillcroft.
Q. subrotunda. *Montg.*	Caithness, in boulder clay; Dalmuir; Cumbrae College; Lochgilp; East Tarbert; Duntroon; Paisley; Garvel Park New Dock, Greenock; Kilchattan Tile-work, Bute; Tangy Glen, near Campbeltown.
Genus ROTALIA. *Lamarck.*	
R. Beccarii. *Linn.*	Caithness, in boulder clay; Paisley; Kilchattan Tile-work, Bute; Tangy Glen, near Campbeltown.
R. orbicularis. *D'Orb.*	Lochgilp.
R. Soldarii. *D'Orb.*	Caithness, in boulder clay.
Genus SPIROLOCULINA. *D'Orbigny.*	
S. limbata. *D'Orb.*	Duntroon.
S. planulata. *Lamk.*	Dalmuir.
Genus TEXTULARIA. *Defrance.*	
T. difformis. *D'Orb.*	Caithness, in boulder clay.
T. sagittula. *Defrance.*	Lochgilp.
T. variabilis. *Will.*	West Tarbert; Duntroon.
Genus TRILOCULINA. *D'Orb.*	
T. oblonga. *Montg.*	Caithness, in boulder clay; Lochgilp; Garvel Park New Dock, Greenock; near Paisley.
T. tricarinata. *D'Orb.*	Garvel Park New Dock, Greenock.
Genus TROCHAMMINA. *Parker & Jones.*	
T. incerta. *D'Orb.*	Caithness, in boulder clay.
T. inflata. *Montg.*	Caithness, in boulder clay; Duntroon.
Genus TRUNCATULINA. *D'Orbigny.*	
T. lobatula. *W. & J.*	Caithness, in boulder clay; Dalmuir; Cumbrae College; Lochgilp; West Tarbert; Crinan; Duntroon; Old Mains, Renfrew; Paisley; Garvel Park New Dock, Greenock; Kilchattan Tile-work, Bute; Tangy Glen, near Campbeltown.
Genus VAGINULINA. *D'Orb.*	
V. legumen. *Linn.*	Caithness, in boulder clay; Tangy Glen, near Campbeltown.
V. linearis. *Montg.*	Caithness, in boulder clay.
Genus VERNEUILINA. *D'Orb.*	
V. polystropha. *Reuss.*	Garvel Park New Dock, Greenock; Kilchattan Tile-work, Bute.

Spongida.

Genus CLIONA. *Grant.*	
C. cælata. *Grant.*	Caithness, in boulder clay; Dalmuir; Old Mains, Renfrew; Paisley; Garvel Park New Dock; Kilchattan Tile-work, Bute; Lochgilp.

Genus GEODIA. *Lamarck.*	
G., sp. ind.	Caithness (*C. W. Peach*).
Genus SPONGILLA. *Lamarck.*	
S. fluviatilis. *Pallas.*	Inter-glacial beds at Cowden Glen, Renfrewshire.

Sub-kingdom: CŒLENTERATA.

Class: Actinozoa.

Order: Zoantharia (Z. sclerodermata).

Genus SPHENOTROCHUS. *Edwards & Haime.*	
S. Wrightii. *Gosse.*	Old Mains, Renfrew.

Sub-kingdom: ANNULOIDA.

Class: Echinodermata.

Ophiuroidea.

Genus OPHIOCOMA. *Agassiz.*	
O. bellis. *Link.*	Dalmuir; Paisley; Garvel Park New Dock, Greenock; the remains of this species consist for the most part of spines and plates.
O. rosula. *Link.*	Caithness, spines in boulder clay.
Genus OPHIURA. *Lamarck.*	
O. albida. *Forbes.*	West Tarbert; Duntroon; Garvel Park New Dock, Greenock; Kilchattan Tile-work, Bute.
O. texturata. *Lamk.*	Garvel Park New Dock, Greenock.
Genus OPHIOLEPIS. *Müller & Troschel.*	
O. gracilis. *Allman.*	Coast, two miles west of Dunbar, in brick-clay (*Allman*); Montrose (*Howden*).

Echinoidea.

Genus ECHINUS. *Linnæus.*	
E. Dröbachiensis. *Müller.*	Dalmuir; Cumbrae College; Lochgilp; East and West Tarbert; Crinan; Duntroon; Old Mains, Renfrew; Paisley; Garvel Park New Dock, Greenock; Kilchattan Tile-work, Bute.
? E. lividus. *Lamk.*	Langbank (*J. Young*).
E. neglectus. *Lamk.*	Caithness, spines in boulder clay.
E. sphæra. *Müller.*	Lochgilp; Garvel Park New Dock, Greenock; Kilchattan Tile-work, Bute.
Genus AMPHIDOTUS. *Agassiz.*	
A., sp. ind.	Kilchattan Tile-work, Bute (*Crosskey & Robertson*).

Holothuroidea.

Genus PSOLUS. *Oken.*	
P. phantopus. *Linn.*	Houston, near Glasgow, in brick clay;* Bute (*Prof. Geikie*).
Genus SIPUNCULUS. *Linnæus.*	
S. Bernhardus? *Forbes.*	Caithness; Dalmuir, remains in *Trophon clathratus;* Garvel Park New Dock, Greenock.

* Owen, *Palæontology*, 1861, p. 41.

Sub-kingdom: ANNULOSA.

Class: Annelida.

Hirudinea.

Genus HÆMOPIS.	
H. sanguisorba.	Jaws and teeth of the "Horse Leech" have been detected in the inter-glacial beds at Cowden Glen, Renfrewshire (*Mahoney*).

Tubicola.

Genus FILOGRANA. *Berkeley.*	
F. implexa. *Berkeley.*	Garvel Park New Dock, Greenock.
Genus PECTINARIA. *Lamarck.*	
? P., sp. ind.	Caithness, in boulder clay (*C. W. Peach*).
Genus SERPULA. *Linnæus.*	
S. triquetra. *Martin.*	Dalmuir; Stevenston (*Landsborough*, fide *Smith*); Lochgilp.
S. vermicularis. *Ellis.*	Caithness, in boulder clay; Dalmuir; Lochgilp; West Tarbert; Garvel Park New Dock; Kilchattan Tile Work, Bute.
Genus SPIRORBIS. *Lamarck.*	
S. carinatus. *Flem.*	Paisley.
S. corrugatus. *Montg.*	Bute (*Smith*).
S. granulatus. *Montg.*	Caithness, in boulder clay.
S. nautiloides. *Lam.*	Dalmuir; Stevenston (*Landsborough*, fide *Smith*). Synonym: *Serpula spirorbis*, Lin.
S. spirillum. *Linn.*	East Tarbert; Duntroon; Paisley; Garvel Park New Dock, Greenock; Arran.

Class: Crustacea.

Cirripedia.

Genus BALANUS. *Lister.*	
B. balanoides. *Linn.*	Paisley; Windmillcroft; Dalmuir (*Smith.*)
B. concavus. *Bronn.*	Aberdeenshire (*Jamieson*, fide *Smith*).
B. crenatus. *Brug.*	Caithness, in boulder clay; Dalmuir; Paisley; Cumbrae College; East and West Tarbert; Crinan; Old Mains, Renfrew; Garvel Park New Dock, Greenock; Kilchattan Tile-work, Bute; Duntroon.
B. Hameri. *Ascanius.*	Lochgilp; Old Mains, Renfrew; Garvel Park New Dock, Greenock; this is the prevailing Cirripede at Uddevalla. ? Synonym: *B. uddevalensis*, Lin.
B. porcatus. *Da Costa.*	Caithness, in boulder clay; Dalmuir; Cumbrae College; Lochgilp; West Tarbert; Crinan; Old Mains, Renfrew; Duntroon; Paisley, attached to large boulders; Garvel Park New Dock, Greenock; Kilchattan Tile-work, Bute; Elie and Errol, Fife. Synonyms: *B. Scoticus* (Brown); ? *B. Costatus* (Smith, *Researches*, p. 48).
Genus VERRUCA. *Schumacher.*	
V. Strömia. *Müller.*	Caithness, in boulder clay; Dalmuir; Cumbrae College; Lochgilp; West Tarbert; Crinan; Duntroon; Old Mains, Renfrew; Paisley; Garvel Park New Dock; Kilchattan Tile-work, Bute; Tangy Glen, near Campbeltown. Synonym: *Creusia verruca* (Lam.)

Ostracoda.

As the subject of Post-tertiary *Ostracoda* is at present undergoing considerable revision at the hands of Messrs. Crosskey, Robertson, and G. S. Brady, it has been thought better to omit any list of these interesting organisms, as a compilation, however carefully executed, must, under existing circumstances, be very imperfect.

Decapoda.

Genus CARCINUS. *Leach.*	
? C. menas. *Linn.*	Bridge of Johnstone, near Paisley (fragments of the carapace, &c., *Prof. Geikie*); Oban (*Prof. Geikie*).
Genus PAGURUS. *Fabricius.*	
P. Bernhardus. *Linn.*	Dalmuir (fragments); Cumbrae College (a claw).

Sub-kingdom: MOLLUSCA.

SECTION: MOLLUSCOIDA.

Class: Polyzoa.

Genus CABEREA. *D'Orbigny.*	
C. Ellisii. *Flem.*	Garvel Park New Dock, Greenock.
Genus CANDA. *Lamouroux.*	
C. reptans. *Linn.*	Duntroon; Paisley.
Genus CELLEPORA. *Fabricius.*	
C. pumicosa. *Linn.*	Caithness, in boulder clay.
Genus CELLULARIA. *Pallas.*	
C. scruposa. *Linn.*	West Tarbert.
Genus CRISIA. *Lamouroux.*	
C. denticulata. *Lam.*	Caithness, in boulder clay.
C. eburnea. *Linn.*	Dalmuir; Lochgilp; Crinan; Duntroon; Paisley; Garvel Park New Dock; Kilchattan Tile-work, Bute; Tangy Glen, near Campbeltown.
Genus DISCOPORELLA. *D'Orb.*	
D. Grignoniensis. *Busk.*	Dalmuir; Duntroon. A crag form.
D. hispida. *Fleming.*	Paisley.
Genus FLUSTRA. *Linnæus.*	
F. avicularia. *Mont.*	Duntroon.
Genus HIPPOTHOA. *Lamouroux.*	
H. catenularia. *Jamieson.*	Caithness, in boulder clay; Dalmuir; Duntroon.
H. divaricata. *Lamx.*	Caithness, in boulder clay.
Genus IDMONEA. *Lamouroux.*	
I. fenestrata. *Busk.*	Garvel Park New Dock, Greenock.
Genus LEPRALIA. *Johnston.*	
L. annulata. *Fabr.*	Do. do. do.
L. concinna. *Busk.*	Garvel Park New Dock, Greenock; Lochgilp.
L. cruenta. *Norman.*	Garvel Park New Dock, Greenock.
L. crystallina. *Norman.*	Do. do. do.
L. hyalina. *Linn.*	Do. do. do.
L. Peachii. *Johnston.*	Caithness, in boulder clay; Cumbrae College; Lochgilp; Dalmuir; Garvel Park New Dock.
L. Peachii var labiosa. *Johnston.*	Caithness, in boulder clay.
L. pertusa. *Esper.*	Dalmuir.
L. simplex. *Johnston.*	Caithness, in boulder clay.

L. tubulosa. *Norman.*	Garvel Park New Dock, Greenock.
L. unicornis. *Fleming.*	Caithness, in boulder clay; Dalmuir, with the central mucro preserved (*Crosskey & Robertson*).
L. verrucosa. *Esper.*	Dalmuir; Duntroon; Garvel Park New Dock, Greenock.
Genus MEMBRANIPORA. *Blainville.*	
M. craticula. *Alder.*	Paisley.
M. Flemingii. *Busk.*	Lochgilp; Garvel Park New Dock, Greenock.
M. unicornis. *Fleming.*	Paisley; Dalmuir; Duntroon.
M., sp. ind.	Caithness, in boulder clay (*C. W. Peach*).
Genus SALICORNARIA. *Cuvier.*	
S., sp. ind.	Caithness, in boulder clay (*C. W. Peach*).
Genus TUBULIPORA. *Lamarck.*	
T. flabellaris. *Fabr.*	Dalmuir.
T. hispida. *Fleming.*	Dalmuir; Caithness, in boulder clay.
T. patina. *Linn.*	Garvel Park New Dock, Greenock.
T. phalangea. *Couch.*	Dalmuir; Duntroon.
T. serpens. *Linn.*	Dalmuir.
?T. verrucaria. *M. Edw.*	Largs (*Landsborough*, fide *Forbes*). This is probably a synonym of *T. phalangea*, Couch.

Class: Brachiopoda.

Genus RHYNCHONELLA. *Fischer.*	
R. (Terebratula) psittacea. *Chem.*	Ayrshire (*Forbes*);* Caithness (*C. W. Peach*).
Genus TEREBRATULA. *Lhwyd.*	
T. caput-serpentis. *Linn.*	Ayrshire (*Smith*).†

SECTION: MOLLUSCA PROPER.

Class: Lamellibranchiata.

Monomyaria.

Genus ANOMIA. *Linnæus.*	
A. ephippium. *Linn.*	Caithness, in boulder clay; Dalmuir; Lucknow Pit, Ayrshire; Paisley; Lochgilp; Garvel Park New Dock, Greenock; West Tarbert; Duntroon; Old Mains, Renfrew; Kilchattan Tile-work, Bute; Gamrie; Cumbrae College.
A. ephippium var aculeata. *Linn.*	Dalmuir; Lochgilp; Paisley; Stevenston (*Smith*); Cumbrae College; Garvel Park New Dock, Greenock; Kilchattan (*Jamieson*).
A. ephippium var squamula. *Linn.*	Paisley; Caithness, in boulder clay; Lochgilp; Garvel Park New Dock, Greenock.
A. patelliformis. *Linn.*	Stevenston (*Smith*); a tertiary fossil of the Clyde Beds (*Jeffreys*). Synonym: *A. undulata*, Gmelin.
Genus OSTREA. *Linnæus.*	
O. edulis. *Linn.*	Lucknow Pit, Ayrshire; Kyles of Bute (*Prof. Geikie*); Gourock (*Prof. Geikie*); Cornton (*Haswell*); Caithness (*Peach*). Messrs. Crosskey and Robertson have failed to discover this species in any of the older glacial beds of the Clyde.

* *Mems. Geol. Survey*, vol. i. p. 406. † *Researches*, p. 55.

Genus Pecten. *Pliny.*	
P. Grœnlandicus. *Sow.*	Elie and Errol, a few feet below the surface; Tangy Glen, near Campbeltown; Montrose, at a depth of from 30 to 40 feet below the surface. Synonym: *P. vitreus.* Gray, [non Chem.]
P. Islandicus. *Müller.*	Dalmuir; Cumbrae College; Paisley; West Tarbert; Duntroon; Old Mains, Renfrew; Kilchattan Tile-work, Bute; Garvel Park New Dock, Greenock; Caithness, in boulder clay; Fort William; Kilmaurs (*J. Young*); Belhelvie; Ellishill; Arran (*Bryce*); Langbank; Lochgilp.
P. maximus. *Linn.*	Kyles of Bute (*Prof. Geikie*); Fairlie (*Prof. Geikie*); Caithness, in boulder clay; Garvel Park New Dock, Greenock.
P. opercularis. *Linn.*	Caithness, in boulder clay; Lucknow Pit, Ayrshire; Kyles of Bute (*Prof. Geikie*); Arran (*Bryce*); Cruden (*Jamieson*).
P. pusio. *Linn.*	Dalmuir (*Smith*). Synonym: *P. sinuosus,* Turton.
P. septemradiatus. *Müller.*	Loch Lomond Beds (*Smith*); Clyde deposits (*Jeffreys*). Synonyms: *P. Danicus,* Chem.; *P. triradiatus,* Müller.
P. similis. *Laskey.*	Fifeshire (*Fleming,* fide *Smith*). A Coralline Crag fossil (*S. Wood*).
P. tigrinus. *Müller.*	Loch Lomond Beds (*Smith*). Synonym: *P. obsoletus,* Pennant.
P. varius. *Linn.*	Dalmuir (*Smith*).

Dimyaria.

Genus Artemis. *Poli.*	
A. lævigata. *Forbes.*	Stevenston (*Landsborough*).
Genus Astarte. *T. Sowerby.*	
A. borealis. *Chem.*	Caithness, in boulder clay; Dalmuir; Lucknow Pit, Ayrshire; Crinan; Stevenston (*Landsborough*); Gamrie; Bute (*Prof. Geikie*); Lochgilp; Gourock (*Prof. Geikie*); Errol; Holy Loch (*Prof. Geikie*); a true arctic form. Synonyms: *A. arctica; A. Islandica.*
A. borealis var semisulcata. *Leach.*	Bute; Wick (*Smith*). Synonym: *Crassina Withami,* Smith.
A. compressa. *Montg.*	Caithness, in boulder clay; Dalmuir; Lochgilp; Gourock; West Tarbert; Duntroon; Old Mains, Renfrew; Garvel Park New Dock, Greenock; Kilchattan Tile-works, Bute; Croftamie, in blue clay; Gamrie; Elie; Paisley; Stevenston (*Landsborough*). Synonyms: *A. propinqua,* Landsborough; *A. multicostata,* Smith; *A. Uddevallensis,* Smith.
A. compressa var globosa. *Müller.*	Paisley; Garvel Park New Dock, Greenock.
A. compressa var striata. *Müller.*	Paisley.
A. crebricostata. *Forbes.*	Dalmuir and Bute (*Smith*). An arctic form.
A. sulcata. *Da Costa.*	Lochgilp; West Tarbert; Old Mains, Renfrew; Garvel Park New Dock, Greenock; Caithness, in boulder clay; Paisley.

A. sulcata var elliptica. *Brown.*	Dalmuir; Lochgilp; Loch Long (*Prof. Geikie*); Paisley; Bute; Croftamie; Belhelvie. Synonyms: *A. gairensis*, Smith; *Crassina elliptica*, Brown; *Crassina ovata*, Brown; *A. semisulcata*, Jeffreys [non Leach].
A. sulcata var Danmoniensis.	Stevenston (*Smith*); Banff (*Smith*).
A. sulcata var Scotica. *M. & R.*	Gamrie (*Prestwich*); Caithness; Clyde Beds (*Smith*).
Genus Axinus. *J. Sowerby.*	
A. ferruginosus. *Forbes.*	Annochie (*Jamieson*). Synonym: *Lucina ferruginosa*, Forbes.
A. flexuosus. *Montg.*	Dalmuir: Lochgilp; East and West Tarbert; Crinan; Duntroon; Cumbrae College.
A. flexuosus var Gouldii. *Phil.*	Lochgilp; Paisley; Kilchattan Tile-work, Bute; Garvel Park New Dock, Greenock. Synonym: *Lucina Gouldii*, Philippi.
A. flexuosus var Sarsii. *Phil.*	Annochie (*Jamieson*); Arran (*Bryce*). Synonym: *Axinus Sarsii*, Philippi.
Genus Cardium. *Linnæus.*	
C. aculeatum. *Linn.*	Stevenston (*Landsborough*).
C. echinatum. *Linn.*	Caithness, in boulder clay; Paisley; Kilchattan Tile-work, Bute; Lochgilp; Gamrie; Cornton, near Bridge of Allan; Belhelvie; Lucknow Pit, Ayrshire.
C. edule. *Linn.*	Caithness, in boulder clay; Dalmuir; Old Mains, Renfrew; Garvel Park New Dock, Greenock; Kilchattan Tile-works, Bute; Cornton, near Bridge of Allan; Lochgilp; Lucknow Pit, Ayrshire.
C. exiguum. *Gmelin.*	Caithness, in boulder clay; Paisley; Lochgilp; Dalmuir (*Smith*); Bute (*Smith*); Garvel Park New Dock, Greenock. Synonym: *C. pygmæum*, F. & H.
C. fasciatum. *Montg.*	Caithness, in boulder clay; Cumbrae College; West Tarbert; Paisley; Garvel Park New Dock, Greenock; Arran; Lochgilp.
C. Grœnlandicum. *Chem.*	Gamrie (*Jamieson*); King-Edward (*Jamieson*).
C. minimum. *Phil.*	Bute (*Smith*). Synonym: *C. Suecicum*, F. & H.
C. Norvegicum. *Spengler.*	Caithness, in boulder clay; Cruden; Cambrae College; Stevenston (*Landsborough*); Lucknow Pit, Ayrshire. Synonym: *C. lævigatum*, Penn et auct. [non Linn.].
Genus Corbula. *Bruguiere.*	
C. gibba. *Olivi.*	East Tarbert; Tangy Glen, near Campbeltown. Synonym: *C. nucleus*, F. & H.
Genus Crenella. *Brown.*	
C. decussata. *Montg.*	Caithness, in boulder clay; Elie.
C. faba. *Müller.*	Errol (*Jamieson*).
Genus Cyamium. *Philippi.*	
C. minutum. *Fabr.*	West Tarbert.
Genus Cyprina. *Lamarck.*	
C. Islandica. *Linn.*	Caithness, in boulder clay; Dalmuir; Cumbrae College; Lochgilp; Lucknow Pit, Ayrshire; Paisley; West Tarbert; Duntroon; Old Mains, Renfrew; Garvel Park New Dock, Greenock; Kilchattan Tile-work, Bute; Gamrie; Elie; Chapelhall; Ellishill; Langbank; Croftamie, Dumbartonshire, in blue clay. Synonym: *Venus Islandicus*, Linn.

Genus DONAX. *Linnæus.*	
D. vittatus. *Da Costa.*	Caithness, n boulder clay; Ayr (*Smith*). Synonym: *D. anatinus*, F. & H.
D. trunculus. *Linn.*	This shell is recorded by Smith * from Stevenston *fide* Landsborough.
Genus GLYCIMERIS. *Lamarck.*	
G. siliqua. *Lamk.*	Caithness, in boulder clay.
Genus KELLIA. *Turton.*	
K. suborbicularis. *Montg.*	St. Fergus (*Jamieson*); a very doubtful determination.
Genus LEDA. *Schumacher.*	
[L. antiqua. *Smith.*]	This is mentioned by Smith in his list, but without any locality; a very doubtful form.
L. arctica. *Gray.*	Elie and Errol; Lucknow Pit, Ayrshire. Montrose; Tyrie. Synonyms: *Leda* (*Nucula*) *truncata*, Brown; *Nucula Portlandica*, Hitchcock. Essentially an arctic species; occurs in the "Leda clay" of Montreal.
L. limatula. *Say.*	King-Edward (*Jamieson*).
L. lucida. *Lovén.*	Do. do.
L. minuta. *Müller.*	Caithness, in boulder clay; Elie and Errol; Bute. Synonyms: *Nucula minuta*, Smith; *Leda caudata*, F. & H.
L. pernula. *Müller.*	Dalmuir; Lochgilp; Duntroon; West Tarbert; Old Mains, Renfrew; Paisley; Crinan; Garvel Park New Dock, Greenock; Tangy Glen, near Campbeltown; Arran; Windmillcroft; Kilmaurs. Synonyms: *Nucula oblonga*, Brown. [Forbes placed *Leda rostrata*, Lamk., recorded from Dalmuir by Smith, as a synonym of *N. oblonga*, Brown.]
L. pernula var baccata. *Steenst.*	Caithness, in boulder clay.
L. pernula var mucilenta. *Steenst.*	Paisley; Garvel Park New Dock; Kilchattan Tile-work, Bute.
L. pygmæa. *Münster.*	Dalmuir; Lochgilp; West Tarbert; Crinan; Duntroon; Old Mains, Renfrew; Paisley; Montrose; Annochie, in brick clay; Caithness, in boulder clay; Tangy Glen, near Campbeltown; Elie and Errol; Windmillcroft; Cumbrae College.
L. pygmæa var gibbosa. *Smith.*	Cumbrae College.
L. pygmæa var lenticula. *Müller.*	Paisley; Garvel Park New Dock; Kilchattan Tile-work, Bute.
Genus LEPTON. *Turton.*	
L. nitidum. *Turton.*	Lochgilp. Occurs in the post-glacial beds of Christiania (*Sars*).
Genus LUCINA. *Bruguière.*	
L. borealis. *Linn.*	Kyles of Bute, and Gourock (*Prof. Geikie*); Caithness, in boulder clay. Synonym: *Venus borealis*, Linn.
L. spinifera. *Montg.*	Caithness, in boulder clay.
Genus LUCINOPSIS. *Forbes & Hanley.*	
L. undata. *Pennant.*	Ayr (*Smith*); Kyles of Bute (*Crosskey*).

* *Researches*, p. 49.

Genus Lutraria. *Lamarck.*	
L. elliptica. *Lamk.*	Kyles of Bute; Lucknow Pit, Ayrshire. Fossil in all our upper tertiaries from the Scotch glacial beds to the coralline crag (*Jeffreys*).
Genus Mactra. *Linnæus.*	
M. solida. *Linn.*	Kyles of Bute (*Prof. Geikie*); Stevenston (*Landsborough*); Forth beds (*Smith*).
M. solida var elliptica. *Brown.*	Stevenston (*Landsborough*); Duntroon; Gamrie.
M. solida var truncata. *Montg.*	Forth beds (*Smith*). This appears to be a very doubtful determination.
M. subtruncata. *Da Costa.*	Cumbrae College; Kilchattan Tile-work, Bute; Forth beds (*Smith*). Synonym: *M. cuneata*, Sow.
M. subtruncata var striata. *Brown.*	Dalmuir.
M. stultorum. *Linn.*	Jeffreys remarks that this shell occurs, "but not commonly, in all our upper tertiary strata, from the Scotch glacial beds (*Smith*) to the coralline crag (*S. Wood*)." *
Genus Modiolaria. *Beck.*	
M. albicostata. *Sow.*	Dalmuir (*Smith*). Perhaps a variety of *M. discors*, Linn.
M. discors. *Linn.*	Paisley. Synonym: *Crenella discors*, F. & H.
M. discors var lævigata. *Gray.*	Dalmuir; Lochgilp; Elie; Errol.
M. nigra. *Gray.*	Kyles of Bute (*Crosskey*); Errol; Garvel Park New Dock, Greenock. Synonyms: *Crenella nigra*, F. & H.; *Modiola nigra*, Gray.
Genus Montacuta. *Turton.*	
M. bidentata. *Montg.*	Dalmuir. Synonym: *Mya bidentata*, Montagu.
M. elevata. *Stimpson.*	Lochgilp; Tangy Glen, near Campbeltown.
M. ferruginosa. *Montg.*	Kilchattan Tile-work, Bute. Synonym: *Mya ferruginosa*, Montagu.
Genus Mya. *Linnæus.*	
M. arenaria. *Linn.*	Bute (*Smith*); Lochgilp, a single valve only.
M. truncata. *Linn.*	Windmillcroft; Dalmuir; Cumbrae College; Lochgilp; East Tarbert; Duntroon; Old Mains, Renfrew; Paisley; Kilchattan Tile-work, Bute; Garvel Park New Dock, Greenock; Wick; Elie (*Jamieson*); Rothesay; King-Edward; Elie.
M. truncata var Uddevallensis. *Forbes.*	Dalmuir; Lochgilp, in situ; Bute and Wick (*Smith*); Rothesay.
Genus Mytilus. *Linnæus.*	
M. edulis. *Linn.*	Dalmuir; Cumbrae College; Lochgilp; Old Mains, Renfrew; Paisley; Garvel Park New Dock, Greenock; Caithness, in boulder clay; Gamrie; Cornton, near Bridge of Allan; Ellishill.
M. modiolus. *Linn.*	Dalmuir; Cumbrae College; Lochgilp; West Tarbert; Old Mains, Renfrew; Paisley; Kilchattan Tile-work, Bute; Caithness, in boulder clay; Crinan; Garvel Park New Dock, Greencck.
Genus Nucula. *Lamarck.*	
N. nitida. *G. B. Sow.*	Paisley (*Crosskey*).

* *British Conchology*, vol. ii. p. 423.

N. nucleus. *Linn.*	Caithness, in boulder clay; Lochgilp; Paisley, Rothesay. Synonym: *Arca nuclea*, Linn.
N. nucleus var tumidula. *Malm.*	Paisley.
N. sulcata. *Bronn.*	Caithness (*Peach*). This appears to be of very doubtful stability.
N. tenuis. *Montg.*	Lochgilp; West Tarbert; Duntroon; Paisley; Garvel Park New Dock, Greenock; Kilchattan Tile-work, Bute; Annochie; Montrose; Cumbrae College.
N. tenuis var expansa. *Reeve.*	Cumbrae College.
N. tenuis var inflata. *Mörch.*	Paisley; Elie and Errol. A decidedly arctic shell.
N. proxima. *Say.*	Recorded from Aberdeenshire by Smith,* *fide* Jamieson. According to Gould this shell is closely allied to, if not identical with, *N. nitida*, Sow.†
Genus PHOLAS. *Lister.*	
P. crispata. *Linn.*	Stevenston (*Landsborough*); Cumbrae College; Gamrie; King-Edward; Kilchattan Tile-work, Bute.
P. dactylus. *Linn.*	Stevenston (*Landsborough*); Ayr (*Smith*).
Genus PSAMMOBIA. *Lamarck.*	
P. Ferröensis. *Chem.*	Kyles of Bute (*Crosskey*).
Genus SAXICAVA. *Fleurian de Bellvue.*	
S. Norvegica. *Spengler.*	Caithness, in boulder clay; Belhelvie; Gourock; Fairlie; Kyles of Bute; Langbank. Synonyms: *Panopœa arctica*, Gould; *P. Bivonœ*, Smith; *Panopœa Norvegica*, F. & H.
S. rugosa. *Linn.*	Caithness, in boulder clay; Elie and Errol; Dalmuir; Lochgilp; Cumbrae College; East and West Tarbert; Old Mains, Renfrew; Garvel Park New Dock, Greenock; Paisley; Kilchattan Tile-work, Bute; Rothesay. Synonym: *S. pholadis*, Linn.
S. rugosa var arctica. *Linn.*	Lochgilp; Dalmuir; East Tarbert; Garvel Park New Dock, Greenock; Kyles of Bute. Synonym: *Mya arctica*, Linn.
S. rugosa var precisa. *Montg.*	Lochgilp, one valve only.
S. sulcata. *Smith.*	Paisley; Montrose; Annochie; Kyles of Bute; Ellishill: Belhelvie; Rothesay.
Genus SCROBICULARIA. *Schumacher.*	
S. alba. *S. Wood.*	Dalmuir; Lochgilp; West Tarbert; Duntroon; Garvel Park New Dock, Greenock. Synonyms: *Ligula Boysii*, Montg.; *Abra alba*, S. Wood; *Syndosmya alba*, F. & H.
S. prismatica. *Montg.*	Lochgilp; Kilchattan Tile-work, Bute; Greenock (*Smith*). Synonyms: *Ligula prismatica*, Montg.; *Syndosmya prismatica*, F. & H.
Genus SOLECURTUS. *De Blainville.*	
S. candidus. *Renier.*	Caithness (*Peach*).

* *Researches*, p. 54.

† Jeffreys, *British Conchology*, vol. ii. p. 150.

Genus Solen. *Linnæus.*	
S. siliqua. *Linn.*	Fragments in Clyde beds (*Prof. Geikie*).
Genus Tapes. *Mühlfeldt.*	
T. decussatus. *Linn.*	Scotch and Irish beds (*Forbes*); Lucknow Pit, Ayrshire.
T. pullastra. *Linn.*	Clyde beds (*Smith*); Lucknow Pit, Ayrshire. Synonym: *Venus pullastra*, Smith.
T. virgineus. *Linn.*	Kyles of Bute (*Prof. Geikie*).
Genus Tellina. *Linnæus.*	
T. Balthica. *Linn.*	Caithness, in boulder clay; Dalmuir; Old Mains, Renfrew; Paisley; Gamrie; King Edward; Belhelvie. This form occurs in the Bridlington and mammaliferous crags, Isle of Man beds, and Scandinavian deposits, &c. Towncroft Farm, near Grangemouth, in muddy sand above a red clay containing seals' bones. Synonym: *T. solidala*, F. & H.
T. calcarea. *Chem.*	Caithness, in boulder clay; Elie; Dalmuir; Lochgilp; Lucknow Pit, Ayrshire; Windmillcroft; Cumbrae College; East Tarbert; Crinan; Duntroon; Old Mains, Renfrew; Paisley; Kilchattan Tile-work, Bute; Cornton, near Bridge of Allan; Chapelhall; Gamrie; Belhelvie; Rothesay. Synonyms: *T. proxima*, Brown; *T. sordida*, Couthouy.
T. crassa. *Linn.*	Ayr (*Smith*).
T. donacina. *Linn.*	Banff (*Smith*).
T. fabula. *Gronovius.*	Lochgilp.
T. Grœnlandica. *Beck.*	Bute (*Forbes*).
T. squalida. *Pulteney.*	Kyles of Bute (*Crosskey*); Dalmuir. Synonym: *T. incarnata*, F. & H.
T. tenuis. *Da Costa.*	Kyles of Bute (*Crosskey*); Gamrie.
Genus Thracia. *Leach.*	
T. myopsis. *Beck.*	Elie; Errol; Greenock (*Jeffreys*). Markedly arctic.
T. papyracea. *Poli.*	Kyles of Bute and Lochgilp (*Prof. Geikie*).
Genus Venus. *Linnæus.*	
V. casina. *Linn.*	Caithness, in boulder clay. From the Clyde beds to the coralline crag (*Jeffreys*).
V. exoleta. *Linn.*	Lucknow Pit, Ayrshire; Clyde beds (*Smith*). Synonyms: *Cythere exoleta*, Lam.; *Artemis exoleta*, Forbes.
V. fasciata. *Da Costa.*	Lucknow Pit, Ayrshire.
V. gallina. *Linn.*	Caithness, in boulder clay. Synonym: *V. striatula*, Donovan.
V. lincta. *Pulteney.*	Caithness, in boulder clay; Dalmuir; Kyles of Bute (*Prof. Geikie*); Clyde beds (*Smith*). Synonym: *Artemis lincta*, F. & H.
V. ovata. *Pennant.*	Caithness, in boulder clay; Tangy Glen, near Campbeltown.
Genus Yoldia. *Möller.*	
Y. hyperborea. *Lovén.*	Errol. ? the young of *Leda arctica*.
Y., sp. ind.	Elie. This is identical with a species found by Dr. Törell at Spitzbergen, in 80° N. Lat. (*Brown*).

Class: Gasteropoda.

Prosobranchiata.

Genus APORRHAIS. *Da Costa.*	
A. pes-pelicani. *Linn.*	Caithness, in boulder clay; Kilchattan Tile-work, Bute; Gourock; King-Edward.
Genus BUCCINUM. *Linnæus.*	
B. ciliatum. *Fabr.*	Bute (*Forbes*).
B. Grœnlandicum. *Chem.*	West Tarbert; Old Mains, Renfrew; Paisley; Dalmuir; Errol. The shells recorded by Prof. Geikie from the Clyde beds as *B. Humphreysianum*, appear to be this species (*Jeffreys*).* According to Messrs. Crosskey and Robertson, this species was recorded in Dr. Thomson's Dalmuir list as *B. striatum*,† Sow. Synonym: *B. cyaneum*, Beck.
B. undatum. *Linn.*	Dalmuir; Cumbrae College; Lochgilp; Paisley; Caithness, in boulder clay; West Tarbert; Duntroon; Gourock; Old Mains, Renfrew; Garvel Park New Dock; Kilchattan Tile-work, Bute; Gamrie; Cornton, near Bridge of Allan. Synonyms: *B. vulgare*, Da Costa; *B. porcatum*, Gmelin.
B. undatum var carinatum. *Turton.*	Bute (*Smith*, fide *Jeffreys*). This is a monstrosity.
Genus CERITHIUM. *Adanson.*	
C. reticulatum. *Da Costa.*	Cumbrae College; Lochgilp; Duntroon.
Genus CERITHIOPSIS. *Forbes & Hanley.*	
C. costulata. *Müller.*	Wick. Synonym: *C. nivea*, Jeffreys.
C. tubercularis. *Montg.*	Jeffreys records this as fossil from the Clyde beds, *fide* Smith.
Genus CHITON. *Linnæus.*	
C. albus. *Linn.*	Fort William (*Jeffreys*).
C. cinereus. *Linn.*	Lochgilp; Garvel Park New Dock, Greenock; Caithness, in boulder clay; Fort William.
C. marmoreus. *Fabr.*	Dalmuir; Lochgilp; Old Mains, Renfrew; Garvel Park New Dock, Greenock; Fort William (*Jeffreys*).
C. ruber. *Linn.*	Dalmuir; Lochgilp; Garvel Park New Dock Greenock; Fort William (*Jeffreys*).
Genus COLUMBELLA. *Lamarck.*	
C. Holböllii. *Möller.*	Fort William (*Jeffreys*).
Genus CYCLOSTREMA. *Marryat.*	
C? costulatum. *Möller.*	Paisley (*Crosskey*); Fort William (*Jeffreys*). Synonym: *Margarita? costulata*, Müller.
Genus DEFRANCIA. *Millet.*	
D. Leufroyii. *Michaud.*	Wick, in boulder clay. Synonyms: *Pleurotoma Leufroyi*, Mich.; *Mangelia Leufroyi*, F. & H.
Genus DENTALIUM. *Linnæus.*	
D. abyssorum. *Sars.*	Caithness, in boulder clay.
D. entalis. *Linn.*	Wick, in boulder clay; Belhelvie; King-Edward; Gamrie.
D. tarentinum. *Lamk.*	Gamrie. Smith makes *D. tarentinum* and *D. dentale*, Linn., synonymous; it is therefore

* *Brit. Conchology*, vol. iv. p. 294.

† *Trans. Geol. Soc. Glasgow*, vol. ii. p. 273.

	doubtful which of the two has been found at this locality.
Genus Fissurella. *Bruguere.*	
F. Græca. *Linn.*	Clyde beds (*Forbes*). Synonym: *F. reticulata*, F. & H.
Genus Fusus. *Bruguere.*	
F. antiquus. *Linn.*	Cumbrae College; West Tarbert; Lochgilp; Paisley (*Prof. Geikie*); Garvel Park New Dock, Greenock; Gourock; Kilchattan Tile-work, Bute; Croftamie, Dumbartonshire, in blue clay; Caithness, in boulder clay.
F. curtus. *Smith.*	Stevenston (*Landsborough*).* Probably a very doubtful species.
F. despectus. *Linn.*	Dalmuir (*Forbes*); Kippet Hills, Loch of Slains, Aberdeenshire (*Jamieson*). Synonyms: *Murex carinatus*, Pen.; *Fusus carinatus*, Lamk.; *F. tornatus*, Gould; *F. carinatus*, Smith.
F. gracilis. *Da Costa.*	Dalmuir.
F. propinquus. *Alder.*	Lochgilp (*Prof. Geikie*); Gamrie; King-Edward.
Genus Helcion. *De Montfort.*	
H. pellucidum. *Linn.*	Dalmuir; Lochgilp; Ayr. Synonym: *Patella pellucida*, Lin.
H. pellucidum var lævis. *Pennant.*	Dalmuir; Banffshire (*Forbes*).
Genus Homalogyra. *Jeffreys.*	
H. atomus. *Phil.*	Dalmuir; Cumbrae College; Lockgilp; West Tarbert; Duntroon; Garvel Park New Dock, Greenock; Old Mains, Renfrew; Paisley; Kilchattan Tile-work, Bute; Fort William. Synonym: *Skenea nitidissima*, F. & H.
Genus Hydrobia. *Hartmann.*	
H. ulvæ. *Pennant.*	Paisley; Dalmuir. Jeffreys says that the males of this species are probably the *Rissoa subumbilicata*, Montg., which under this name is recorded by Smith from Dalmuir.
Genus Lacuna. *Turton.*	
L. divaricata. *Fabr.*	Dalmuir; Cumbrae College; Lcchgilp; West Tarbert; Duntroon; Old Mains, Renfrew; Paisley; Garvel Park New Dock, Greenock; Kilchattan Tile-work, Bute; Caithness; Gamrie; King-Edward; Fort William. Synonym: *L. vincta*, F. & H.
L. divaricata var quadrifasciata. *Montf.*	Dalmuir. Synonym: *Turbo quadrifasciatus*, Montf.
L. pallidula. *Da Costa.*	Dalmuir; Lochgilp. Synonyms: *Turbo pallidulus*, Turton; *Nerita pallidula*, Da Costa.
L. pallidula var neritoidea. *Gould.*	Dalmuir; Fort William.
L. puteola. *Turton.*	Dalmuir; Cumbrae College; Paisley; Garvel Park New Dock, Greenock; Fort William. Synonym: *Turbo puteolus*, Turton.

* Smith, *Researches*, p. 49.

Genus LITORINA. *Ferussac.*	
L. limita. *Lovén.*	Dalmuir; Cumbrae College; Lochgilp; Paisley; East and West Tarbert; Garvel Park New Dock. Synonyms: *L. palliata*, Say; *L. arctica*, Müller; *Turbo expansus*, Brown.
L. litorea. *Linn.*	Dalmuir; Cumbrae College; East and West Tarbert; Paisley; Crinan; Old Mains, Renfrew; Kilchattan Tile-works, Bute; Caithness; Cornton, near Bridge of Allan; Croftamie, Dumbartonshire, in blue clay; Windmillcroft; Lochgilp; Lucknow Pit.
[L. neritoides. *Linn.*]	Recorded by Prof. Geikie from certain of the Clyde beds. Some doubt regarding the correctness of this is expressed by Jeffreys.*
L. obtusata. *Linn.*	Dalmuir; Cumbrae College; Lochgilp; East and West Tarbert; Duntroon; Paisley; Old Mains, Renfrew; Garvel Park New Dock, Greenock; Kilchattan Tile-work, Bute; Caithness, in boulder clay; Lucknow Pit, Ayrshire. Synonym: *L. littoralis*, F. & H.
L. obtusata var neritiforme. *Brown.*	East Tarbert.
L. rudis. *Maton.*	Dalmuir; Lochgilp; Paisley; Garvel Park New Dock, Greenock; Kilchattan Tile-work, Bute; Cornton, near Bridge of Allan; Rothesay.
L. rudis var patula. *Jeffreys.*	Dalmuir; Bute (*Smith*).
L. rudis var sexatilis. *Johnston.*	Garvel Park New Dock, Greenock.
L. squalida. *B. & S.*	Fort William; Ellishill; Invernettie; Railway cutting between Drymen and Gartness, Stirlingshire, in gravel with other arctic shells (*Jamieson*); Paisley.
Genus MARGARITA. *Leach.*	
M. cinerea. *Couthouy.*	Bute (*Prof. Geikie*); Rothesay (*Smith*).
M. olivacea. *Brown.*	Clyde beds (*Jeffreys*).
Genus MENESTHO. *Müller.*	
M. albula. *Fabr.*	Paisley.
Genus MÖLLERIA. *Jeffreys.*	
M. costulata. *Müller.*	Old Mains, Renfrew; Paisley; Fort William.
Genus MUREX. *Linnæus.*	
M. erinaceus. *Linn.*	Dalmuir (*Smith*).
Genus NASSA. *Lamarck.*	
N. incrassata. *Ström.*	Caithness, in boulder clay; King-Edward; Kyles of Bute (*Prof. Geikie*); Dalmuir (*Smith*); Lochgilp. Synonym: *N. macula*, Montg.
N. reticulata. *Linn.*	Duntroon.
Genus NATICA. *Adanson.*	
N. affinis. *Gmelin.*	Dalmuir; Cumbrae College; Lochgilp; West Tarbert; Duntroon; Old Mains, Renfrew; Paisley; Garvel Park New Dock, Greenock; Caithness, in boulder clay; Kilchattan Tile-work, Bute; Gamrie (*Chambers*); Gourock; King-Edward; Rothesay. Synonym: *N. clausa*, B. & S.
N. Alderi. *Forbes.*	Caithness, in boulder clay; King-Edward.

* *Brit. Conchology*, vol. iii. p. 362.

	Synonyms: *N. intermedia* et *Marochiensis*, Philippi; *N. nitida*, F. & H.
N. catena. *Da Costa.*	Paisley; Bute; Gourock. This species is mentioned in Dr. Thomson's Dalmuir list as *N. glaucinoides*, Sow.* Synonyms: *N. monilifera*, F. & H.; *N. glaucina.*
[N. fragilis. *Smith.*]	Dalmuir. Considered by E. Forbes to be a much decayed *N. monilifera.*
[N. glaucinoides. *Sow.*]	Recorded from King-Edward by Smith in Jamieson's list of Aberdeenshire glacial shells,† and also by Landsborough from Stevenston.‡ It is a crag-fossil. Forbes also regarded this fossil as identical with *N. monilifera.*
N. Grœnlandica. *Beck.*	Elie and Errol; Old Mains, Renfrew; Garvel Park New Dock, Greenock; Kilchattan Tile-work, Bute; Kilmaurs (*J. Young*). Synonym: *N. pusilla*, F. & H.
N. Islandica. *Gmelin.*	Bute; Caithness, in boulder clay; Gamrie; King-Edward. Synonyms: *N. helicoides*, F. & H.; *N. pallida*, B. & S. Jeffreys remarks that it is difficult to determine whether *N. pallida*, B. & S., is a synonym of *N. Grœnlandica* or *N. Islandica.*§ Under the designation of *N. pallida* a species is recorded from Dalmuir; Paisley; Elie and Errol; Gamrie; King-Edward; Kilchattan Tile-work (*Jamieson*); Caithness.
N. Montacuti. *Forbes.*	Bute (*Smith*); Clyde beds (*Smith*). Synonym: *Natica Montagui*, Forbes.
N. Smithii. *Brown.*	Ardincaple, near Helensburgh (*Smith*). Synonyms: *Bulbus Smithii*, Brown; also probably *N. flava*, Gould; *N. aperta*, Lovén.
N. sordida. *Phil.*	Caithness, in boulder clay (*Peach*).
Genus Odostomia. *Fleming.*	
O. acicula. *Phil.*	Caithness, in boulder clay.
O. albella. *Lovén.*	Do. do.
O. conoidea. *Brocchi.*	Lochgilp.
O. pallida. *Montg.*	Do.
O. Lukisi. *Jeffreys.*	Garvel Park New Dock, Greenock.
O. spiralis. *Montg.*	Lochgilp; Dalmuir.
O. turrita. *Hanley.*	Lochgilp.
O. unidentata. *Montg.*	Lochgilp; Garvel Park New Dock; Kilchattan Tile-work, Bute.
Genus Patella. *Lister.*	
P. vulgata. *Linn.*	Paisley; Caithness, in boulder clay; Fort William (*Jeffreys*); Lucknow Pit, Ayrshire.
Genus Pleurotoma. *Lamarck.*	
P. nebula. *Montg.*	Caithness (*Jamieson*). Synonym: *Mangelia nebula*, F. & H.; *P. ginnaniana*, Phil.
P. pyramidalis. *Ström.*	Caithness; Gamrie; King-Edward; Fort William (*Jeffreys*); Kilchattan Tile-work, Bute; Dalmuir; Cumbrae College; Lochgilp; West Tarbert; Crinan; Duntroon; Paisley. Synonyms: *Defrancia Vahlii*, Beck; *Mangelia pyramidalis*, Ström.

* Crosskey and Robertson, *Trans. Geol. Soc. Glasgow*, vol. ii. p. 273.
† *Researches*, p. 57. ‡ *Pro. Geol. Soc.*, vol. iii. p. 444. § *Brit. Con.*, vol. iv. p. 218.

[P. rufa. *Montg.*]	Recorded from Kyles of Bute (*Prof. Geikie*); King-Edward and Gamrie (*Jamieson*). Jeffreys has only recognised this species as a glacial shell from the Belfast deposit,* those from the localities just quoted probably being the preceding form, *P. pyramidalis.*
P. Trevelyana. *Turton.*	Wick (*Peach*); West Tarbert; King-Edward and Gamrie (*Jamieson*); Kilchattan Tile-work, Bute; Hebrides (*Jeffreys*). Synonyms: *Mangelia Trevelyana*, F. & H.; *Fusus decussatus*, Couthouy.
P. turricula. *Montg.*	Cumbrae College; West Tarbert; Duntroon; Paisley; Dalmuir (*Smith*); Garvel Park New Dock, Greenock; Lochgilp; Oban (*Prof. Geikie*); Kilchattan Tile-work, Bute; Caithness; Gamrie; King-Edward. Synonyms: *Mangelia turricula*, F. & H.; *Fusus discrepans*, Brown.
P. violacea. *Migh & Ad.*	Dalmuir; Cumbrae College; Lochgilp; Old Mains, Renfrew; Garvel Park New Dock, Greenock; Kilchattan Tile-work, Bute.
Genus PUNCTURELLA. *R. T. Lowe.*	
P. Noachina. *Linn.*	Dalmuir; Lochgilp; Cumbrae College; Old Mains, Renfrew; Garvel Park New Dock, Greenock; Fort William. Synonym: *Cremoria Flemingiana*, Leach.
Genus PURPURA. *Bruguière.*	
P. lapillus. *Linn.*	Dalmuir; Cumbrae College; East and West Tarbert; Crinan; Paisley; Kilchattan Tile-work, Bute; Loch Long (*Prof. Geikie*); Caithness; Lochgilp; Lucknow Pit, Ayrshire.
Genus RISSOA. *Fréminville.*	
R. cancellata. *Da Costa.*	Lochgilp. Synonym: *R. crenulata*, F. & H.
R. costata. *Adams.*	Largs (*Landsborough*).
R. inconspicua. *Alder.*	Crinan.
R. inconspicua var ventrosa. *Montg.*	Dalmuir (*Smith*).
R. membranacea. *Adams.*	Bute (*Smith*). Synonym: *R. labiosa*, F. & H.
R. parva. *Da Costa.*	West Tarbert; Duntroon; Paisley (*Jamieson*); Garvel Park New Dock, Greenock; Fort William.
R. parva var interrupta. *Adams.*	Dalmuir; Caithness; Cumbrae College; Lochgilp; West Tarbert; Duntroon; Old Mains, Renfrew; Paisley; Garvel Park New Dock, Greenock; Kilchattan Tile-work, Bute. Synonym: *Turbo interruptus*, Adams.
R. proxima. *Alder.*	Lochgilp. Synonym: *R. striatula*, Jeffreys.
R. reticulata. *Adams.*	Lochgilp.
R. soluta. *Phil.*	Paisley. Synonym: *R. globosa*, Martin.
R. striata. *Adams.*	Dalmuir; Cumbrae College; Lochgilp; West Tarbert; Duntroon; Old Mains; Garvel Park New Dock; Greenock; Kilchattan Tile-work, Bute. Synonym: *Turbo semicostatus*, Montg.
R. striata var arctica. *Lovén.*	Garvel Park New Dock, Greenock.

* *Loc. cit.*, p. 494.

R. striata var saxitilis. *Möller.*	Paisley.
R. violacea. *Desmarets.*	Lochgilp.
Genus SCALARIA. *Lamarck.*	
S. Grœnlandica. *Chem.*	Fairlie; King-Edward (*Jamieson*).
Genus SKENEA. *Fleming.*	
S. planorbis. *Fabr.*	Dalmuir; Windmillcroft; Cumbrae College; Lochgilp; West Tarbert; Duntroon; Old Mains, Renfrew; Paisley; Kilchattan Tile-work, Bute; Garvel Park New Dock.
Genus TECTURA. *Cuvier.*	
T. virginea. *Müller,*	Dalmuir; Cumbrae College; Lochgilp; Old Mains, Renfrew; Paisley; Garvel Park New Dock, Greenock; Kilchattan Tile-work, Bute; Gamrie. Synonyms: *Acmœa virginea,* F. & H.; *Lottia virginea,* Alder; *Patella virginea,* Müller.
Genus TRICHOTROPIS. *Broderip & Sowerby.*	
T. borealis. *Brod. & Sow.*	Garvel Park New Dock, Greenock; Bute (*Smith*).
Genus TROCHUS. *Rondeletius.*	
T. cinerarius. *Linn.*	Stevenston (*Landsborough*); Garvel Park New Dock; Kilchattan Tile-work, Bute; Lochgilp (*Prof. Geikie*); Lucknow Pit, Ayrshire.
T. cinerus. *Couthouy.*	Clyde beds (*Jeffreys*). Synonym: *Margarita striata,* B. & S.
T. Grœnlandicus. *Chem.*	Dalmuir; Cumbrae College; Lochgilp; East Tarbert; Old Mains, Renfrew; Paisley; Garvel Park New Dock, Greenock; Caithness; Fort William; Rothesay; Errol. Synonyms: *Margarita undulata,* G. B. Sow; *Turbo incarnatus,* Couthouy; *Trochus inflatus,* Brown.
T. helicinus. *Fabr.*	Dalmuir; Cumbrae College; East Tarbert; Paisley; Garvel Park New Dock, Greenock; Oban; Fort William.
[T. lineatus. *Da Costa.*]	Paisley (*Smith*). Jeffreys doubts the correctness of this determination.*
T. magus. *Linn.*	Clyde beds (*Smith*); Lucknow Pit, Ayrshire.
T. millegranus. *Phil.*	Fort William (*Jeffreys*).
T. tumidus. *Montg*	Dalmuir; Lochgilp; Old Mains, Renfrew; Garvel Park New Dock, Greenock; Kilchattan Tile-work, Bute; Fort William.
T. Vahlii. *Möller.*	Paisley; Caithness. Synonym: *Margarita Vahlii.*
T. Zizyphinus. *Linn.*	Caithness, in boulder clay.
Genus TROPHON. *De Montfort.*	
T. clathratus. *Linn.*	Dalmuir; Lochgilp: West Tarbert; Duntroon; Old Mains, Renfrew; Paisley; Garvel Park New Dock; Kilchattan Tile-work, Bute; Caithness; Gamrie; Belhelvie; King-Edward; Rothesay. Synonyms: *Fusus imbricatus,* Smith; *Fusus scalariformis,* Gould; *F. Peruvianus,* Sow.
T. clathratus var Gunneri. *Lovén.*	Dalmuir; Duntroon; Old Mains, Renfrew; Paisley; Garvel Park New Dock, Greenock; Gamrie; King-Edward.

* *Brit. Con.,* vol. iii. p. 319.

T. truncatus. *Ström.*	Dalmuir; Cumbrae College; Lochgilp; West Tarbert; Duntroon; Garvel Park New Dock; Old Mains, Renfrew; Paisley; Kilchattan Tile-work, Bute; Caithness; Gamrie; King-Edward. Synonym: *Murex Bamffius*, Donovan.
Genus TURRITELLA. *Lamk.*	
T. erosa. *Couthouy.*	Elie. Synonym: *T. polaris*, Beck. A markedly arctic form.
T. reticulata. *Migh. & Ad.*	King-Edward (*Jamieson*). Synonyms: *Mesalia reticulata*, Migh. and Ad.; *T. lactea*, Möller.
T. terebra. *Linn.*	Caithness, in boulder clay; Gourock (*Prof. Geikie*); King-Edward and Auchleuchries (*Jamieson*). Synonym: *T. communis*, F. & H.
Genus VELUTINA. *Fleming.*	
V. lævigata. *Pennant.*	Dalmuir; Crinan; Garvel Park New Dock, Greenock; Kilchattan Tile-work, Bute.
V. undata. *Smith.*	Old Mains, Renfrew; Dalmuir; Garvel Park New Dock, Greenock; Paisley (*Jamieson*). Synonym: *V. zonata*, Gould.

Opisthobranchiata.

Genus ACTÆON. *De Montfort.*	
A. tornatilis. *Linn.*	Caithness, in boulder clay; Lochgilp. Synonym: *Tornatella fasciata*, F. & H.
Genus CYLICHNA. *Lovén.*	
C. alba. *Brown.*	Dalmuir; Lochgilp; Duntroon; Paisley; Garvel Park New Dock, Greenock; Annochie; Gamrie. Synonym: *Volvaria alba*, Brown.
C. cylindracea. *Pennant.*	Lochgilp; Paisley; ? Bute; ? St. Fergus. Synonym: *Bulla cylindracea*, Pen.
C. obstricta. *Gould.*	Dalmuir; Lochgilp.
Genus SCAPHANDER. *De Montfort.*	
S. lignarius. *Linn.*	Greenock (*Robertson*, fide *Jeffreys*).
Genus TORNATELLA. *Lamk.*	
[T. pyramidata.]	Aberdeenshire (*Smith*, fide *Jamieson*).
Genus UTRICULUS. *Brown.*	
U. hyalinus. *Turton.*	Dalmuir; Cumbrae College; Duntroon; Paisley; Garvel Park New Dock; Kilchattan Tile-work, Bute. Synonym: *Bulla hyalina*, Turton.
U. mammillatus. *Phil.*	Dalmuir. Synonym: *Cylichna mammillata*, F. & H.
U. obtusus. *Montg.*	Cumbrae College; Lochgilp; West Tarbert; Duntroon; Paisley; Garvel Park New Dock, Greenock; Kilchattan Tile-work, Bute; ? St. Fergus. Synonym: *Cylichna obtusa*, F. & H.
U. truncatulus. *Brug.*	Duntroon. Synonym: *Cylichna truncatula*, F. & H.

Note.—In a paper on the "Palæontology of the Post-glacial Drifts of Ireland" (*Geol. Mag.*, vol. x., 1873, p. 447), Mr. A. Bell mentions having obtained the fry of living Mediterranean forms from glacial clay got near Greenock, viz., *Conus Mediterraneus* and *Cardita trapezia*.

VERTEBRATA.

Class: Pisces.

Fish bones have been found in the brick-clay of Invernettie, near Aberdeen,* and in the boulder clay of Caithness.† Fish vertebræ are recorded by Messrs. Crosskey and Robertson from the following localities:—Paisley; Kilchattan Tile-work, Bute; Garvel Park New Dock, Greenock;‡ &c. From the last-named locality *Otolites* have also been obtained.

Class: Aves.

The skeleton of a bird is recorded by Jamieson from brick-clay on the side of the river Dee near Aberdeen,§ and bones from the Paisley deposit by Messrs. Crosskey and Robertson. In the Appendix to his paper "On the Glacial Drift of Scotland," ‖ Prof. Geikie mentions the fact that the furculum of a gull was found in brick-clay at the Bridge of Johnston, near Paisley. Smith describes this as the "fourchette of a diver" (*Researches*, p. 14), and adds that the bed was 54 feet above sea-level.

Class: Mammalia.

Section: PLACENTALIA.

Ungulata.

Genus Bos. *Linnæus.*	
B. primigenius. *Boj.*	In clay forming the bed of the Clyde opposite Jordanhill (*Scouler*);¶ in clay in Rothesay Bay (*Prof. Geikie*); in the inter-glacial beds of Cowden Glen, Renfrewshire.**
Genus Cervus. *Linnæus.*	
C. alces. *Linn.*	Recorded by the late Mr. Smith, of Jordanhill, from the marl beds of Perthshire; the exact locality, according to Dr. J. A. Smith, is Airleywight, in marl underlying moss. Dr. Smith likewise records numerous other instances of the occurrence of this species in the post-tertiary beds of Scotland, the most important of which are—a marl pit in Forfarshire; in clay and gravel at Strath Halladale, Sutherlandshire; in a peat-moss on the edge of Williestruther Loch, valley of the river Slitrig, Roxburghshire, with the skull of *Bos longifrons;* in a peat-bog at Oakwood, near Selkirk. Synonym: *Alces malchis*, Gray. (J. A. Smith, *Pro. Soc. Ant. Scotland*, 1871.)
? C. dama. *Linn.*	Recorded by Smith from Kilmaurs, in boulder clay.
? C. elaphus. *Linn.*	Do. do. do.
C. tarandus. *Linn.*	Bed of the Clyde opposite Jordanhill (*Scouler*); Croftamie, Dumbartonshire, with shells in blue clay, about 18 feet from the surface.

* Jamieson, *Quart. Journ.*, vol. xiv. p. 518.
† Peach, *Pro. Roy. Phys. Soc. Ed.*, vol. iii. p. 403.
‡ *Trans. Geol. Soc. Glasgow*, vols. iii. and iv.
§ *Loc. cit.* p. 510.
‖ *Trans. Geol. Soc. Glasgow*, vol. i.
¶ *Ed. New Phil. Journ.*, vol. lii. p. 135.
** J. Geikie, *Geol. Mag.*, vol. v. p. 393.

Genus EQUUS. *Linnæus.*
E. caballus. *Linn.* — At Cowden Glen, with the remains of *Bos primigenius.*[*]

Genus MEGACEROS. *Owen.*
M. Hibernicus. *Owen.* — Inter-glacial beds at Cowden Glen, Renfrewshire.[†] Synonym: *Cervus Megaceros*, Hart. (The remains of this extinct elk were found in the shell marl of an old silted-up loch in the parish of Maybole, Ayrshire,[‡] with those of *Cervus elaphus* and *Bos primigenius;* with the antlers of *Cervus capreolus*, in a deposit of gravel, earth, and large boulders, probably a river accumulation, at Coldingham, Berwickshire.[§] It seems most likely, however, that in the latter instance the remains were those of the true elk, *Cervus alces.*[‖])

Proboscoidea.

Genus ELEPHAS. *Linnæus.*
E. primigenius. *Blum.* — At Chapelhall, near Airdrie, the bone of an elephant was found at a height of 350 feet above the sea in laminated sand underlying till; Greenhill Quarry, Kilmaurs, in a peaty deposit underlying sand, in which were glacial shells, and overlaid by about 40 or 50 feet of till; a tusk was discovered during the excavation of the line of the Union Canal between Edinburgh and Falkirk; remains found at Cliftonhall 15 to 20 feet from the surface, in boulder clay; Bishopbriggs, near Glasgow.

Carnivora.

Genus PAGOMYS.
P. fœtidus. *Gray.* — Remains of seals have been found in red clay at Westfield of Auchmacoy, near Aberdeen (*Jamieson*);[¶] in laminated clay at Montrose; in brick-clay at Springfield, near Stratheden, Fife,[**] 150 feet above sea level, and about 16 feet from surface; a pelvis of a seal was obtained from brick-clay at Tyrie, 30 feet above high-water mark; and at a depth of about 19 feet;[††] at Portobello the remains of a seal were likewise found in brick-clay, 20 feet above high-water mark, at a depth of 15 feet from the surface.[‡‡] At Camelon, in a bed of clay 90 feet above the level of the Frith of Forth; portions of a skeleton were found in a brick deposit at Errol; lastly, in sinking a pit on Towncroft Farm, Grangemouth, seal bones were found in a red clay, 80 feet from the surface, and about 68 feet below sea level. The majority of the remains mentioned in this list have been examined by

* R. Craig, *Trans. Geol. Soc. Glasg.*, vol. iv. p. 18.
† *Loc. cit.*
‡ *New Statistical Acc. Scot.*, 1845, vol. v. p. 353.
§ J. Hardy, *Trans. Berwicksh. Nat. Hist. Soc.*, vol. i. p. 247.
‖ Dr. J. A. Smith, *Proc. Soc. Ant. Scot.*, 1871, p. 325.
¶ *Quart. Journ. Geol. Soc.*, vol. xiv. p. 514.
** Page, *Geologist*, vol. i. p. 538.
†† Allman, *Ed. New Phil. Journ.*, 1858, vol. viii. p. 147.
‡‡ *Loc. cit.*, 1859, vol. x.

Prof. Turner, who pronounced them to belong to the small arctic seal (*Pagomys fœtidus*, Gray), and not, as has usually been stated, to *Phoca vitulina*, Linn. (Turner, *Pro. Roy. Soc. Ed.*, 1869–70, pp. 105–114.)

Note.—The remains of *Bos longifrons* and *Bos primigenius* were obtained in an old river-gravel in Glasgow; the first in Rutherglen Loan, the latter in Greendyke Street. (J. Bennie, *Trans. Geol. Soc. Glasgow*, vol. ii. p. 152.)

EXPLANATORY NOTES REGARDING CERTAIN OF THE LOCALITIES MENTIONED IN THE FOREGOING LIST.

Windmillcroft.—When compared with glacial deposits of other localities in the Clyde district the clay at Windmillcroft is peculiar from the very scanty proportion of animal remains found in it, no doubt arising from the fact, that the conditions of deposition were less favourable than those of some other localities.*

Dalmuir.—On the banks of the Dalmuir burn, about eight and half miles from Glasgow. This locality was extensively explored and described by Dr. T. Thomson, Mr. James Smith, and again by Messrs. Crosskey and Robertson. From the examination of the latter investigators, it appears there are two shell beds, an upper clay bed containing the mass of the shells, and a lower more sandy deposit, in which the shells are found in the best state of preservation, and of larger size. The sandy bed is underlaid by stiff blue boulder clay without shells, and the whole capped by water-worn gravel.†

Cumbrae College.—A shell-bearing sand bed near the College, Isle of Cumbrae, described by Messrs. Crosskey and Robertson, who consider that it appertains to the older Glacial deposits of the West of Scotland. It is peculiarly remarkable for the very large proportion of sand which enters into its composition.‡

Lochgilp.—Near the bridge crossing the Crinan Road to Lochgilphead. At this locality the boulder clay is immediately overlaid by the shell clay, the usual laminated clay of the Clyde beds being absent. There is no transition from one to the other, the line of demarcation between the boulder clay and shell clay being sharp and distinct. The former does not contain any organic remains. The shell bed in addition to its fauna, which is very abundant, contains a few boulders, less striated, and smaller in size than those in the underlying boulder clay, and often with *Serpulæ* attached. At one part of the deposit *Mya truncata*, and its variety *M. Uddevalensis* occur in their natural position, with the valves united.§

Lucknow Pit.—Ardeer Ironworks Ayrshire. There is evidence to show that here there was either a shell bed containing an admixture of true arctic and temperate shells, or else two beds, one with purely arctic forms, the other with temperate. Messrs. Crosskey and Robertson point out that a commingling of arctic and temperate forms occurs in several Norwegian post-tertiary beds. This was the first locality in the west of Scotland at which *Leda arctica* was found.‖

* Rev. C. H. Crosskey, *Trans. Geol. Soc. Glasgow*, vol. ii. p. 115.
† Crosskey and Robertson, *Trans. Geol. Soc. Glasgow*, vol. ii. p. 267.
‡ *Ibid.*, vol. iii. p. 113. § *Ibid.*, p. 118. ‖ *Ibid.*, p. 127.

East Tarbert.—A bed of clay containing arctic shells in the Black burn, at the north-east corner of Tarbert Loch. The mollusca are few in number both specifically and numerically. From the water-worn condition of many of the remains, it would appear that this deposit contains a large percentage of transported organisms. *Trochus helicinus* is very characteristic.*

West Tarbert.—A shell-clay on the south side of Tarbert Loch, near its head. Many of the shells in this bed, such as *Buccinum undatum*, and *Pecten Islandicus* attain a considerable size.†

Crinan.—A thin shell-bearing stratum occurs on a small plateau on a promontory at the north side of No. 11 Lock, on the Crinan Canal. The shells are scarce and very fragmentary.‡

Duntroon.—A stiff brown clay exposed at high-water mark a little to the south of Duntroon Castle. Amongst an abundant fauna, *Pleurotoma pyramidalis* is very characteristic.§

Old Mains.—A shell bed exposed in a tramway cutting, between the Houston Pit, No. 5, and old Mains Farm, Renfrew. The deposit consists of brown sand and earth with a large number of stones covered with *Balani*.‖

Paisley.—Throughout the Paisley beds, the organic remains are chiefly confined to the lower half of the deposit, but they may occur in any part of the section from the boulder clay to the summit. Messrs. Crosskey and Robertson remark that hitherto the boulder clay of the Paisley district has not yielded any vestiges of life, but occasionally beds containing arctic shells have been found beneath the boulder clay. The same writers observe "that the laminated clay, which has been regarded as the unfossiliferous base of the shell clay, is not unfossiliferous, but contains Foraminifera," such as *Polystomella striato-punctata*, &c.¶

Garvel Park New Dock, Greenock.—The fossiliferous clay here lies in a trough in boulder clay, and besides the usual shells of arctic type, characteristic of the Clyde beds, contains one or two additional forms of mollusca, not hitherto plentifully found in similar beds of the neighbourhood. Much speculation has been caused by the very confused appearance presented by this deposit generally. Unfavourable opinions as to its genuineness have been advanced, but Messrs. Crosskey and Robertson have arrived at the conclusion that the conditions of the deposit, both as regards the matrix and organic contents, are perfectly reconcileable with established facts.**

Kilchattan Tile-work, Bute.—A deposit of muddy sand containing shells is here superimposed on the usual laminated clay resting on reddish boulder clay, at the north-west side of Kilchattan Bay. The characteristic shells are *Tellina calcarea*, *Axinus flexuosus*, *Scrobicularia prismatica*, *Cyprina Islandica*, *Mya truncata*, and *Utriculus obtusus*. In the interior of many of the *Mya* valves, thick patches of the muddy sand have become so firmly indurated, as to be scarcely removable. These patches appear to consist of a strong calcareous base. This deposit has been well described by both Mr. Jamieson, and Professor Geikie.††

Tangy Glen.—About six miles from Campbeltown, on the Tarbert road. At this locality may be seen the rarer phenomenon of the fossiliferous clay *overlaid* by boulder clay of a dark reddish brown. Organic remains are rare, especially the Mollusca. Both the latter and the Ostracoda more nearly

* *Trans. Geol. Soc. Glasgow*, vol. iii. p. 321. † *Ibid.*, p. 324. ‡ *Ibid.*, p. 327.
§ *Ibid.*, p. 328. ‖ *Ibid.*, p. 331. ¶ *Ibid.*, pp. 334-341.
** *Ibid.*, vol. iv. pp. 32-45. †† *Ibid.*, pp. 128-133.

resemble those found in the fossiliferous beds of the east than of the west of Scotland, the former of which are generally considered to be much more arctic in character. The prevailing shell is *Leda pygmæa.* Two other forms rare in Scotch glacial beds are met with here, *Pecten Grælandicus* and *Montacuta elevata.**

Caithness.—The boulder clay of Caithness rises in many places to a height of two hundred feet above the sea-level, and varies in thickness from sixty to eighty feet. It is a tough and compact mass, with numerous striated and polished boulders, and resembles in general appearance the till which contains no shells. The fragmentary remains of mollusca are scattered without order through the general mass, many of them striated, some few with portions of the epidermis still remaining and hardly any in a good state of preservation, certainly in the case of bivalves, never with the valves united. Here and there nests and pockets of sand are met with.†

Montrose.—In a dull red to greenish grey laminated clay, forty feet above the sea-level, occupying the estuary valley of the South Esk, arctic shells, many pieces of chalk, and the bones of the seal have been found.‡

Cowden Glen.—In a section on the Crofthead and Kilmarnock Railway, in Cowden Glen, Neilston, Renfrewshire, are exposed a series of clay, sand, gravel and peaty beds, in hollows between two beds of till. From these beds, evidently of lacustrine origin, have been obtained numerous genera and species of *Diatomaceæ, Desmideæ, Entomostraca,* mosses, and the remains of many plants. In addition to the foregoing, a portion of the skull of *Bos primigenius,* bones of the great Irish elk and the horse have also been obtained from these beds.

Kilmaurs.—Sandstone quarry of Woodhill, near Kilmaurs. At various times during the past half century, the remains of the mammoth and reindeer have been found in a peaty layer between two thin beds of sand and gravel, overlaid by till or boulder clay, and resting directly on the sandstone rock of the quarry. From their position, it has been contended by some writers, that these remains are of *pre*-glacial origin, but an extended examination of the neighbourhood does not bear out this view. It appears that the bed of sand and gravel in which the remains were found, is only one of numerous similar intercalated deposits. The bed rock rises up here and there into prominences giving it a very irregular surface, and it was probably against one of these that the remains were washed, a lower boulder clay or till appearing in the deeper irregularities of the rocky bed.§

Chapelhall.—Near Airdrie. In a laminated sand underlying till, a bone of the mammoth was found about three hundred and fifty feet above the sea-level. Professor Geikie remarks that the ground has been so altered in mining operations, that it is impossible to fix the exact condition under which this bone was found.‖ *Tellina calcarea* was found at this locality in brick-clay between two beds of till, at a height of 510 feet above sea-level. (Smith, *Researches,* p. 141.)

Stevenston.—The shells recorded from this locality were obtained from a bed of blue clay nine feet thick, underlying thirty to forty feet of sand.¶

* *Trans. Geol. Soc. Glasgow,* vol. iv. pp. 134-137.
† C. W. Peach, *Pro. Roy. Phys. Soc. Ed.,* vol. iii. pp. 38, 396.
‡ J. C. Howden, *Trans. Geol. Soc. Ed.,* vol. i. p. 141.
§ Geikie, *Mems. Geol. Survey, Expl.* 22, *Scotland.*
‖ *Glacial Drift of Scotland. Trans. Geol. Soc. Glasgow,* vol. i.
¶ Landsborough, *Pro. Geol. Soc.,* vol. iii. p. 444.

Cornton.—Near Bridge of Allan. A clay was discovered in a burn behind Christie's Brick-work, from which were obtained some years ago the bones of a whale, at a depth of nine feet from the surface, together with pieces of the bark of the alder and nuts of the hazel. This deposit is most probably a post-glacial or recent accumulation.*

Croftamie.—Dumbartonshire. At about one hundred feet above the level of the sea the horn of a reindeer was discovered in blue clay, seven feet in thickness, under till, accompanied by arctic shells, in all about eighteen feet from the surface.†

Annochie.—Five miles north of Peterhead, on the Aberdeenshire coast. The shells occur in a bed of fine clay a few feet above the sea-level, passing underneath the beach; they are entire, with the epidermis preserved but very much decayed.‡

Belhelvie.—A clay pit five miles north of Aberdeen, and from thirty to forty feet above high-water mark. The shells are fragmentary and occur in a black stratum in laminated clay which rests on boulder clay.§

Ellishill.—Railway cutting three miles west of Peterhead. The shells are found in a red clay at an elevation of one hundred and twenty feet above the sea.‖

Gamrie.—Seven miles south of Banff. Shells occur in a thin sand bed at an elevation of about one hundred and fifty feet. The mass of the deposit, consisting of fine sand and clay, extends to a height of three hundred feet.¶

King-Edward.—Five miles south-south-east of Banff. Shells found in silt and gravel at an elevation of from one hundred and fifty to two hundred feet, some of them entire and in situ.

Tyrie.—Near Kinghorn, Fife. The shells at this locality were first made known by the late Dr. Fleming.

Invernettie.—A brick-work, one mile south of Peterhead.**

Auchleuchries.—Twenty miles north of Aberdeen. Shells occur as broken fragments in a thick mass of gravel, about three hundred feet above the sea-level.††

Portobello.—At the brick and tile works near Portobello, brick-clays are well exhibited, constantly containing *Foraminifera, Entomostraca*, and occasionally shells. Hugh Miller and Prof. Geikie record the occurrence of *Scrobicularia piperata*, nuts of the hazel, branches of the oak, beech, thorn, and other trees.‡‡ In Abercorn brick-field *S. piperata* occurs in its natural position. Pebbles of chalk and flint and nests of sand are frequently met with. The latter also contain pieces of chalk, flint, grit, together with *Foraminifera, Entomostraca*, and shelly débris, which appear to be different from those in the clay itself. These deposits are probably in part post-glacial.

Elie.—On the Fife coast, eleven miles south of St. Andrews. Shells are contained in a bed of clay passing out to sea, and are not in a good state of preservation. The fauna of this bed is decidedly arctic in character, more so in fact than in most other localities.§§

Errol.—Clay pit on the north side of the river Tay, eight miles east of Perth. The clay from which the shells are obtained is about forty-five feet above sea-level. *Leda arctica* is obtained here.‖‖

* J. Haswell, *Geol. Mag.*, vol. ii. p. 182. † J. A. Smith, *Pro. Roy. Phys. Ed.*, vol. i. p 247.
‡ Jamieson, *Quart. Journ. Geol. Soc.*, vol. xxi. p. 196. § *Ibid.*, p. 196.
‖ *Ibid.*, p. 196. ¶ *Ibid.*, p. 196. ** *Ibid.*, p. 197. †† *Ibid.*, p. 196.
‡‡ Geikie, *Mems. Geol. Survey, Memoir* 32, *Scotland*, p. 128.
§§ Jamieson, *Quart. Journ. Geol. Soc.*, vol. xxi. p. 196. ‖‖ *Ibid.*

Towncroft Farm.—Near Grangemouth. At this place a shaft was sunk through alternating beds of sand, mud, gravel, clay, and till, to a depth of one hundred and fifty-five feet. At a depth of about eighty feet, the bones of the small arctic Seal (*Pagomys fœtidus*, Gray) were found, accompanied by two species of Foraminifera; *Polymorphina compressa*, D'Orb, and *Nonionina asterizans*, F. and M. In the same shaft, at about a depth of only four feet from the surface, a portion of a horn of a large red deer was met with in blue mud and sand.*

Rothesay.—In the course of an excavation for a gasometer in the town of Rothesay, a shelly-clay was cut, underlying a bed of stratified sea-sand, which was again overlaid by a stratum of moss.†

NOTE H.

Map showing the Principal Direction of the Glaciation of Scotland.

The earliest general sketch-map of Scotland, showing the direction of the glaciation, is that which accompanies my brother's paper "On the Glacial Phenomena of Scotland," published in the *Transactions of the Glasgow Geological Society*. Since the date of that publication our knowledge has greatly increased. A large part of the midland and southern districts of the country has been surveyed by the Geological Survey, and to their maps I must refer those who desire detailed information. I have not found it possible on the present small map to indicate all the places where striæ have been observed by the Survey, for in many cases they are so closely set that a very large scale would be required for the purpose. Those who have been engaged on the work of the Survey in the districts referred to are Professor Geikie, Dr. (now Professor) Young, and Messrs. B. N. Peach, H. M. Skae, R. L. Jack, J. Horne, D. R. Irvine, and myself. Another observer, Mr. Jolly, has also published a short account of some part of the Merrick district in Galloway. In the north of Scotland the chief additions to our knowledge have been made by Mr. Jamieson, who has traced the glaciation of Caithness, and noted a number of localities on the borders of the Moray Frith, and along the coasts of the north-west Highlands. Mr. C. W. Peach has described the glaciation of the Shetlands; Dr. Howden has ascertained the direction of the glaciation in the Montrose district, and I have given some account of the glaciation of Lewis, and have mapped the striæ in a number of the western sea-lochs, and along the borders of several of the highland lakes. I append here a few notes supplementary to the general account of the glaciation of Scotland given in chapters vi. and vii. I have stated at p. 101 that the Scotch ice coalesced on the floor of what is now the North Sea with the Scandinavian mer de glace. I believe Mr. Croll was the first to indicate that such must have been the case. He pointed to the direction of the striæ on the south coasts of the Moray Frith, in Caithness, and in Shetland, and showed how they could not have been produced by ice streaming out from Scotland in a regular way. He reasoned that at the climax of glacial cold the Scandinavian ice-sheet and the Scottish glaciers attained so great a

* Turner, *Proc. Ed. Roy. Soc.*, 1869-70, pp. 105-114.
† Hugh Miller, *Sketch Book Pop. Geol.* App., p. 324.

thickness that they could not possibly have floated off in the sea that separates the one country from the other. And he explained the abnormal direction of the striæ in the Shetland Islands, in Caithness, and the north-east of Aberdeen, by inferring that the Scandinavian ice-sheet pressed up against the Scotch ice, and thus partially deflecting it, compelled it to over-ride Caithness from south-east to north-west. As another indication that some great obstruction opposed the progress of the Scotch ice as it crept outwards upon the bed of the German Ocean, I would point to the direction of the glaciation in the lower reaches of the Tweed valley. In this valley the ice flowed from west to east at Kelso, but for no apparent reason it gradually turned away to south-east, as it continued on its course, until at last it overflowed Holy Island, and the district round Belford, in a direction parallel to the trend of the coast. This is certainly not the direction which a glacier that filled the Tweed valley would naturally take; but seeing that the Scandinavian ice-sheet cannot fail to have occupied the bed of the German Ocean, we have in its presence a simple explanation of the curious trend of the striæ referred to. The Scotch ice, impeded in its outward course, was compelled to hug and overflow the English coast in its progress towards the south.

A similar great deflection of the Scotch mer de glace took place off the north coast of Ireland. In that region the Irish and Scottish ice-sheets coalesced—one portion flowing westwards into the Atlantic, and the other south-west and south into the Irish Sea. The fact of this deflection is attested by the position of the striæ in the south of Scotland (Wigtonshire), by the direction of the fiord-basin of the Sound of Jura (see map of Physiography of West Coast, &c., and Note E), and the deep hollow in front of Rathlin Island: by the deflection of the glacial striæ in the north of Ireland, and by the fact that the Scotch ice overflowed the Isle of Man, as is shown by the composition of the till in that island, and the direction of the glacial striæ, which my colleague, Mr. Horne, assures me point from north-east to south-west, *i.e.* parallel to the backbone of the island.

The direction of the glaciation in the basin of the Clyde is very striking. From a study of the map it will be observed that the ice which streamed down the vale of the Leven and the Gareloch, instead of flowing out to sea by the Frith of Clyde, was forced away to south-east and east, eventually crossing the whole breadth of Scotland, and doubtless coalescing at last with the Scandinavian ice-sheet. From this we may imagine how great must have been the accumulation of glacier-ice that filled up the lower reaches of the Frith of Clyde, and poured southwards over the bed of the Atlantic towards the south-west coast of Scotland and the north of Ireland. It will be noticed, moreover, that at various places the striæ cross and recross, and there is sometimes an apparent confusion, especially on the high-grounds that extend from Renfrewshire into Lanarkshire. These curious appearances are due—first, to the meeting of the two great opposing streams from north and south; and second, to the removal of obstructions to the natural flow of the ice during the retreat of the great mer de glace. The two colossal ice-streams met somewhere above Hamilton. Certain appearances would even lead one to infer that the highland stream sometimes reached as far south as Lesmahagow, for. Mr. B. N. Peach got scattered fragments of mica-schist, gneiss, and other typical highland rocks in the till of that district. But it is evident that the region between Lesmahagow and Cambuslang was a kind of debatable ground upon which the rival ice-streams were liable to occasional deflections. The general trend of the striæ west and south-west of Strathavon, however, clearly indicates that the high-grounds in that district were glaciated by the ice that streamed outwards from the Highlands. If we draw an undulating line from the sea-coast near Ayr north-east to the valley of the Irvine, and thence across the watershed into the Avon, and east to Lesmahagow, then down the valley of the Clyde to Carluke, sweeping it away to the east by Wilsontown, and thereafter continuing it along the crest of the Pentlands and the northern slopes of the Lammermuir Hills, by Reston and Ayton to the sea, we shall roughly indicate the meeting-place of the two great ice-streams. All along this line we have a "debatable ground" of variable breadth throughout which we find a commingling in the till of stones which have come from the north and south. South of it characteristic highland stones do not occur, and north of it stones derived from the south are similarly absent. When the ice-sheets were melting back, and so breaking up into a series of gigantic local glaciers, the direction of flow occasionally became modified. The highland glaciers, no longer hampered in their course by the ice piled up upon the low-grounds, were enabled to follow the natural slope of their beds; and this gave rise to another set of striæ. Examples of such cross-hatching of striæ occur, as my colleague, Mr. Jack, has proved, very abundantly in the basin of the Clyde. During the climax of glacial cold the lower reaches of the Frith of Clyde were choked with ice descending from the mountain-glens of Argyleshire, and hence the ice that streamed down the Gareloch and the valley of the Leven was forced to overflow the high-grounds behind Greenock, and so to continue on its way *up* the valley of the Clyde towards Glasgow. Hereabouts the pressure of the ice-sheet advancing from the south began to be felt, and a portion of the highland ice-stream became deflected, flowing first south-east, then south, and last south-west, while another portion continued on its easterly course across the whole breadth of the country to the North Sea.

In a preceding Note (E) I have spoken of the "undertow" of the ice-sheet, and endeavoured to show how the lower strata of the massive ice-sheet might be deflected, while the upper portions of the stream flowed on in one continuous direction. There is abundant proof to show that this was the case in a less or greater degree in almost every part of the country. The striæ are deflected by isolated hills and projecting bosses, while the general direction of glaciation remains unchanged. Nor is there wanting evidence that would lead one to infer that here and there the lower strata of ice followed one route, while the upper strata were moving in almost an exactly opposite direction.

[Since this Note was in type No. 116 of the Geological Society's Journal has appeared. It contains a paper by Mr. J. F. Campbell ("Notes on the Glacial Phenomena of the Hebrides") in which the direction of the glacial striæ in some of the islands of the Outer Hebrides is given. These directions I have transferred to my map. Mr. Campbell considers that the islands have been glaciated from *north-west* to *south-east*. My own observations in the Outer Hebrides compel me, on the contrary, to believe that the overflowing ice came from the mainland of Scotland, *i.e.*, from *south-east* to *north-west*. See *Quart. Journ. Geol. Soc.*, vol. xxix. p. 532.]

INDEX.

ABERDEENSHIRE, boulder clay beds of, 212; brick-clays of, 260; sand and gravel in, 238
Aberdeen, submerged peat at, 321
Abies picea in Orkney peat, 318
Acts of Scottish Parliament and state of forests, 334
Adhémar, M., on disturbance of earth's centre of gravity by polar ice-cap, 216
Admiralty charts of Scottish sea-lochs, 297
Æneas Silvius on want of woods in Scotland, 334
Agassiz, Professor, cited, 31, 42, 274, 423
Age of palæolithic deposits, 467; of cave deposits, 473
Airdrie, shelly clay at, 201
Alaska, mammaliferous deposits of, 496
Alluvial deposits of Scotland, 340
Alpine glaciers, 52
Alternations of cold and warm climates, 133
America (North), superficial deposits of, 410; glaciation of, 411; lakes of, 412; unmodified drift of, 413; pre-glacial beds in, 413; fossils in drift of, 414; scratched pavements in till of, 415; marine fossils in boulder clay of, 415; fossiliferous beds in unmodified drift, 416; inter-glacial forest-bed in, 417; succession of changes during glacial epoch in, 419; glacial lakes in, 424; shelly clays of, 424; terminal moraines in, 424; mammaliferous deposits of, 499
Andrews, Professor E., on Illinois drift, 415
Antarctic ice-sheet, 101; regions, climate of, 138
Anticlinal arches, 288
Aosta, glacial deposits at mouth of valley of, 525
Arago, M., on eccentricity of earth's orbit, 134
Arctic regions, former changes of climate in the, 103
Åsar of Sweden, 385; theories of, 391
Asia, ancient glaciers in, 379
Atlas mountains, former glacial action in, 379
Atmospheric denudation in recent times, 340
Auvergne, 378

BACK, Captain, description of glaciated region by, 412; on sand-hills of Barrens, 421
Baer, Von, on action of frost in Nova Zembla, 52
Bald on mammalian remains at Woodhill Quarry, 163
Barren Grounds, 197, 411
Bear, 368, 376, 471, 488
Beaver, 346, 368, 376, 417, 526
Beds in boulder clay, 209
Belcher, Sir E., on pine-tree at Wellington Channel, 497
Bell, Mr. A., on Wexford beds, 374
Bell, Mr. D., on glacial lake near Glasgow, 185
Bennie, Mr. J., on chalk fragments in brick-clay at Portobello, 268
Bison, 376, 409, 481, 494
Black Forest, ancient glaciers of, 378
Blair Drummond moss, 311
Blown sands, 315
Boar, 488
Bœthius cited, 331
Borings through glacial, &c., deposits, 151, 177, 183, 187
Bos longifrons, 346
Bos primigenius, 162, 346, 405, 451

Bos aurus, 407
Boulder clay beds of Scotland, 205
Boulder clay, lower, of Lancashire, 361; upper, of ditto, 362; of East Anglia, 369
Brick-clays, &c., distribution of, in Scotland, 258; height of, above sea, 260; general character of, 261
Bridlington shell-bed, 370
Britain, continental condition of, in post-glacial times, 325
British Columbia, glaciation of, 380
Brittany, submerged peat of, 324
Bronze age, 434
Brown, Dr. R., on glaciation of British Columbia, 380
Buffalo, 499
Buried forests of Scotland, 318
Buried ravines in Scotland, 169, 174, 175, 176, 179, 182, 187

CALEDONIA, early historical notices of, 328
Campbell, Mr. J. F., on glaciation of the Hebrides, 565
Canoes in Scottish raised beaches, 312
Carinthia, glacial and inter-glacial deposits of, 407
Carse lands of Scotland, 311
Cave-bear, 405, 444, 471
Cave deposits of England, 440
Caves of England, how formed, 438
Caucasus, ancient glaciers of, 379
Cervus elaphus, 405
Chalk fragments in Portobello brick-clays, 265
Chalmers, Mr., on ancient forests of Scotland, 331
Chambers, R., on parallel roads of Lochaber, 274; on raised beaches at Trinity, 309; on Swedish åsar, 395
Change in direction of ice-flow (Scotland), 217; in America, 419
Change in the earth's axis of rotation, 108
Changes of climate in arctic regions, evidence of former, 103
Changes in distribution of sea and land, 109
Changes in physical features of south England during palæolithic times, 447
Channel Islands, submerged peat of, 324
Chantre, M., on Rhone glacier, 401
China, traces of former glacial action in, 379
Climate, alternations of, 106; changes of, by varying eccentricity of earth's orbit, 133; cosmical changes of, 107; hypothetical causes of change of, 107; influenced by varying distribution of land and sea, 119; of Britain, to what due, 112; of Deception Island, 138; of inter-glacial periods in Scotland, 196; of inter-glacial period in Switzerland, 404; of palæolithic period in Britain, 450
Climates, former mild, in arctic regions, 103
Climates of early geological ages, well-marked, 106
Clyde, inter-glacial deposits in valley of, 189
Clyde Trustees, borings made for, 187
Cones of sand and gravel, 228
Configuration of Scottish mountains, &c., 22, 83
Continental Britain, climate of, 326
Contorted beds accompanying Scotch till, 158, 166
Contorted brick-clays, &c., 265
Contorted drift at Cromer, 368
Cowdon Burn, fossiliferous beds in till of, 162, 561
Crag and tail, 19, 97
Crevasses in glaciers, 47
Croll, Mr. J., on glacier motion, 40; on effects of eccentricity of earth's orbit, 135; on ocean currents and their effect on climate of glacial epoch, 140; on periods of great eccentricity of the orbit, 147; on buried course of Kelvin Water, 184; on effect of polar ice-cap on earth's centre of gravity, 216; on deflection of ice-sheet in north of Scotland, 563
Cromer drift, 367
Cross-hatched striæ, 91, 217
Currents (ocean), their effect on climate of glacial epoch, 140, 145; indirect effect of varying eccentricity of the orbit upon, 140; cause of, 145

DANA, Professor, on Connecticut ice-sheet, 420
Darwin, Mr., on former glacial action in South America, 518; on alternations of cold and warm periods, 518
Dawkins, Mr. W. Boyd, cited, 450, 451, 456, 461
Dawson, Professor, on fossiliferous bed below till, 413; on marine fossils in boulder clay of St. Lawrence, 416; on traces of former glacial action in silurian and carboniferous strata, 512, 513

Deas, Mr., on borings at Glasgow, 187
Debatable ground between ice-flows in Scotland, 91
Decay of Scottish peat-mosses, 339
Deflection-basins on bed of Scottish seas, 522
Deflection of ice-markings, 91, 95
De Laske, Mr. J., on erratics of Mount Desert Island, 423
Denudation in post-glacial times, 342; in last inter-glacial period, 342
Denuded surface of till below sand and gravel, 227, 242
De Rance, Mr., on glacial deposits of north-west of England, 361
Destruction of inter-glacial beds by ice-sheet, 190, 197
Diagonal bedding in kames, 231, 247
Dispersal of erratics by colossal glaciers, 225, 252
Disturbance of drainage-systems by glaciers, 191
Dog, 488
Dora Baltea, 525
Dora Riparia, 525
Drift-dammed lakes, 284
Drift deposits of Scotland (lower), 6; ditto (upper), 218
Drums of till in Scotland, 18, 97
Duration of glacial and inter-glacial periods, 199
Dürnten inter-glacial beds, 404

EAGLESHAM, near Glasgow, high-level terraces at, 233, 248
Earth, movements of the, 123; eccentricity of orbit of the, 124; supposed change in axis of rotation of the, 108
East Anglia, glacial deposits of, 358, 366
Eccentricity of the earth's orbit, 124; effects of variations in the, 133; periods of high, 148
Eisblink glacier, 57; submarine bank in front of, 57, 61
Elephant, 368, 405, 414, 417, 451, 469, 499, 526
Elk, Irish, 162, 346, 368, 376, 481
Elk or moose deer, 346
Emys, 526
England, glacial deposits of, 355, 426
English Channel, submerged peat in, 323
Eocene, erratics in, 514
Equinoxes, precession of the, 126
Erdmann, Professor, on Scandinavian drifts, 385
Erie clays of North America, 418
Erratics, 220, 221; position of, 223; conditions under which dispersed, 224, 252, 255, 257; at higher levels than rocks from which derived, 223, 257; on tops and slopes of kames, 232
Etheridge (jun.), Mr. R., on traces of former glacial action in palæozoic rocks of Australia, 513; list of fossils from Scotch glacial beds, 533
Etheridge, Mr., on fossils of Italian tertiaries, 528
Europe, ancient glaciers of, 378; general succession of glacial deposits in, 427
Evans, Mr., on archæological changes, 448; on stone implements in high- and low-level gravels, 469

FALLOW deer, 346
Falsan, M., on Rhone glacier, 401
False-bedding in kames, 231, 247
Faraday, Professor, on fracture and regelation, 36
Faults in Scotch coal-fields, 289
Favre, M., on fresh-water bed between glacial deposits, 405
Felis Caffer, 451
Fiord-basins of Scotland, 520
Floating ice during accumulation of Scotch boulder clay, 217
Forbes, David, on ancient glacial action in Cordillera of South America, 518
Forbes, J. D., on glacier motion, 36; on the first appearance of stones on the surface of glaciers, 257
Forbes, Ed., on climate, &c., of post-glacial period, 349
Forest-bed of Cromer, 367
Forests of post-glacial Britain, 326; causes of the decay of, 327; destruction of by the Romans, 329; historical notices of Scottish, 330
Formations, Scottish superficial, 7
Forster, Capt., cited, 138
Fossils in Scottish sand and gravel series, 245, 254; paucity of, in high level brick-clays, 279; in Erie clay-beds, 280
Fox, 346, 376
Fracture and regelation of ice, 37
Frith of Forth, submerged rock-basin in, 519
Frost, action of, on rocks, 52, 341

GASTALDI, Professor, on miocene glacial deposits, 514, 530; on succession of deposits on plains of Piedmont, 526
Geikie, Professor, on scratched pavements in Scotch till, 164; on fossils

in till, 159, 160; on Kilchattan brick-clay, 263
German Ocean, once filled with glacier ice, 99, 564; submerged peat of, 324, fresh-water shells in, 324; site of ancient river in bed of, 325
Glacial action, traces of former, in older formations, 511
Glacial epoch, date and duration of, 148
Glacial erosion, effects of, 85
Glacial erosion of rock-basins, 294
Glacial lake at Neidpath, 181; near Glasgow, 186; in Lochaber, 274
Glacial period in southern hemisphere, 518
Glacial rivers, turbid water of, 50, 77
Glacial series of East Anglia, 369
Glaciation of Scotland, 22, 75, 90, 519, 562; of England, 357; of Ireland, 373; of Scandinavia, 381; of Switzerland, 398; of Carinthia, 407; of North America, 411
Glacier, general character of an alpine, 46
Glacier-ice, formation of, 34
Glacier motion, theories of, 36
Glacier of the Leven valley, 188
Glaciers, gigantic local, during formation of Scotch boulder clay, 216
Glaciers, temperature of, 42
Glaciers, smoothed and striated rocks below, 49
Glaciers of Greenland, 56; icebergs shed from them, 60; scarcity of surface moraines upon, 61; sometimes terminate on land and give rise to rivers, 63
Glutton, 451
Goat, 346, 488, 526
Godwin-Austen, Mr., on ancient river of North Sea area, 325; on age of cave mammals, 473; on erratics in French carboniferous, 513; in new red sandstone of Devonshire, 514; in chalk at Croydon, 514
Greenland, its glacial aspect, 54
Grund-moränen in Switzerland, 402
Gulf-stream, effect of, on climate of Europe, 112; origin of, 140, 463

HAAST, Dr., on glaciers of New Zealand, 518
Hare, 376
Harkness, Professor, on Irish glacial deposits, 374
Harmer, Mr. F. W., on English drifts, 366
Hayes, Dr., on open polar sea, 66; expedition of upon Greenland mer de glace, 66
Heat transmitted through ice, 41, through glaciers, 43
Hector, Dr., on glaciers of New Zealand, 518
Heer, Professor, on Swiss inter-glacial beds, 404
Height above sea reached by Scotch till, 94, 96; by boulder clay, 207, 216; by morainic débris and erratics of southern uplands of Scotland, 218; by kames, 252
Height of land during formation of Scotch till, 99
Herschel, Sir J., on eccentricity of earth's orbit, 133
Highland valleys, kames opposite mouths of, 238
Hind, Professor, on drift deposits of Moisie river, 414
Hippopotamus, 368, 451
Höfer, Professor, on Carinthian superficial deposits, 407
Holmström, Mr., on Swedish inter-glacial beds, 384
Hooker, Dr., on pine-tree from Wellington Channel, 498
Hörbye, M., on glaciation of Norway, 381
Horne, Mr. J., on glacial striæ in Isle of Man, 564
Horse, 162, 368, 376, 481, 488, 494
Howden, Dr., on gravels at Montrose, 272: on striæ in Forfarshire, 563
Hull, Professor, on Irish glacial deposits, 374
Human relics in raised beaches (Scotland), 310
Humboldt glacier, 56, 71
Huxley, Professor, on inter-glacial fossils, 245
Hyæna, 451, 454, 489

ICEBERG deposits, 79; of North America, 417
Iceberg, effect of stranded, 72
Icebergs of arctic seas, 70; stones in and upon, 60, 71
Icebergs, origin of, 58
Ice-flow, deflection of, in central valley of Scotland, 85; direction of indicated by stones in till, 90
Ice-flutings and groovings deflected by hills, 95
Ice-foot, of arctic shores, 67; in Scotland during formation of boulder clay, 214
Ice-rafts with stones and rubbish, 96
Ice-sheet, advance and retreat of, during deposition of boulder-clay,

214; Scottish, coalescent with Scandinavian ice-sheet, 101, 563
Ice, thickness of, formed by direct freezing on sea, 67
Inter-glacial deposits of Scotland, 157, 179, 182, 194; chiefly of fresh-water origin, 181; destruction of, by ice-sheet, 187, 197
Inter-glacial marine deposits (Scotland), 163, 201
Inter-glacial beds of England, 362, 369; of Scandinavia, 384; of Switzerland, 404; of North America, 415; of Italy, 526
Inter-glacial periods, climate of, in Scotland, 196
Inter-glacial period, the last, 397, 428
Ireland, reached by Scotch land-ice, 99, 522; glacial deposits of, 373, 426
Irish elk, 162, 346, 368, 481
Irish Sea filled with land-ice, 84
Iron age, 431
Irvine, Mr. D. R., on erratics of Rinns of Galloway, 219
Islands of Scottish coast striated from mainland, 98
Isochimenal lines in Europe, 114, 462
Isothermal lines in Europe, 461
Italy, glacial and inter-glacial deposits of, 525

JACK, Mr. R. L., on Scotch till, 11; on kames of Fintry hills, 252; on boulder clay beds of Stirlingshire, 212; on fault near Loch Lomond, 295; on cross striæ in basin of the Clyde, 563
Jamieson, Mr. J. F., on boulder clay beds of north of Scotland, 212; on height of erratics in Highlands, 220; on sand and gravel of Highland valleys, 238; on absence of kames from Caithness, 240; on river-like bedding of gravel in Highland valleys, 241; on formation of kames, 247; on stratified deposits at high levels, 249; on brick-clays, &c., of Aberdeenshire, 272; on parallel roads of Glen Roy, 274; on raised beaches, 311; on blown sands, 316
Judd, Mr., on traces of former ice-action in Scotch jurassic strata, 514

KAMCHATKA, shattered rock-surfaces of, 53
Kames, 228, 234, 246; of denudation, 246; origin of certain, 246; greatest height above sea of, 252
Kane, Dr., on Mary Minturn River, 63; on open polar sea, 66; on musk ox in ice-foot, 69
Keilhau, Mr., on erratics of Trondhjemfiord, 382
Kendall, Lieut., on climate of Deception Island, 318
Kennedy's Sound, 65
Kent's Cavern, 440
Kinahan, Mr., on sand deposits in Ireland, 375

LACUSTRINE beds in till, 181, 195
Lake district of England, glacial deposits of, 358
Lakes and sea-lochs of Scotland, 282
Lakes of Italy, moraines at foot of, 529
Lakes, silted-up, 344
Land, elevation and depression of, 109; effect of massing of, under equator and round the poles, 110, 116
Lebanon, moraines of, 379
Leda clays of North America, 424
Lemming, 451
Lewis, glaciation of, 99, 523; boulder clay of, 209; succession of glacial deposits in, 211; absence of kames from, 240; lakes of, 284
Lignites of Italy, 526; of Switzerland, 404
Lion, 451, 454, 489
Lisbon, winter temperature of, 113
Loch Lomond, map and sections of, 518
Lochs of Scotland: Lochs Broom, Long, Fyne, Sunart, 297; Loch Etive, 300, 520; Lochs Torridon, Cairn-Bahn, Lydoch, Rannoch, Fannich, Luichart, 302; Loch Doon, 219, 271, 292; Loch Langabhat, 240; Loch Skene, 269; Loch Leven, 284; Loch Trool, 292; Loch Lomond, 295, 518; Lochs Shin, Linnhe, 296; Loch Alsh, 323, 521; Loch of Forfar, 347; Lochs Hourn, Carron, Eishart, Craignish, Crinan, 521; Lochs Shell, Bhracadail, Eynort, Bhreatal, Scavaig, Nevis, 523
Lower drifts of Scotland, 6
Lubbock, Sir J., on commingling of mammalian remains in river-gravels, 469
Lyell, Sir C, on changes of climate, 109; on marl-beds of Forfar, 345; on English caves, 439

Machairodus latidens, 451
Macclesfield beds, 362
Mackintosh, Mr., on upper boulder clay of Lancashire, 362

Maclaren, Mr. on erratics of Pentland Hills, 222
Mälar lake, åsar in basin of, 393
Mammalian remains in till, 16, 164; in inter-glacial deposits, 162, 163, 165, 405, 407, 419, 526; in post-glacial deposits, 346, 376, 409, 481, 485, 487, 488; in palæolithic deposits, 431, 450, 467, 486
Mammaliferous deposits of foreign countries, 486
Mammoth, 162, 163, 325, 368, 409, 451, 494
Manor Water, changes in course of, 182; buried rock-basin at head of, 303
Marine drift, high-level, 232, 248, 252, 362, 375
Markinch, kames at, 229, 238
Marl deposits in Scotland, 343
Marmot, 451
Martins, M. Ch., on Swedish åsar, 395
Mary Minturn River, 63
Mastodon, 414, 417, 419, 499
Mediterranean, climatal influence of, 461
Melrose, beds below till at, 155
Mer de glace of Greenland, 55; of Scotland, 83; of England, 360; of Ireland, 375; of Scandinavia, 381; of Switzerland, 401; of North America, 419
Merrick mountains, 271
Michelotti, Sig., on fossils in pliocene of northern Italy, 528
Middle sand and gravel of Lancashire, &c., 362; of Ireland, 374
Moel Tryfan, 362
Molecular theory of glacier motion, 43
Moorfoot Hills, angular gravel and terraces of, 241, 248
Moose-deer, 346, 453
Moray Frith, 277, 316
Moraine profonde or till, 88
Moraines, 51, 76; rare in Greenland, 61; of latest glacial period in Scotland, 268; of English lake district, 363; of Scandinavia, 368; of Rhone glacier, &c., 401; of Carinthia, 407; of "second period," why better developed than those of first ice period in Switzerland, 407; of North America, 424; of northern Italy, 525
Morainic débris of Scotland, 219, 220, 242; conditions under which deposited, 220; below sand and gravel, 242; denudation of, by large rivers, 243; of north-west of England and Wales, 360, 363; of Ireland, 373; of Norway, 381
Morainic lakes, 284
Morlot, M., on two glacial periods in Switzerland, 407
Mörschweil, inter-glacial beds at, 404
Moseley, Canon, on glacier motion, 38
Motion of glaciers, 36; rate of, 38
Mounds of sand and gravel, 228, 363, 374, 385, 390, 420
Mud and silt derived from glaciers, 50, 77
Musk sheep, 451

Nathorst, Mr., on Swedish inter-glacial deposits, 384
Neidpath, Peebles, beds below till at, 157, 181
Neolithic deposits, their distribution, 480; of Switzerland, 487; of Italy, 488
Neolithic period, 435
Ness, sand and gravel in Vale of the, 238
Newberry, Dr., on drift deposits of Ohio, 417, 419; on boulder in coal, 513
New Kilpatrick, inter-glacial deposits at, 183
Newport, raised beaches at, 308
Nielston, fossiliferous beds in till near, 162
Nithsdale, gravel terraces of, 248
Nith, valley of the, 84
Norfolk coast sections, 366
Normandy, submerged peat of, 322
Norrie's Law, kame of, 230
Northern drift, 390
North Minch, filled with glacier ice, 99; depth of, 298, 326
North Sea, 280
North Uist, submerged peat of, 323
North-west districts of England, glacial deposits of, 357
Norway, glaciation of, 381
Norwich Crag, 366
Nova Zembla, disintegrated rocks of, 52; climate of, 114
Nutation, 128

Oak, range of, 320
Oban, submerged peat at, 323
Obliquity of the ecliptic, 131; effect of increased, on climate, 142
Ochil Hills, striated rocks of, 80; overflowed by highland ice-sheet, 94; erratics on, 222; kames at base of, 237; height of till on, 94
Orkneys, submerged peat and trees in, 321

Oscillations of level during glacial epoch, 201, 203, 365, 395, 417, 425
Oscillations of level produced by weight of polar ice-cap, 216

PABBAY (Hebrides), submerged peat of, 323
Packard, Dr. A. S., on drift of Labrador, 414; on glacial lakes of the White Mountains, 424
Paisley hills, 80, 185, 277
Palæolithic deposits, their geological age, 467; their distribution, 477; of Italy, 488
Palæolithic, neolithic, and mammaliferous deposits of foreign countries, 486
Palæolithic period, 435; climate of, 450
Paucity of fossils in high-level brick-clays of Scotland, 279
Peach, Mr. B. N., on till of Ochils, 94; on till of Lesmahagow, 564
Peat-moss, composition of, 317
Peat-mosses, submerged, 314-325; origin of, 337; decay of, 339
Peccary, 499
Peeblesshire, sand and gravel in valleys of, 249
Peebles, terraces at, 306
Pengelly, Mr., on Kent's Cavern, 440
Pentland Hills, striated rocks of, 80; till on, 94; erratics on, 222, 223
Perched blocks. (See Erratics.)
Petaries, grants of, in Scotland, 333
Peterhead, boulder clay beds at, 212
Physical condition of Scotland during formation of till, 98
Physical condition of Scotland during formation of boulder clay, 215
Physical condition of Scotland during formation of clays with arctic shells, 277
Physical condition of Scotland in post-glacial times, 348
Physiography of western Scotland as it would appear upon an elevation of 600 ft., 519
Pine, Scotch, range of, 319
Pliocenic alluvia of Northern Italy, 526
Portobello brick-clays, 265
Post-glacial deposits of Scotland, 304, 317, 341; of England and Ireland, 376; of Scandinavia, 388; of Switzerland, 409, 487; of Italy, 488
Post-glacial period in Scotland, summary, 348
Precession of the equinoxes, 128; effect of, on climate, during high eccentricity of the earth's orbit, 139
Pre-glacial lakes, 182
Prestwich, Mr., on river-gravels, 468; on shells in river-gravels, 469; on river-drifts at Hoxne, 474; on changes of drainage-systems in England, 477
Pumpelly, Mr., on rock-basins in northern Asia, 380
Pyrenees, 378

QUEEN Charlotte Islands, glaciation of, 380

RACOON, 499
Raised beaches, 304; human relics in, 310, 312
Ramsay, Professor, on distribution of lakes, 283; on rock-basins, 286; on glaciers of North Wales, 364; on age of cave-deposits, 473; on glacial period in the Permian, 514
Rathlin Island, submerged rock-basin north of, 522
Ravines, pre-glacial and post-glacial, in Scotland, 168
Recent deposits of Scotland, 304, 316, 340
Red deer, 346, 481
Reindeer, 162, 163, 346, 451, 453
Reston (Berwickshire), highland erratics in till at, 225
Rhinoceros, 368, 407, 409, 451, 494
Rhone glacier, 399; coalescent with Rhine glacier, &c., 401; moraine profonde of, in low-grounds of France, 401
Richardson, Dr., on glaciated aspect of the Barren Grounds, 411; on absence of mammalian remains east of Rocky Mountains, 497
Ridges of sand and gravel, 228
Rinns of Galloway, erratics of, 219
River deposits below till, 154; in till, 158, 193
River-gravels of southern England, 444, 468
River ravines, pre-glacial and post-glacial, in Scotland, 168
Rivers, glacial, 50, 77
Rivers, large, during deposition of sand and gravel series of Scotland, 243
Rock-basins, 283, 285; absence of, from highly faulted districts, 289; double in some valleys, 302; silted up, 303; submerged, round Scottish coasts, 519; in fiords, 520; opposite islands, 522

Rocks, striated, below till, 20, 80; below glacier, 49
Roebuck, 346, 488
Rome, Rev. J. L., on East Anglian drifts, 366
Ross, Sir J. C., on antarctic ice, 102
Rum, island of, rock-basin north of, 523
Rye Water, 191

SAHARA, a sea during glacial epoch, 402
St. Mary's Loch, 303
Saltpans, destruction of timber for, 333
Sand and gravel on denuded till, 227; distribution of in Scotland, 227; stratified down valleys, 240; pass into angular débris, 241; older glacial deposits denuded below, 242, 403; mode of origin of, 243, 249; fossils in, 245
Sand dykes, 175
Scandinavian superficial deposits, 381; general succession of, 427
Scotland, superficial formations of, 6; hills and mountains of, rounded, 22; aspect of, during climax of glacial cold, 101; physical aspect of, during Roman period, 329
Scottish sea-lochs, 296
Sea-lochs and submerged rock-basins, 300, 519
Selwyn, Mr., on traces of former glacial action in palæozoic rocks of Australia, 513
Siberia, climate of, 114, 460; mammaliferous deposits of, 494
Sierra Famatina, height of snow-line on, 33
Sierra Guadarrama, ancient glaciers of, 379
Sierra Nevada, ancient glaciers of, 379
Sierra Nevada (California), ancient glaciers of, 380, 422
Sheep, 488
Shells, broken and striated in boulder clay, 206, 209, 210, 212
Shore deposits in fiords, 251
Skae, Mr. H. M., on high-level terraces of Nithsdale, 248
Skene, Loch, moraines at, 269, 284
Slitrig Water, fossiliferous beds in till at, 159
Smith, Mr., of Jordanhill, cited, 258
Smith's Sound, 65
Snow-line, 32; height of, 33
Solinus, his description of the Orkneys, 330
Solway Frith, confluent glaciers flowing into, 84
Southern hemisphere, why it has a cooler mean annual temperature than the northern, 142
Spey, sand and gravel in valley of, 238
Stag, 488
Steinbock, 407
Stevenston, blown sand of, 316
Stinchar, valley of the, boulder clay in, 207; raised beaches at mouth of, 308
Stoddard, Professor O. N., on scratched pavements in till of Miami, 415
Stone age, 434
Stone period, newer, in Scotland, 347
Stones, in Scotch till, 14; lines of, in till, 23; below glaciers, whence derived, 50, 85; in till, whence derived, 86; commingling of, in till, from separate districts, 91; in till indicate direction of ice-flow, 91; in brick-clay beds, 263, 264
Stratified deposits at high levels, 248
Striæ, direction of, in Scotland, 80, 84, 95; height of in Highlands, 83; direction of, in England, 360; in Scandinavia, 381; of North America, 419
Striated rocks below till, 20, 80; origin of, 81
Striated stones, scarcity of in terminal moraines, 76
Striated pavements in till, 164, 415
Submarine bank in front of Eisblink glacier, 57, 61
Submerged trees and peat, 314, 321
Submerged rock-basins off west coast of Scotland, 297, 519
Submergence, extent of, in Scotland, during formation of kames, 254; during inter-glacial periods, 202; during formation of boulder clay, 215
Succession of glacial deposits, tables of, 425
Summary of geological changes in Scotland in post-tertiary times, 351, 425
Summer of polar regions, 37
Superficial formations of Scotland, general distribution of, 6
Sutherland, range of Scotch pine in, 320
Sweden, glaciation of, 381; upper and lower till of, 383; inter-glacial deposits of, 384; åsar of, 385; erratics on åsar, 387; shelly clays of, 387; post-glacial beds of, 388; moraines of, 388; succession of

changes in, during glacial epoch, 389; erratics derived from, 390; origin of åsar of, 392
Switzerland, superficial deposits of, 398; erratics of, 399; till or moraine-profonde of, 402; ancient alluvium of, 403; inter-glacial beds of, 404; gravel and erratics overlaying inter-glacial beds of, 405, 406; terminal moraines of last glacial period in, 406; moraines of first and second period in, 407; post-glacial beds of, 409
Sylva Caledoniæ, 329
Synclinal troughs, 288

TABLE of sedimentary strata, 511
Table of quaternary deposits of Britain, and their foreign equivalents, 516, 517
Tables of general succession of glacial deposits, 425
Tain, boulder clay at, 212
Talla Linnfoots, buried rock-basin at, 303
Tallert Bank, 61
Tay, basin of, submerged peat and trees in, 322
Tents Moor, blown sand of, 316
Terraces of gravel and sand at Eaglesham, 232, 248
Teviot, kames in valley of, 247
Thickness of Scottish mer de glace, 83; of Scandinavian, 381; of Swiss, 400; of North American, 420
Tiddeman, Mr., on ice-sheet of north-west England, 360; on human remains in Victoria Cave, 510
Tiger, 451
Till of Scotland, 8; its general character, 10; its stones, 14, 75; its local character, 16, 93; its distribution, 16, 88, 95; its terraced aspect in Southern Uplands, 17, 96; its drums or sowbacks, 19, 97; smoothed and broken rocks under, 20, 80; its nests and patches of sand and gravel, 23; early theories of, 27; origin of, 87; stones of, whence derived, 86; striated pavements in, 164; contorted beds in and below, 166; beds accompanying, 150, 165, 179; river-deposits in and below, 193; lacustrine deposits in, 181, 185, 195
Till of the Ochil Hills, the Pentlands, Paisley and Kilmarnock highgrounds, Tinto Hill, &c., 94
Till of England, 359; of Ireland, 373; of Scandinavia, 383; of Switzerland, 402; of North America, 413
Tillon Burn, pre-glacial course of, 176
Tiree, submerged peat of, 323
Torfæus on the Orkneys, 331
Törnebohm, Mr., on erratics of Jemtland, 382; on Swedish till, 383; on mild inter-glacial period in Sweden, 384; on elevation reached by the åsar, 387; on moraines of Lake Wener, 388; his theory of formation of åsar, 392
Traigh Chrois, Lewis, boulder clay at, 209
Trees below peat, 318
Trees below silt of carse lands, 314
Trinity, raised beaches at, 309
Tweed, glacier of, 84; drums of till in valley of, 97; stratified deposits below till of, 155, 157; changes in course of, since glacial times, 182
Tyndall, Dr., his theory of glacier motion, 36
Tyrie brick-clays, 266
Tytler, Mr., on old Scottish forests, 332

UDDINGSTON, shelly clay at, 260
Ulleswater, 364
Undertow of Scottish ice-sheet, 523, 565
Upernavik, 70
Upper drifts of Scotland, 218
Urus, 162, 346, 405, 451
Utznach inter-glacial beds, 404

VALLAY, submerged peat of, 323
Vancouver's Land, 380
Viscous theory of glacier motion, 36
Vosges Mountains, ancient glaciers of, 378

WALES, glacial deposits of, 358
Ward, Mr. J. C., on Lake district, 252
Wark, kames at, 229
Washington Territory, 380
Waste of mountains, &c., 52, 87
Water below Greenland glaciers, 63
Watson, Mr., on range of trees, 319
Waves of translation, 28
Weathering of rocks, 340
Wetzikon inter-glacial beds, 404
Wexford gravels, 374
Whitaker, Mr. W., on Thames gravels, 468; on correlation of boulder clays, 475
Whiteadder Water, kames in valley of, 237

White Mountains, glacial lakes of, 424
Whitney, Professor J. D., on driftless region of Wisconsin, 500
Whittlesey, Mr., on drift of North America, 413, 415, 416
Wigtonshire, boulder clay of, 213
Wild boar, 346
Wild cat, 346
Wilkes, Commodore, cited, 101
Winds, effect of variations in eccentricity of the earth's orbit upon, 140
Winter temperature, lines of equal, in Europe, 114
Wishaw, pre-glacial river-courses in neighbourhood of, 176
Wolf, 346, 376
Wooded condition of post-glacial Britain, 326
Woodhill Quarry, Ayrshire, 162
Wood in till, 16, 164
Wood, Mr. Nicholas, on buried river-course, 181
Wood, Mr. S. V., on glacial deposits of East Anglia, 366, 369
Woods, destruction of, in Scotland, 333

YARROW, ice-flutings in valley of the, 95
Young, Professor, cited, 152
Ythan, sand and gravel in valley of the, 238

THE END.

VIRTUE AND CO., PRINTERS, CITY ROAD, LONDON.

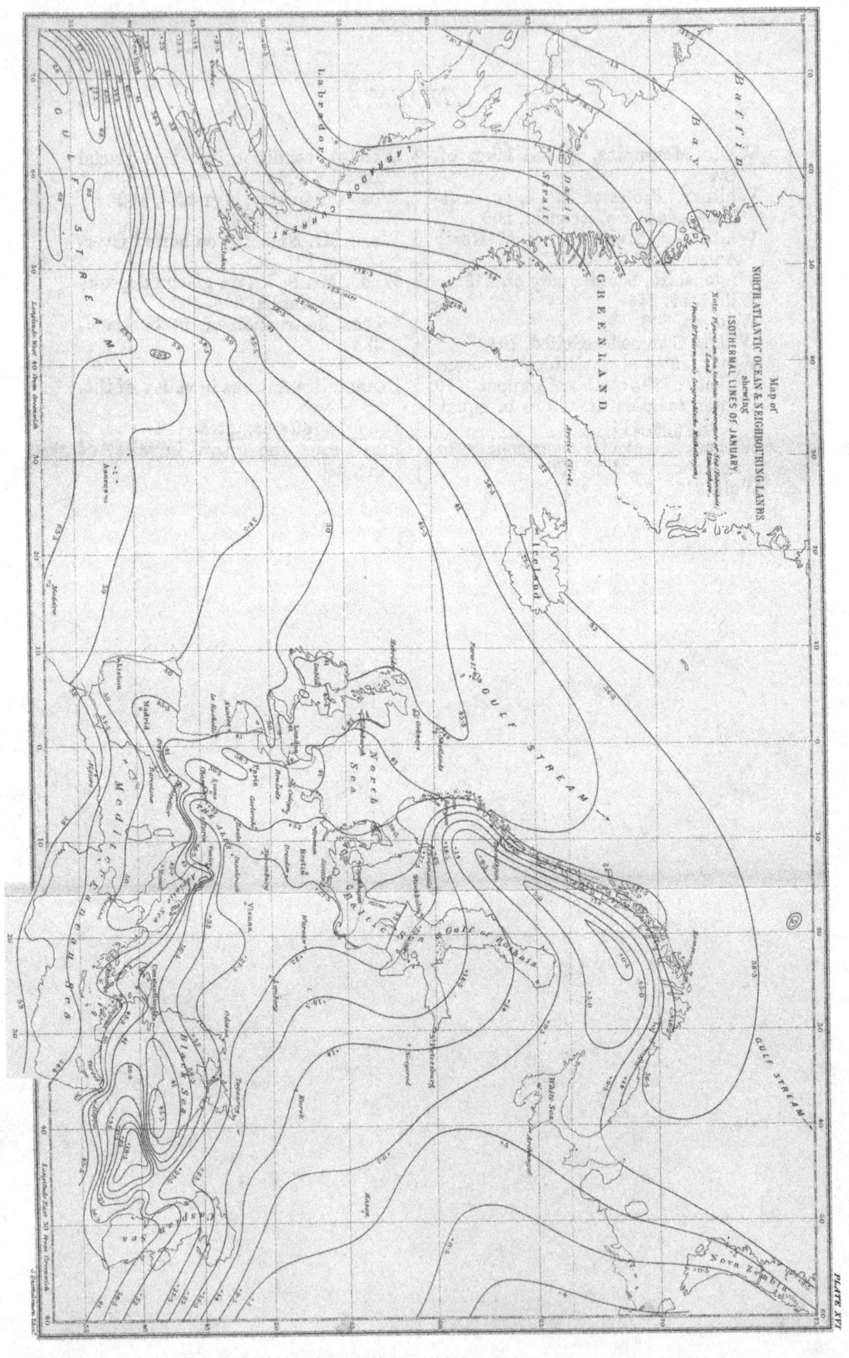

The material originally positioned here is too large for reproduction in this reissue. A PDF can be downloaded from the web address given on page iv of this book, by clicking on 'Resources Available'.

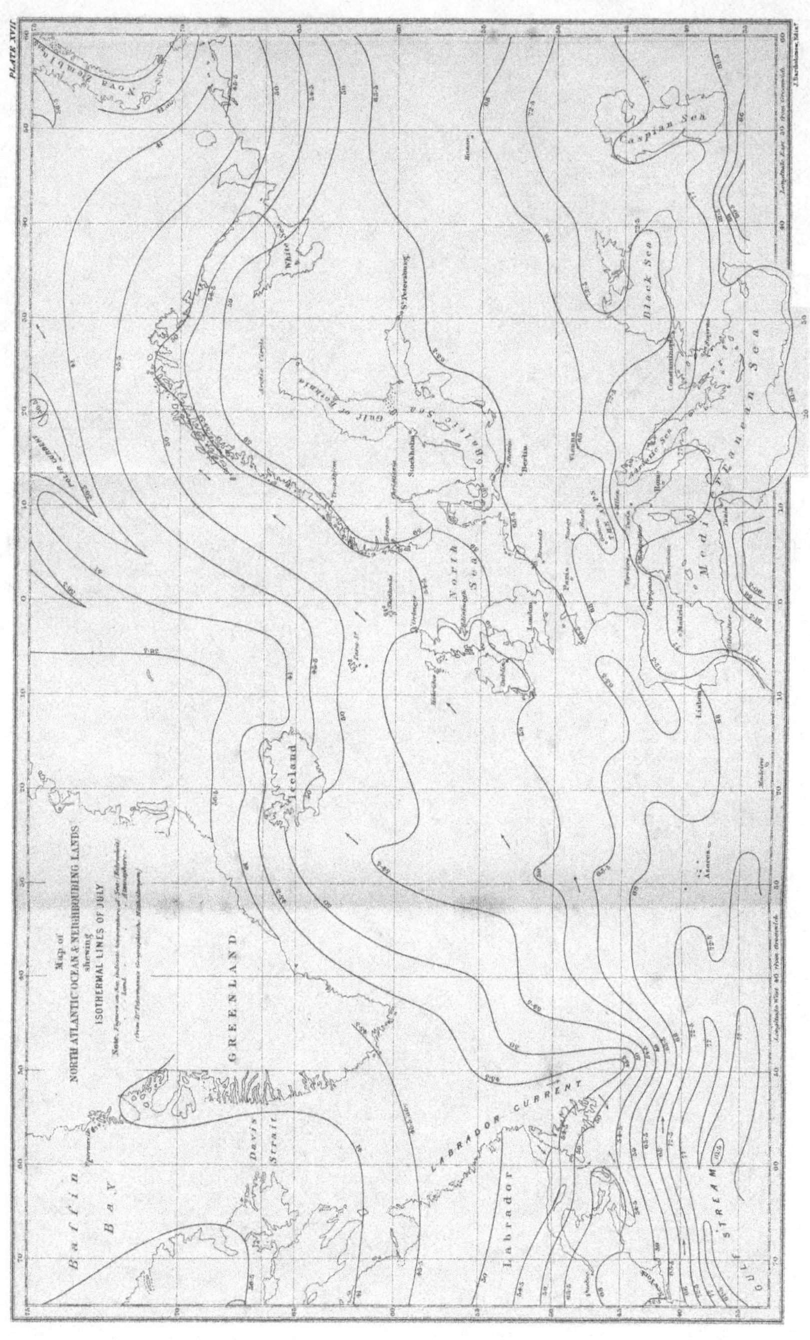

The material originally positioned here is too large for reproduction in this reissue. A PDF can be downloaded from the web address given on page iv of this book, by clicking on 'Resources Available'.

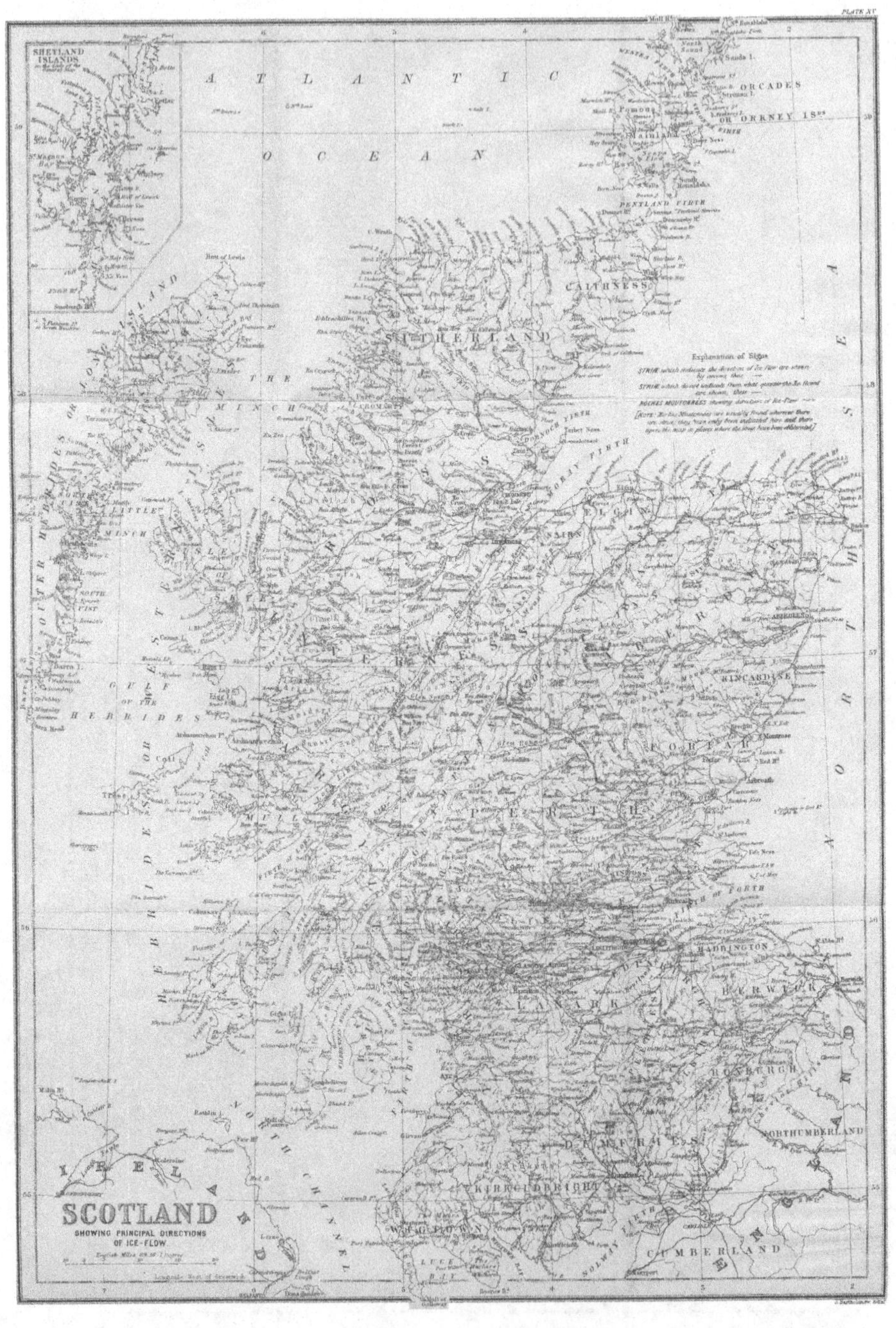

The material originally positioned here is too large for reproduction in this reissue. A PDF can be downloaded from the web address given on page iv of this book, by clicking on 'Resources Available'.

Printed in the United States
By Bookmasters